Jie Jack Li

Name Reactions

A Collection
of Detailed Reaction Mechanisms

Second Edition

Springer

Jie Jack Li, Ph. D.
Pfizer Global Research and Development
Chemistry Department
2800 Plymouth Road
Ann Arbor, MI 48105
USA
e-mail: jack.li@pfizer.com

ISBN 3-540-40203-9 Springer-Verlag Berlin Heidelberg New York

Cataloging-in-Publication Data applied for
Bibliographic information published by Die Deutsche Bibliothek
Die Deutsche Bibliothek lists this publication in the Deutsche Nationalbibliografie;
detailed bibliographic data is available in the Internet at <http://dnb.ddb.de>.

Springer-Verlag Berlin Heidelberg New York
a member of BertelsmannSpringer Science+Business Media GmbH

http://www.springer.de

© Springer-Verlag Berlin Heidelberg 2003
Printed in Germany

Typesetting: Dataconversion by author
Cover-design: design & production, Heidelberg
Printed on acid-free paper   2 / 3020 xv - 5 4 3 2 1 0 -

To Vivien

Preface to the second edition

The second edition includes five points of improvement: (a) Additional 16 name reactions have been supplemented; (b) I have corrected typos and a few dubious mechanisms in the first edition. I wish to thank Prof. Rick L. Danheiser of Massachusetts Institute of Technology and Mr. Yiqian Lian of Michigan State University for invaluable comments and suggestions. I have also incurred many debts of gratitude to Prof. Brian M. Stoltz of California Institute of Technology and his students, Eric Ashley, Doug Behenna, Dan Caspi, Neil Garg, Blake Greene, Jeremy May, Sarah Spessard, Uttam Tambar, Raissa Trend, and Ryan Zeidan for proofreading the final draft of the second edition; (c) The references are expanded and updated; (d) A more thorough index has been implemented so the reader may navigate through the book more easily; (e) The short descriptions of name reactions given as mnemonics seem to be helpful to both novices and veterans. As a result, I added the descriptions for most reactions. Finally, I am grateful for permission to use the postage stamps on the inner covers from respective postal authorities, who still retail the copyrights of those stamps.

Jack Li
Ann Arbor, Michigan, May 2003

Preface to the first edition

What's in a name? That which we call a rose by any other name would smell as sweet.[1] Contrary to Shakespeare's claim, *name reactions* in organic chemistry and the corresponding mechanisms are nevertheless fascinating for their far-reaching utilities as well as their insight into organic reactions. Understanding their mechanisms greatly enhances our ability to solve complex synthetic problems. As a matter of fact, some name reactions are the direct results of a better understanding of the mechanisms as exemplified by the Barton–McCombie reaction.[2] In addition, our knowledge of how reactions work can shed light on side reactions and by-products. When a reaction does not give the "desired" product, the mechanism may provide clues to where the reaction has gone awry.

I started collecting named and unnamed organic reactions and their corresponding mechanisms while I was a graduate student. It occurred to me that many of my fellow practitioners are doing exactly the same, and that these efforts could be made easier through a monograph tabulating interesting and useful mechanisms of name reactions. To this end, I have updated my collection with many *contemporary* name reactions and added more recent references, especially up-to-date review articles. In reflecting the advent of asymmetric synthesis, relevant name reactions in this field have been included to the repertoire. Since the step-by-step mechanisms delineated within are mostly self-explanatory, detailed verbal explanations are not offered, although some important jargons entailing the types of transformations are highlighted. Short descriptions of name reactions are given as mnemonics rather than accurate definitions. With regard to the references, the first one is generally the original article, whereas the rest are related articles and review articles. Readers interested in in-depth coverage of name reactions are encouraged to follow up with the references as well as relevant books.[3-7]

I would like to express my grateful thanks to Profs. Brian J. Myers of Ohio Northern University, Jeffrey N. Johnston of Indiana University and Christian M. Rojas of Barnard College, who read the manuscript and offered many invaluable comments and suggestions. Special thanks are due to Profs. Gordon W. Gribble of Dartmouth College, Louis S. Hegedus of Colorado State University and Thomas R. Hoye of University of Minnesota for their critique of the drafts. In addition, I am very much indebted to Nadia M. Ahmad, John (Jack) Hodges, Michael D. Kaufman, W. Howard Roark, Peter L. Toogood and Kim E. Werner for proofreading the manuscript. Any remaining errors are, of course, solely my own. I am also grateful to Ms. Ann Smith of Merck & Co., Inc. for her helpful communications and discussions. Last but not the least, I wish to thank my wife, Sherry Chun-hua Cai, for her understanding and support throughout the project.

Jack Li
Ann Arbor, Michigan, November 2001

References

1. William Shakespeare, *"Romeo and Juliet"* Act II, Scene ii, **1594–1595**.
2. Derek H. R. Barton, *"Some Recollections of Gap Jumping"* American Chemical Society, Washington, DC, **1991**.
3. Mundy, B. R.; Ellerd, M. G. *Name Reactions and Reagents in Organic Synthesis* John Wiley & Sons, New York, **1988**.
4. Laue, T.; Plagens, A. *Named Organic Reactions* John Wiley & Sons, New York, **1999**.
5. *"Organic Name Reactions"* section, *The Merck Index* (13th edition), **2001**.
6. Smith, M. B.; March, J. *"Advanced Organic Chemistry"* (5th edition), Wiley, New York, **2001**.
7. Hassner, A.; Stumer, C. *Organic Synthesis Based on Named Reactions* Pergamon, **2002**.

Table of Contents

XVI

Abbreviations and Acronyms

Ac	acetyl
AIBN	2,2'-azobisisobutyronitrile
Alpine-borane®	*B*-isopinocamphenyl-9-borabicyclo[3.3.1]-nonane
B:	generic base
9-BBN	9-borabicyclo[3.3.1]nonane
BINAP	2,2'-bis(diphenylphosphino)-1,1'-binaphthyl
Boc	*tert*-butyloxycarbonyl
t-Bu	*tert*-butyl
Cbz	benzyloxycarbonyl
m-CPBA	*m*-chloroperoxybenzoic acid
CuTC	copper thiophene-2-carboxylate
DABCO	1,4-diazabicyclo[2.2.2]octane
dba	dibenzylideneacetone
DBU	1,8-diazabicyclo[5.4.0]undec-7-ene
DCC	1,3-dicyclohexylcarbodiimide
DDQ	2,3-dichloro-5,6-dicyano-1,4-benzoquinone
DEAD	diethyl azodicarboxylate
Δ	solvent heated under reflux
(DHQ)$_2$-PHAL	1,4-bis(9-*O*-dihydroquinine)-phthalazine
(DHQD)$_2$-PHAL	1,4-bis(9-*O*-dihydroquinidine)-phthalazine
DIBAL	diisobutylaluminum hydride
DMA	*N,N*-dimethylacetamide
DMAP	*N,N*-dimethylaminopyridine
DME	1,2-dimethoxyethane
DMF	dimethylformamide
DMS	dimethylsulfide
DMSO	dimethylsulfoxide
DMSY	dimethylsulfoxonium methylide
DMT	dimethoxytrityl
dppb	1,4-bis(diphenylphosphino)butane
dppe	1,2-bis(diphenylphosphino)ethane
dppf	1,1'-bis(diphenylphosphino)ferrocene
dppp	1,3-bis(diphenylphosphino)propane
E1	unimolecular elimination
E2	bimolecular elimination
E1cb	2-step, base-induced β-elimination *via* carbanion
Eq	equivalent
HMPA	hexamethylphosphoric triamide
Imd	imidazole
LAH	lithium aluminum hydride
LDA	lithium diisopropylamide
LHMDS	lithium hexamethyldisilazane
LTMP	lithium 2,2,6,6-tetramethylpiperidine

M	metal
Mes	mestyl
MVK	methyl vinyl ketone
NBS	*N*-bromosuccinimide
NCS	*N*-chlorosuccinimide
NIS	*N*-iodosuccinimide
NMP	1-methyl-2-pyrrolidinone
Nu	nucleophile
PCC	pyridinium chlorochromate
PDC	pyridinium dichromate
SET	single electron transfer
S_NAr	nucleophilic substitution on an aromatic ring
S_N1	unimolecular nucleophilic substitution
S_N2	bimolecular nucleophilic substitution
TBAF	tetrabutylammonium fluoride
TBDMS	*tert*-butyldimethylsilyl
TBS	*tert*-butyldimethylsilyl
Tf	trifluoromethanesulfonyl (triflyl)
TFA	trifluoroacetic acid
TFAA	trifluoroacetic anhydride
TFP	tri-*o*-furylphosphine
THF	tetrahydrofuran
TIPS	triisopropylsilyl
TMEDA	*N,N,N',N'*-tetramethylethylenediamine
TMP	tetramethylpiperidine
TMS	trimethylsilyl
Tol	toluene or tolyl
Tol-BINAP	2,2'-bis(di-*p*-tolylphosphino)-1,1'-binaphthyl
Ts	tosyl

Abnormal Claisen rearrangement

Further rearrangement of the normal Claisen rearrangement product with the β-carbon becoming attached to the ring.

References

1. Hansen, H.-J. In *Mechanisms of Molecular Migrations;* vol. 3, Thyagarajan, B. S., ed.; Wiley-Interscience: New York, **1971**, pp 177–200. (Review).
2. Shah, R. R.; Trivedi, K. N. *Curr. Sci.* **1975**, *44*, 226.
3. Kilenyi, S. N.; Mahaux, J. M.; Van Durme, E. *J. Org. Chem.* **1991**, *56*, 2591.
4. Nakamura, S.; Ishihara, K.; Yamamoto, H. *J. Am. Chem. Soc.* **2000**, *122*, 8131.
5. Schobert, R.; Siegfried, S.; Gordon, G.; Mulholland, D.; Nieuwenhuyzen, M. *Tetrahedron Lett.* **2001**, *42*, 4561.
6. Puranik, R.; Rao, Y. J.; Krupadanam, G. L. D. *Indian J. Chem., Sect. B* **2002**, *41B*, 868.

Alder ene reaction

Addition of an enophile to an alkene *via* allylic transposition.

enophile

ene

reaction

References

1. Alder, K.; Pascher, F.; Schmitz, A. *Ber. Dtsch. Chem. Ges.* **1943**, *76*, 27.
2. Oppolzer, W. *Pure Appl. Chem.* **1981**, *53*, 1181. (Review).
3. Oppolzer, W. *Angew. Chem.* **1984**, *96*, 840.
4. Mackewitz, T. W.; Regitz, M. *Synthesis* **1998**, 125–138.
5. Johnson, J. S.; Evans, D. A. *Acc. Chem. Res.* **2000**, *33*, 325. (Review).
6. Stratakis, M.; Orfanopoulos, M. *Tetrahedron* **2000**, *56*, 1595–1615.
7. Mikami, K.; Nakai, T. In *Catalytic Asymmetric Synthesis;* 2nd ed.; Ojima, I., ed.; Wiley-VCH: New York, **2000**, 543–568. (Review).
8. Leach, A. G.; Houk, K. N. *Chem. Commun.* **2002**, 1243.
9. Lei, A.; He, M.; Zhang, X. *J. Am. Chem. Soc.* **2002**, *124*, 8198.
10. Shibata, T.; Takesu, Y.; Kadowaki, S.; Takagi, K. *Synlett* **2003**, 268.

Aldol condensation

Condensation of a carbonyl with an enolate or an enol. A simple case is addition of an enolate to an **aldehyde** to afford an alco**hol**, thus the name **aldol**.

References

1. Wurtz, W. A. *Bull. Soc. Chim. Fr.* **1872**, *17*, 436.
2. Nielsen, A. T.; Houlihan, W. J. *Org. React.* **1968**, *16*, 1–438. (Review).
3. Mukayama, T. *Org. React.* **1982**, *28*, 203–331. (Review).
4. Mukayama, T.; Kobayashi, S. *Org. React.* **1994**, *46*, 1–103. (Review on Tin(II) enolates).
5. Saito, S.; Yamamoto, H. *Chem. — Eur. J.* **1999**, *5*, 1959–1962. (Review).
6. Johnson, J. S.; Evans, D. A. *Acc. Chem. Res.* **2000**, *33*, 325–335. (Review).
7. Denmark, S. E.; Stavenger, R. A. *Acc. Chem. Res.* **2000**, *33*, 432–440. (Review).
8. Palomo, C.; Oiarbide, M.; Garcia, J. M. *Chem. — Eur. J.* **2002**, *8*, 36–44. (Review).
9. Alcaide, B.; Almendros, P. *Eur. J. Org. Chem.* **2002**, 1595–1601. (Review).
10. Wei, H.-X.; Hu, J.; Purkiss, D. W.; Paré, P. W. *Tetrahedron Lett.* **2003**, *44*, 949.

4

Allan–Robinson reaction

Synthesis of flavones or isoflavones.

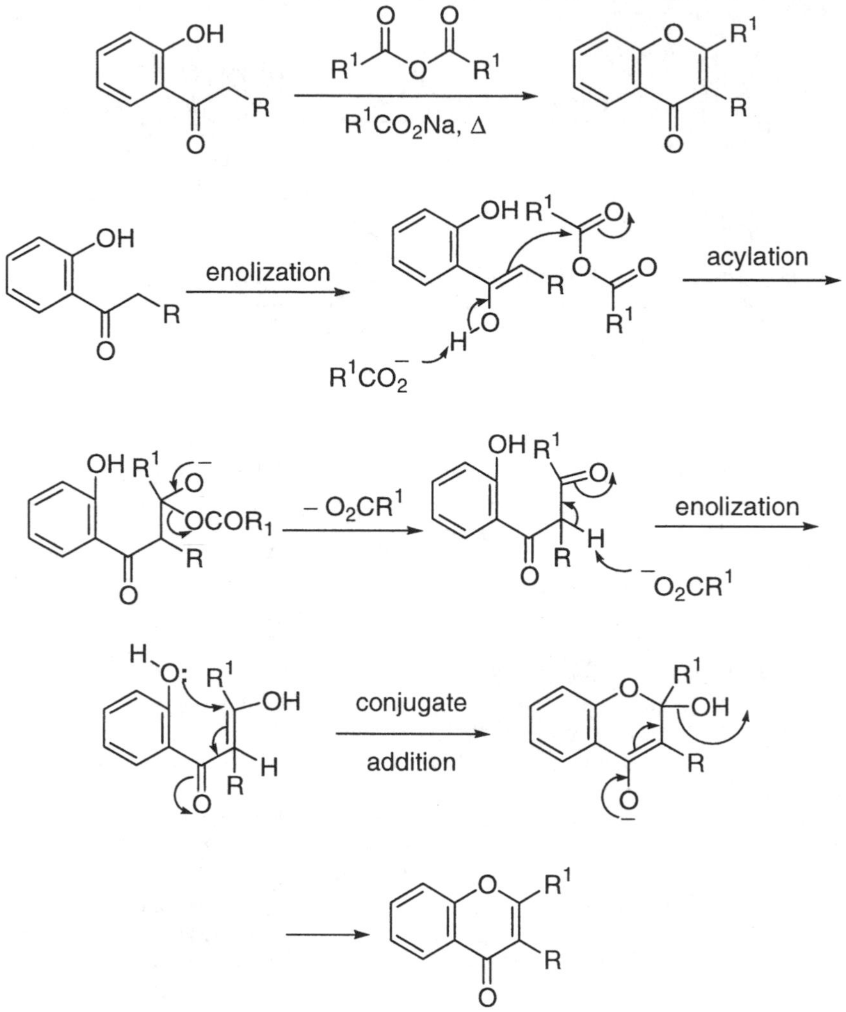

References

1. Allan, J.; Robinson, R. *J. Chem. Soc.* **1924**, *125*, 2192.
2. Szell, T.; Dozsai, L.; Zarandy, M.; Menyharth, K. *Tetrahedron* **1969**, *25*, 715.
3. Wagner, H.; Maurer, I.; Farkas, L.; Strelisky, J. *Tetrahedron* **1977**, *33*, 1405.
4. Dutta, P. K.; Bagchi, D.; Pakrashi, S. C. *Indian J. Chem., Sect. B* **1982**, *21B*, 1037.
5. Patwardhan, S. A.; Gupta, A. S. *J. Chem. Res., (S)* **1984**, 395.
6. Horie, T.; Tsukayama, M.; Kawamura, Y.; Seno, M. *J. Org. Chem.* **1987**, *52*, 4702.

7. Horie, T.; Tsukayama, M.; Kawamura, Y.; Yamamoto, S. *Chem. Pharm. Bull.* **1987**, *35*, 4465.
8. Horie, T.; Kawamura, Y.; Tsukayama, M.; Yoshizaki, S. *Chem. Pharm. Bull.* **1989**, *37*, 1216.

Alper carbonylation

Palladium-catalyzed ring expansion-carbonylation of azirines.

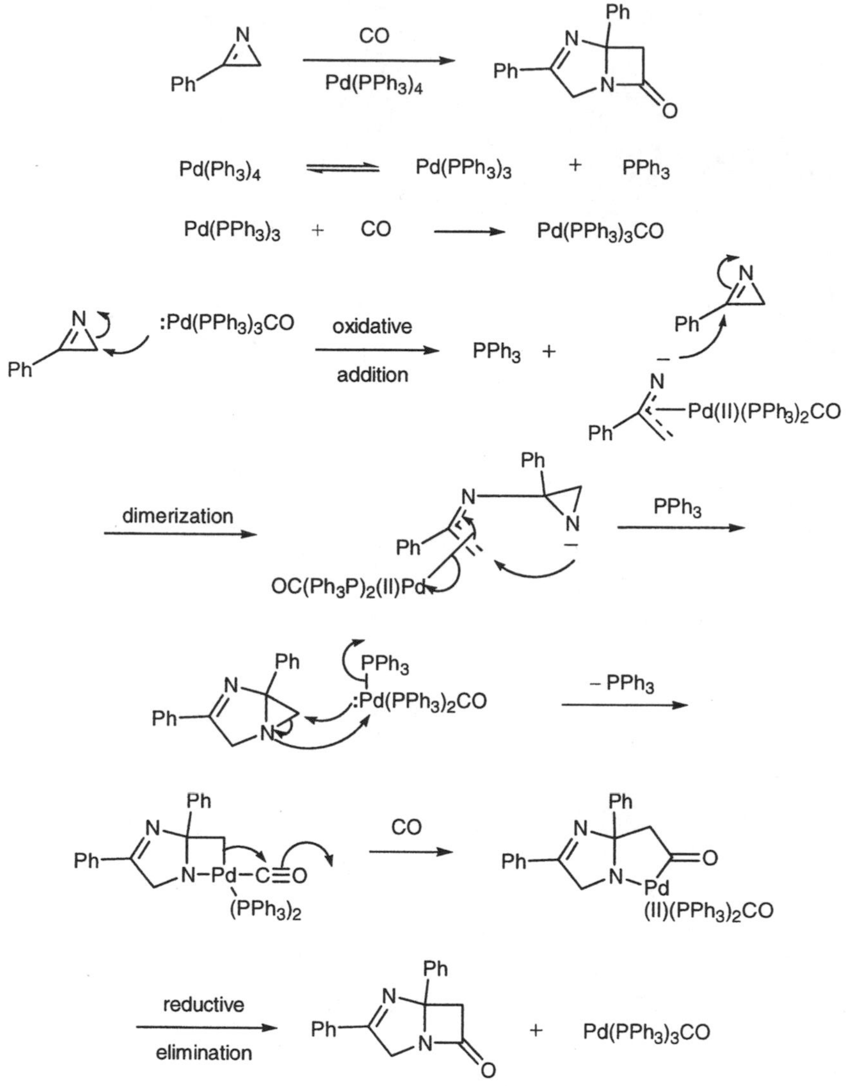

An alternative mechanism:

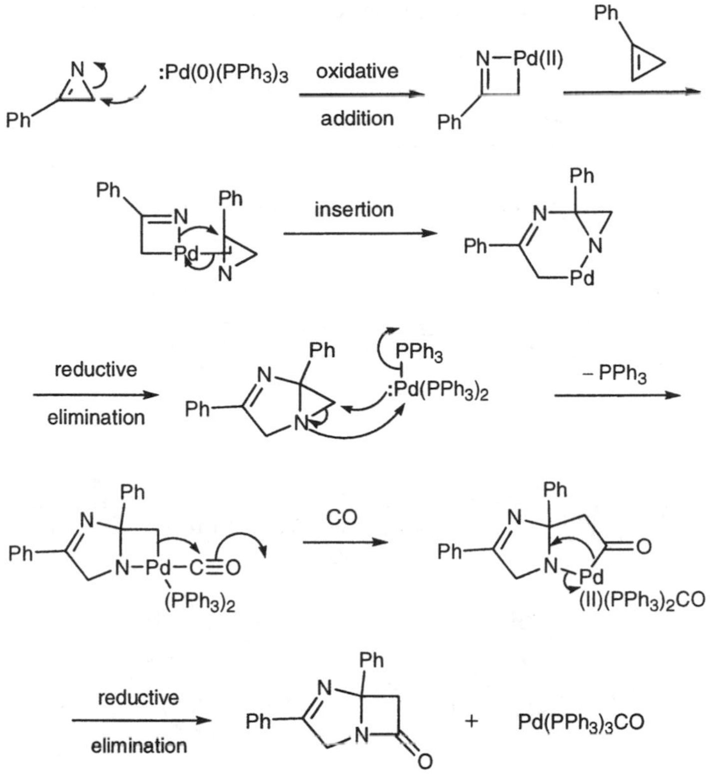

References

1. Alper, H.; Perera, C. P.; Ahmed, F. R. *J. Am. Chem. Soc.* **1981**, *103*, 1289.
2. Alper, H.; Perera, C. P. *Organometallics* **1982**, *1*, 70.
3. Alper, H.; Hamel, N. *Tetrahedron Lett.* **1987**, *28*, 3237.
4. Alper, H. *Aldrichimica Acta* **1991**, *24*, 3. (Review).
5. Jia, L.; Ding, E.; Roberts, J. E.; Anderson, W. R. *Abstr. Pap.-Am. Chem. Soc.* (**2000**), 220th INOR-265.
6. Butler, D. C. D.; Inman, G. A.; Alper, H. *J. Org. Chem.* **2000**, *65*, 5887.

8

Amadori rearrangement

Reversible conversion of aldosylamine to the corresponding ketosylamine.

glycosylamine

1-amino-1-deoxy-2-ketose

tautomerization

References

1. Amadori, M. *Atti Accad. Nazl. Lincei* **1925**, *2*, 337.
2. Hodges, J. E. *Adv. Carbohydrate Chem.* **1955**, *10*, 169. (Review).
3. Simon, H.; Kraus, A. *Fortschr. Chem. Forsch.* **1970**, *14*, 430.
4. Yaylayan, V. A.; Huyghues-Despointes, A. *Carbohydr. Res.* **1996**, *286*, 187.
5. Wrodnigg, T. M.; Stutz, A. E.; Withers, S. G. *Tetrahedron Lett.* **1997**, *38*, 5463.
6. Kadokawa, J.-I.; Hino, D.; Karasu, M.; Tagaya, H.; Chiba, K. *Chem. Lett.* **1998**, 383.
7. Turner, J. J.; Wilschut, N.; Overkleeft, H. S.; Klaffke, W.; Van Der Marel, G. A.; Van Boom, J. H. *Tetrahedron Lett.* **1999**, *40*, 7039.
8. Cremer, D. R.; Vollenbroeker, M.; Eichner, K. *Eur. Food Res. Technol.* **2000**, *211*, 400.
9. Liu, Z.; Sayre, L. M. *Chem. Res. Tox.* **2003**, *16*, 232.

Angeli–Rimini hydroxamic acid synthesis

Hydroxamic acid formation from aldehyde and *N*-sulfonylhydroxylamine.

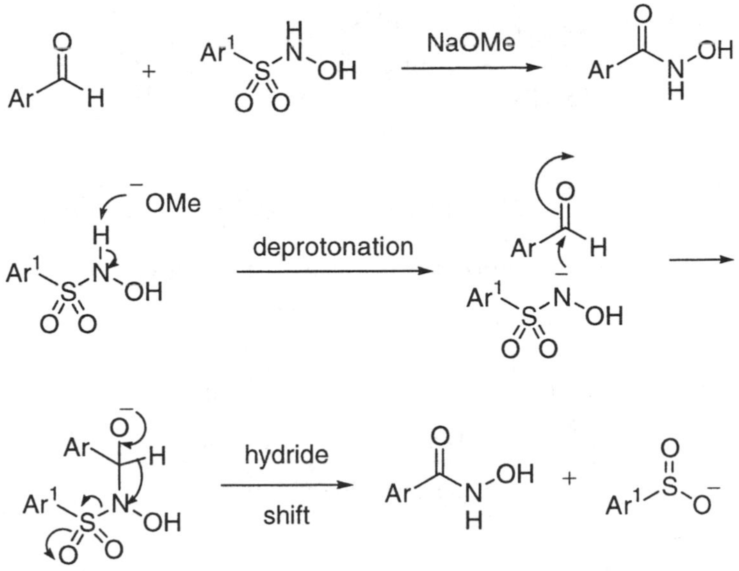

References

1. Angeli, A. *Gazz. Chim. Ital.* **1896**, *26(II)*, 17.
2. Balbiano, L. *J. Chem. Soc.* **1913**, *102(I)*, 474.
3. Yale, H. L. *Chem. Rev.* **1943**, *33*, 228.
4. Hassner, A.; Wiederkehr, E.; Kascheres, A. J. *J. Org. Chem.* **1970**, *35*, 1962.
5. Zhou, S.; Xie, F.; Xu, Z.; Ni, S. *Huaxue Shiji* **2001**, *23*, 154.

ANRORC mechanism

Addition of Nucleophiles, Ring Opening and Ring Closure.

$$N^* = N^{15}$$

Chichibabin amination → ring opening

common nitrile intermediate

6-*exo-dig* ring closure → tautomerization →

5-*endo-trig* ring closure → − 2 H oxidation →

References

1. Lont, P. J.; Van der Plas, H. C.; Koudijs, A. *Recl. Trav. Chim. Pays-Bas* **1971**, *92*, 207.
2. Lont, P. J.; Van der Plas, H. C. *Recl. Trav. Chim. Pays-Bas* **1973**, *92*, 449.
3. Van der Plas, H. C. *Acc. Chem. Res.* **1978**, *11*, 462. (Review).
4. Kost, A. N.; Sagitulin, R. S. *Tetrahedron* **1981**, *37*, 3423.
5. Rykowski, A.; Van der Plas, H. C. *J. Org. Chem.* **1987**, *52*, 71.
6. Briel, D. *Pharmazie* **1999**, *54*, 858.
7. Rykowski, A.; Wolinska, E.; Van der Plas, H. C. *J. Heterocycl. Chem.* **2000**, *37*, 879.
8. Buscemi, S.; Pace, A.; Pibiri, I.; Vivona, N.; Spinelli, D. *J. Org. Chem.* **2003**, *68*, 605.

Arndt–Eistert homologation

One carbon homologation of carboxylic acids using diazomethane.

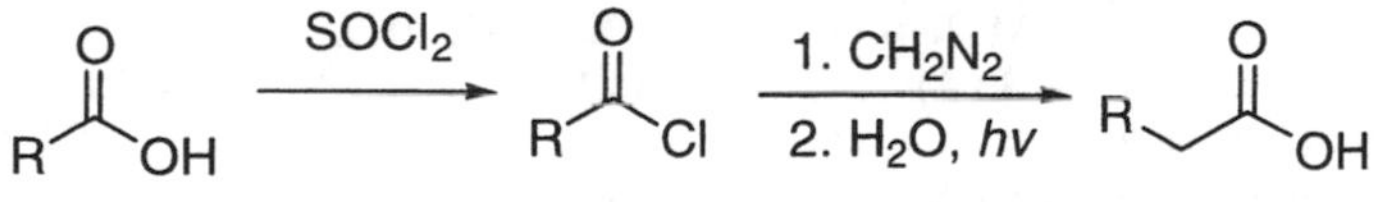

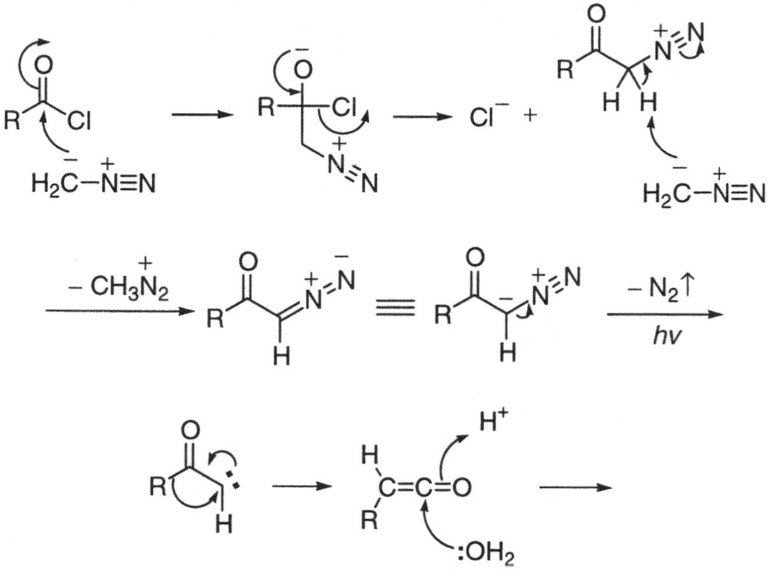

α-ketocarbene intermediate   ketene intermediate

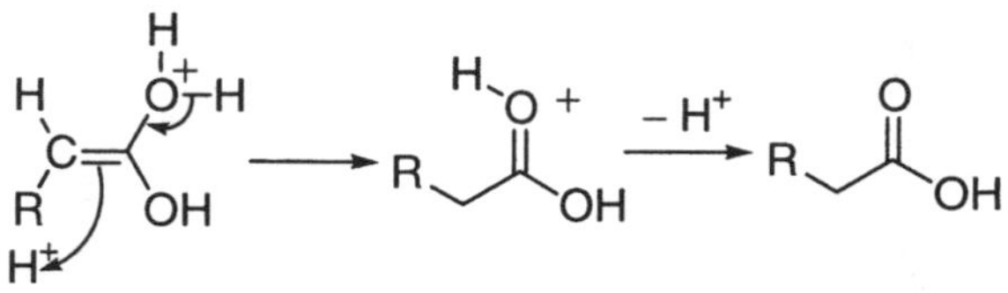

side reaction:

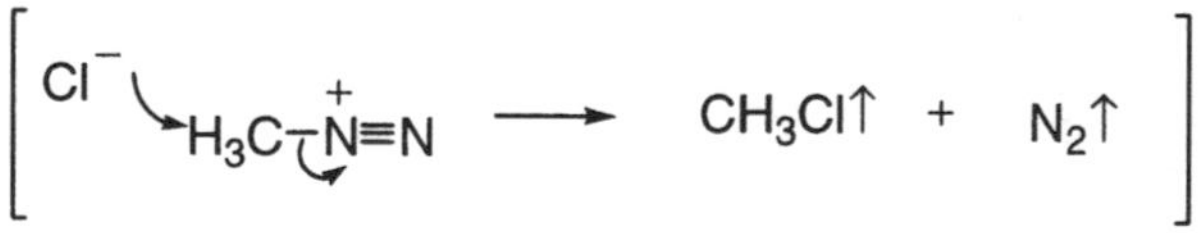

References

1. Arndt, F.; Eistert, B. *Ber. Dtsch. Chem. Ges.* **1935**, *68*, 200.
2. Kuo, Y. C.; Aoyama, T.; Shioiri, T. *Chem. Pharm. Bull.* **1982**, *30*, 899.

12

3. Podlech, J.; Seebach, D. *Angew. Chem., Int. Ed. Engl.* **1995**, *34*, 471.
4. Matthews, J. L.; Braun, C.; Guibourdenche, C.; Overhand, M.; Seebach, D. *Enantiose-lective Synthesis of β-Amino Acids* **1997**, pp105–126.
5. Katritzky, A. R.; Zhang, S.; Fang, Y. *Org. Lett.* **2000**, *2*, 3789.
6. Cesar, J.; Sollner Dolenc, M. *Tetrahedron Lett.* **2001**, *42*, 7099.
7. Katritzky, A. R.; Zhang, S.; Mostafa Hussein, A. H.; Fang, Y.; Steel, P. J. *J. Org. Chem.* **2001**, *66*, 5606.
8. Vasanthakumar, G.-R.; Babu, V. V. S. *Synth. Commun.* **2002**, *32*, 651.
9. Chakravarty, P. K.; Shih, T. L.; Colletti, S. L.; Ayer, M. B.; Snedden, C.; Kuo, H.; Tyagarajan, S.; Gregory, L.; Zakson-Aiken, M.; Shoop, W. L.; Schmatz, D. M.; Wyvratt, M.J.; Fisher, M. H.; Meinke, P. T. *Bioorg. Med. Chem. Lett.* **2003**, *13*, 147.

Auwers reaction

Conversion of coumarones to flavonols by treatment of 2-bromo-2-(α-bromobenzyl)coumarones with alcoholic alkali.

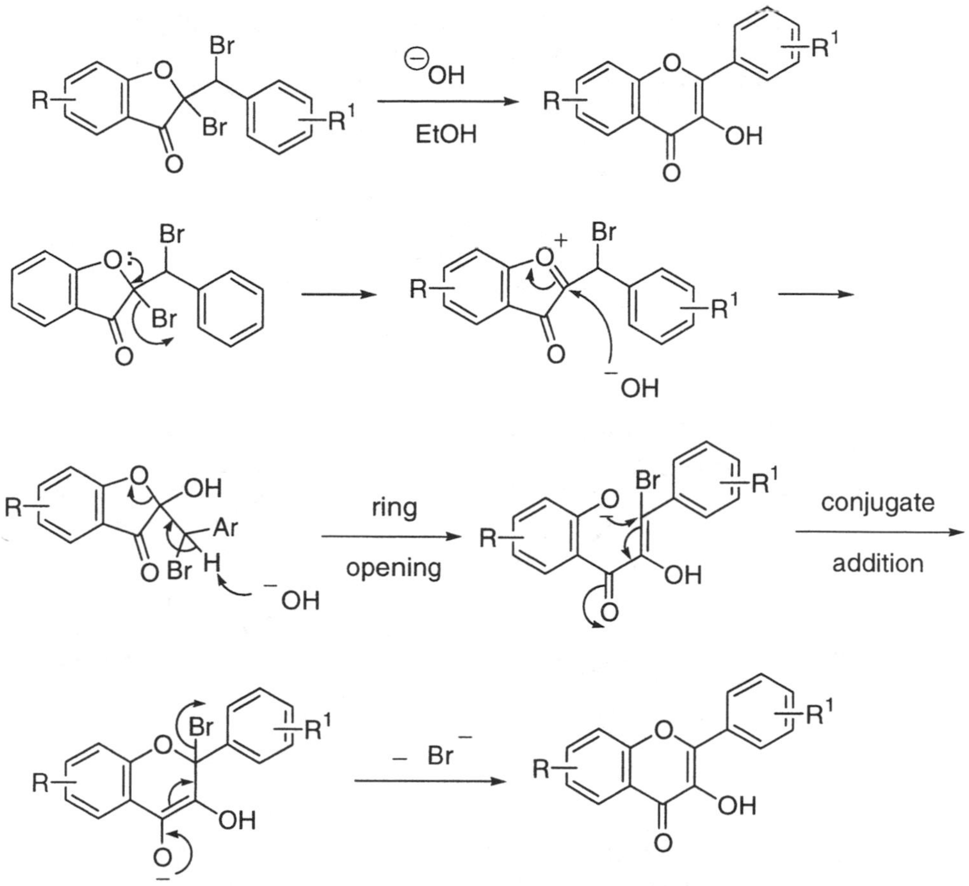

References

1. Auwers, K. *Ber. Dtsch. Chem. Ges.* **1908**, *41*, 4233.
2. Minton, T. H.; Stephen, H. *J. Chem. Soc.* **1922**, *121*, 1598.
3. Ingham, B. H.; Henry, S.; Ronald, T. *J. Chem. Soc.* **1931**, 895.
4. Wawzonek, S. *Heterocyclic Compounds* **1951**, *2*, 245.
5. Philbin, E. M.; O'Sullivan, W. I. A.; Wheeler, T. S. *J. Chem. Soc.* **1954**, *245*, 4174.
6. Bird, C. W.; Cookson, R. C. *J. Org. Chem.* **1959**, *24*, 441.
7. Rozenberg, V. I.; Nikanorov, V. A.; Svitan'ko, Z. P.; Bakhmutov, V. I.; *Z. Org. Khimii* **1981**, *17*, 2009.

14

Baeyer–Drewson indigo synthesis

Applicable for the detection of *o*-nitrobenzaldehyde.

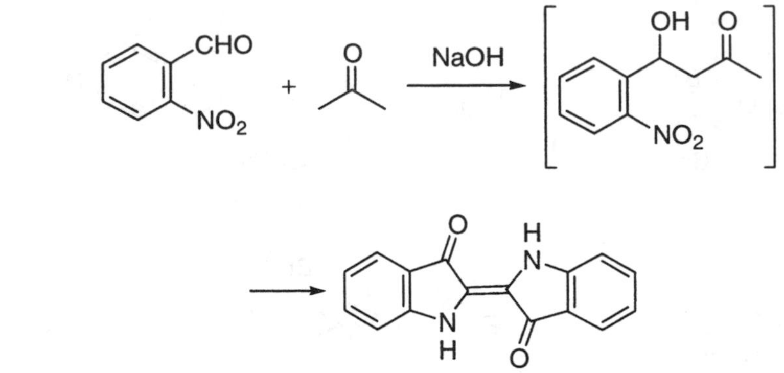

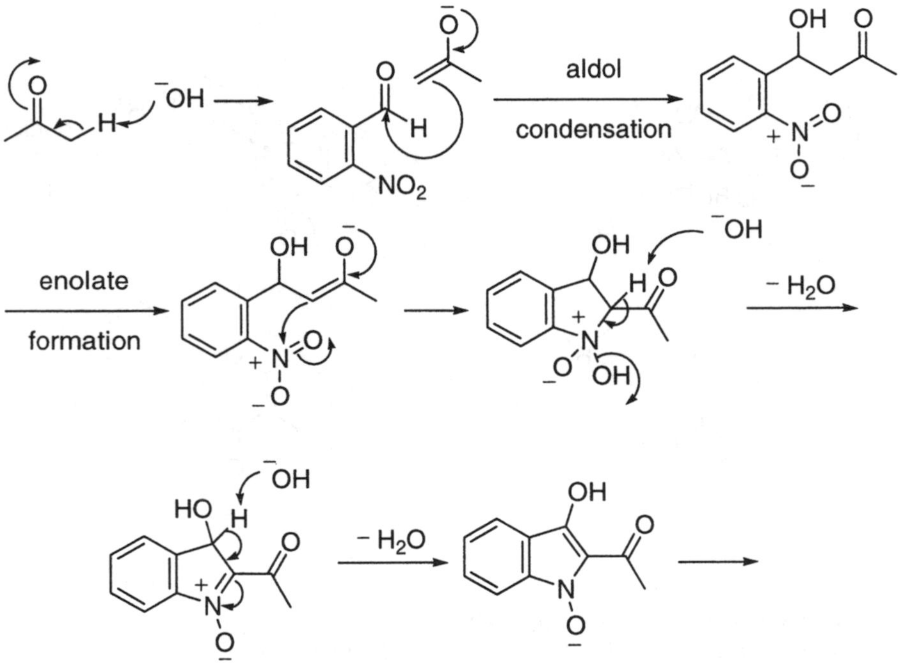

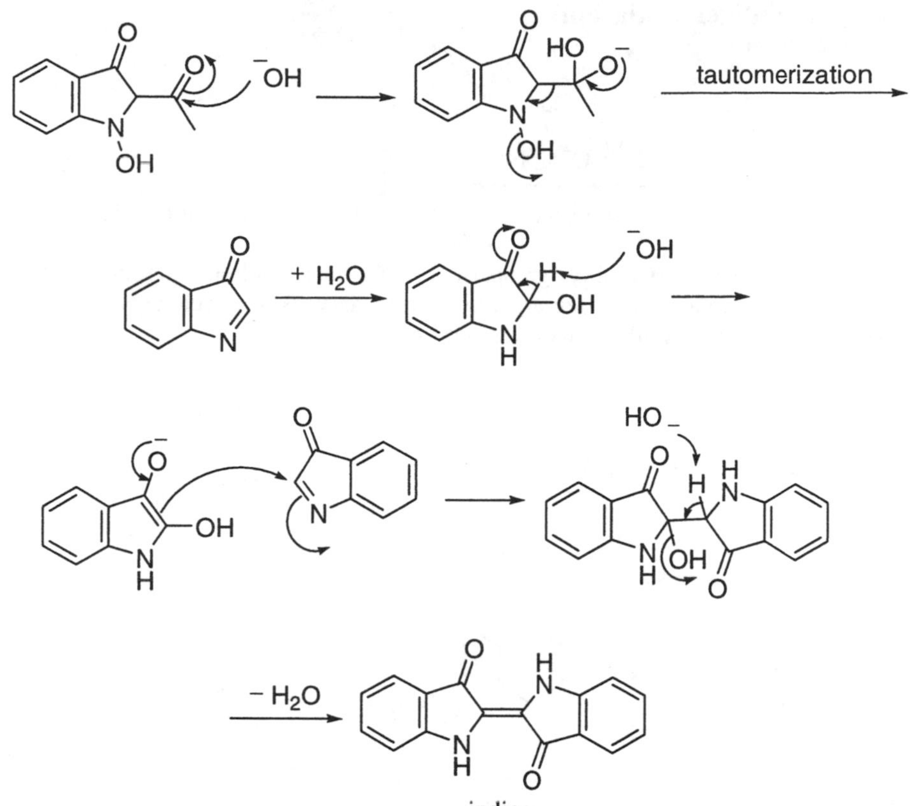

15

indigo

References

1. Baeyer, A.; Drewson, V. *Ber. Dtsch. Chem. Ges.* **1882**, *15*, 2856.
2. Friedlander, P.; Schenck, O. *Ber. Dtsch. Chem. Ges.* **1914**, *47*, 3040.
3. Hinkel, L. E.; Ayling, E. E. *J. Chem. Soc.* **1932**, 985.
4. Hassner, A.; Haddakin, M. *J. Tetrahedron Lett.* **1962**, 975.
5. Sainsbury, M. In *Rodd's Chemistry of Carbon Compounds IVB*, **1977**, 346. (Review).
6. Torii, S.; Yamanaka, T.; Tanaka, H. *J. Org. Chem.* **1978**, *43*, 2882.
7. Voss, G; Gerlach, H. *Chem. Ber.* **1989**, *122*, 1199.
8. McKee, J. R.; Zanger, M. *J. Chem. Educ.* **1991**, *68*, A242.

Baeyer–Villiger oxidation

General scheme:

The most electron-rich alkyl group (more substituted carbon) migrates first. The general migration order: tertiary alkyl > secondary alkyl > cyclohexyl > benzyl > phenyl > primary alkyl > methyl >> H

e.g.:

References

1. v. Baeyer, A.; Villiger, V. *Ber. Dtsch. Chem. Ges.* **1899**, *32*, 3625.
2. Know, G. R. *Tetrahedron* **1981**, *37*, 2697.
3. Krow, G. R. *Org. React.* **1993**, *43*, 251. (Review).
4. Renz, M.; Meunier, B. *Eur. J. Org. Chem.* **1999**, *4*, 737.
5. Bolm, C.; Beckmann, O. *Compr. Asymmetric Catal. I-III* **1999**, *2*, 803. (Review).
6. Crudden, C. M.; Chen, A. C.; Calhoun, L. A. *Angew. Chem., Int. Ed.* **2000**, *39*, 2851.
7. Fukuda, O.; Sakaguchi, S.; Ishii, Y. *Tetrahedron Lett.* **2001**, *42*, 3479.
8. Watanabe, A.; Uchida, T.; Ito, K.; Katsuki, T. *Tetrahedron Lett.* **2002**, *43*, 4481.
9. Kobayashi, S.; Tanaka, H.; Amii, H.; Uneyama, K. *Tetrahedron* **2003**, *59*, 1547.

Baker–Venkataraman rearrangement

Base-catalyzed acyl transfer reaction that converts α-acyloxyketones to β-diketones.

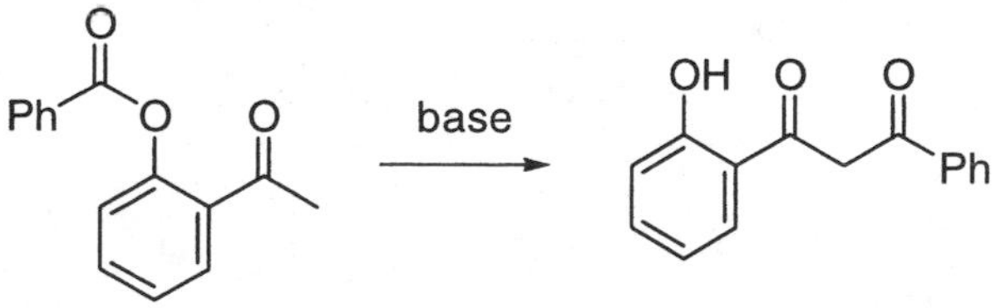

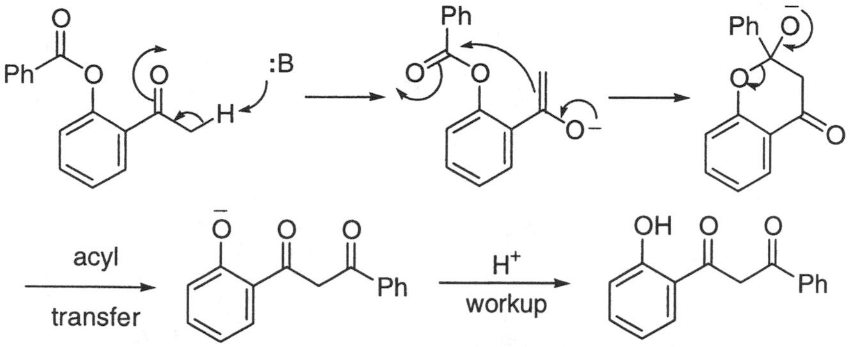

References

1. Baker, W. *J. Chem. Soc.* **1933**, 1381.
2. Kraus, G. A.; Fulton, B. S.; Wood, S. H. *J. Org. Chem.* **1984**, *49*, 3212.
3. Bowden, K.; Chehel-Amiran, M. *J. Chem. Soc., Perkin Trans. 2* **1986**, 2039.
4. Makrandi, J. K.; Kumari, V. *Synth. Commun.* **1989**, *19*, 1919.
5. Reddy, B. P.; Krupadanam, G. L. D. *J. Heterocycl. Chem.* **1996**, *33*, 1561.
6. Kalinin, A. V.; Snieckus, V. *Tetrahedron Lett.* **1998**, *39*, 4999.
7. Pinto, D. C. G. A.; Silva, A. M. S.; Cavaleiro, J. A. S. *New J. Chem.* **2000**, *24*, 85.
8. Thasana, N.; Ruchirawat, S. *Tetrahedron Lett.* **2002**, *43*, 4515.

Bamberger rearrangement

Acid-mediated rearrangement of *N*-phenylhydroxylamine to 4-aminophenol.

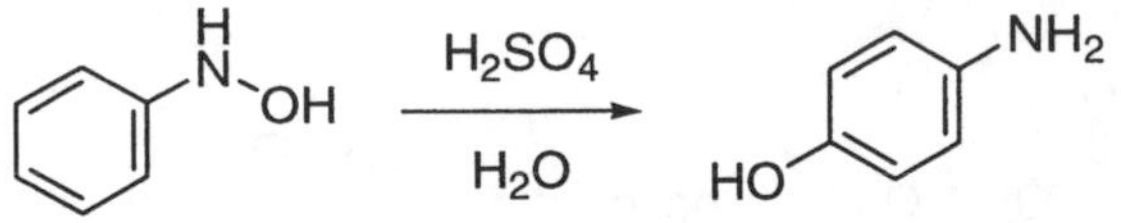

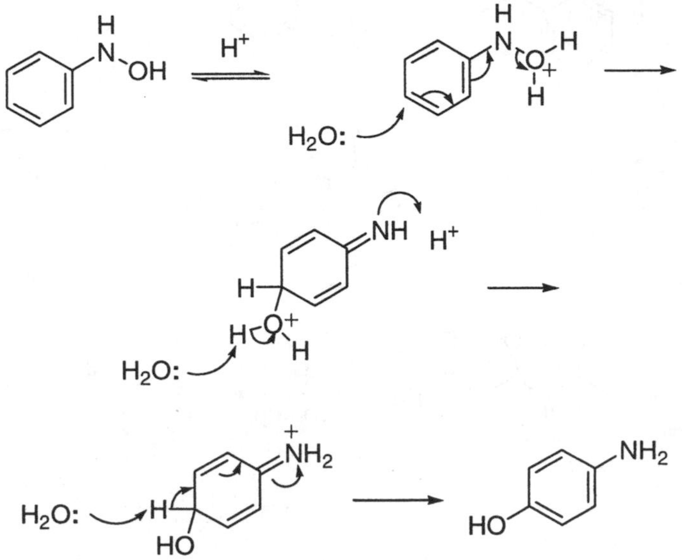

References

1. Bamberger, E. *Ber. Dtsch. Chem. Ges.* **1894**, *27*, 1548.
2. Shine, H. J. In *Aromatic Rearrangement;* Elsevier: New York, **1967**, pp 182–190. (Review).
3. Sone, T.; Tokuda, Y.; Sakai, T.; Shinkai, S.; Manabe, O. *J. Chem. Soc., Perkin Trans. 2* **1981**, 298.
4. Fishbein, J. C.; McClelland, R. A. *J. Am. Chem. Soc.* **1987**, *109*, 2824.
5. Zoran, A.; Khodzhaev, O.; Sasson, Y. *J. Chem. Soc., Chem. Commun.* **1994**, 2239.
6. Fishbein, J. C.; McClelland, R. A. *Can. J. Chem.* **1996**, *74*, 1321.
7. Naicker, K. P.; Pitchumani, K.; Varma, R. S. *Catal. Lett.* **1999**, *58*, 167.
8. Pirrung, M. C.; Wedel, M.; Zhao, Y. *Synlett* **2002**, 143.

Bamford–Stevens reaction

The Bamford–Stevens reaction and the Shapiro reaction share a similar mechanistic pathway. The former uses a base such as Na, NaOMe, LiH, NaH, NaNH$_2$, *etc.*, whereas the latter employs bases such as alkyllithiums and Grignard reagents. As a result, the Bamford–Stevens reaction furnishes more-substituted olefins as the thermodynamic products, while the Shapiro reaction generally affords less-substituted olefins as the kinetic products.

In protic solvent:

In aprotic solvent:

References

1. Bamford, W. R.; Stevens, T. S. M. *J. Chem. Soc.* **1952**, 4735.
2. Casanova, J.; Waegell, B. *Bull. Soc. Chim. Fr.* **1975**, *3–4(Pt. 2)*, 922.
3. Shapiro, R. H. *Org. React.* **1976**, *23*, 405. (Review).
4. Adlington, R. M.; Barrett, A. G. M. *Acc. Chem. Res.* **1983**, *16*, 55. (Review).
5. Sarkar, T. K.; Ghorai, B. K. *J. Chem. Soc., Chem. Commun.* **1992**, *17,* 1184.
6. Nickon, A.; Stern, A. G.; Ilao, M. C. *Tetrahedron Lett.* **1993**, *34*, 1391.
7. Olmstead, K. K.; Nickon, A. *Tetrahedron* **1998**, *54*, 12161.
8. Olmstead, K. K.; Nickon, A. *Tetrahedron* **1999**, *55*, 7389.
9. Khripach, V. V.; Zhabinskii, V. N.; Kotyatkina, A. I.; Lyakhov, A. S.; Fando, G. P.; Govorova, A. A.; van de Louw, J.; Groen, M. B.; de Groot, A. *Mendeleev Commun.* **2001**, *4,* 144.
10. May, J. A.; Stoltz, B. M. *J. Am. Chem. Soc.* **2002**, *124*, 12426.

Bargellini reaction

Synthesis of hindered morpholinones and piperazinones from acetone and 2-amino-2-methyl-1-propanol or 1,2-diaminopropanes.

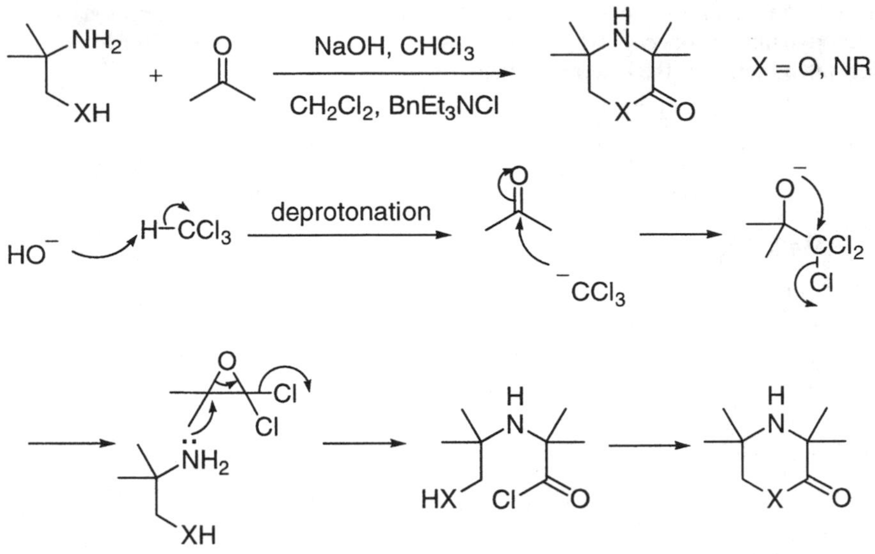

References

1. Bargellini, G. *Gazz. Chim. Ital.* **1906**, *36*, 329.
2. Lai, J. T. *J. Org. Chem.* **1980**, *45*, 754.
3. Lai, J. T. *Synthesis* **1981**, 754.
4. Lai, J. T. *Synthesis* **1984**, 122.
5. Lai, J. T. *Synthesis* **1984**, 124.
6. Rychnovsky, S. D.; Beauchamp, T.; Vaidyanathan, R.; Kwan, T. *J. Org. Chem.* **1998**, *63*, 6363.

Bartoli indole synthesis

7-Substituted indoles from the reaction of *ortho*-substituted nitroarenes and vinyl Grignard reagents.

nitroso intermediate

References

1. Bartoli, G.; Leardini, R.; Medici, A.; Rosini, G. *J. Chem. Soc., Perkin Trans. 1* **1978**, 892.
2. Bartoli, G.; Bosco, M.; Dalpozzo, R.; Todesco, P. E. *J. Chem. Soc., Chem. Commun.* **1988**, 807.
3. Bartoli, G.; Palmieri, G.; Bosco, M.; Dalpozzo, R. *Tetrahedron Lett.* **1989**, *30*, 2129.
4. Bosco, M.; Dalpozzo, R.; Bartoli, G.; Palmieri, G.; Petrini, M. *J. Chem. Soc., Perkin Trans. 2* **1991**, 657.
5. Bartoli, G.; Bosco, M.; Dalpozzo, R.; Palmieri, G.; Marcantoni, E. *J. Chem. Soc., Perkin Trans. 1* **1991**, 2757.
6. Dobson, D. R.; Gilmore, J.; Long, D. A. *Synlett* **1992**, 79.
7. Dobbs, A. P.; Voyle, M.; Whittall, N. *Synlett* **1999**, 1594.
8. Dobbs, A. *J. Org. Chem.* **2001**, *66*, 638.
9. Pirrung, M. C.; Wedel, M.; Zhao, Y. *Synlett* **2002**, 143.
10. Garg, N. K.; Sarporg, R.; Stoltz, B. M. *J. Am. Chem. Soc.* **2002**, *124*, 13179.

Barton decarboxylation

Radical decarboxylation *via* the corresponding thiocarbonyl derivatives of the carboxylic acids.

Barton ester

2,2'-azobisisobutyronitrile (AIBN)

$CO_2\uparrow$ + R•⌒⌒H⌢SnBu$_3$ ⟶ R^{-}H + Bu$_3$Sn •

References

1. Barton, D. H. R.; Crich, D.; Motherwell, W. B. *J. Chem. Soc., Chem. Commun.* **1983**, 939.
2. Barton, D. H. R.; Zard, S. Z. *Pure Appl. Chem.* **1986**, *58*, 675.
3. Barton, D. H. R.; Bridon, D.; Zard, S. Z. *Tetrahedron* **1987**, *43*, 2733.
4. Magnus, P.; Ladlow, M.; Kim, C. S.; Boniface, P. *Heterocycles* **1989**, *28*, 951.
5. Barton, D. H. R. *Aldrichimica Acta* **1990**, *23*, 3.
6. Gawronska, K.; Gawronski, J.; Walborsky, H. M. *J. Org. Chem.* **1991**, *56*, 2193.
7. Eaton, P. E.; Nordari, N.; Tsanaktsidis, J.; Upadhyaya, S. P. *Synthesis* **1995**, 501.
8. Crich, D.; Hwang, J.-T.; Yuan, H. *J. Org. Chem.* **1996**, *61*, 6189.
9. Elena, M.; Taddei, M. *Tetrahedron Lett.* **2001**, *42*, 3519.
10. Materson, D. S.; Porter, N. A. *Org. Lett.* **2002**, *4*, 4253.

Barton–McCombie deoxygenation

Deoxygenation of alcohols by means of radical scission of their corresponding thiocarbonyl derivatives.

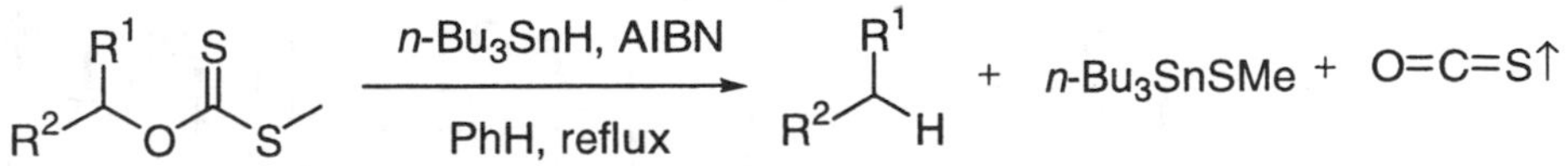

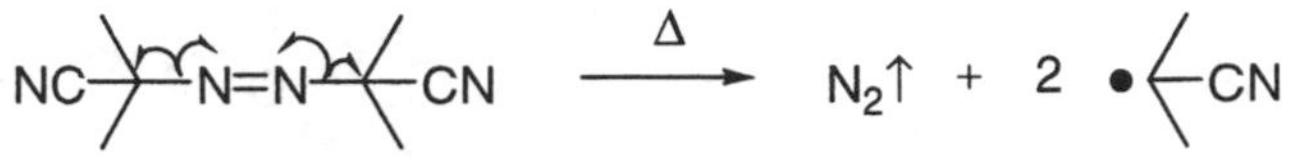

2,2'-azobisisobutyronitrile (AIBN)

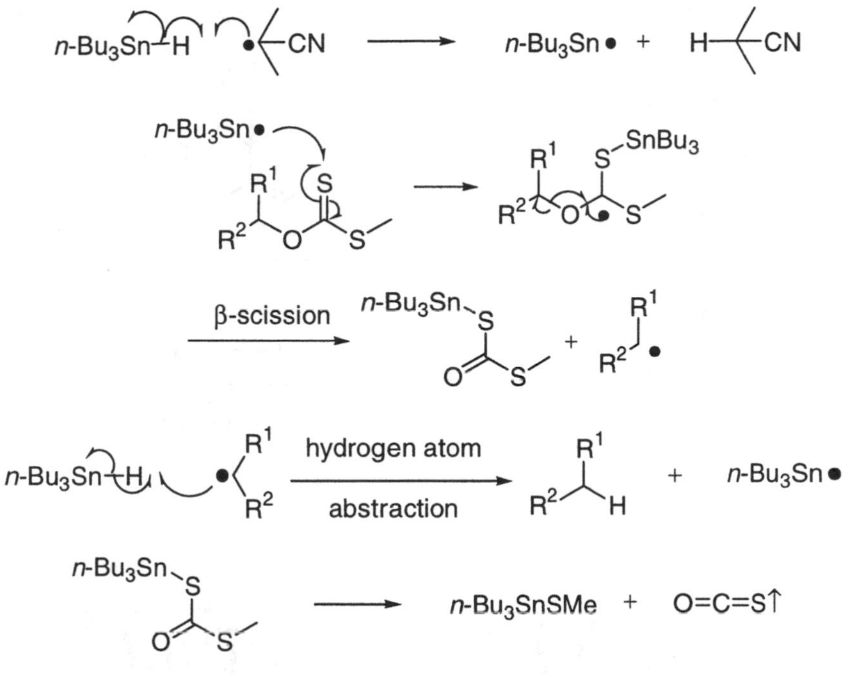

β-scission

hydrogen atom abstraction

References

1. Barton, D. H. R.; McCombie, S. W. *J. Chem. Soc., Perkin Trans. 1* **1975**, 1574.
2. Zard, S. Z. *Angew. Chem., Int. Ed. Engl.* **1997**, *36*, 672.
3. Lopez, R. M.; Hays, D. S.; Fu, G. C. *J. Am. Chem. Soc.* **1997**, *119*, 6949.
4. Hansen, H. I.; Kehler, J. *Synthesis* **1999**, 1925.
5. Cai, Y.; Roberts, B. P. *Tetrahedron Lett.* **2001**, *42*, 763.
6. Clive, D. L. J.; Wang, J. *J. Org. Chem.* **2002**, *67*, 1192.
7. Rhee, J. U.; Bliss, B. I.; RajanBabu, T. V. *J. Am. Chem. Soc.* **2003**, *125*, 1492.

Barton nitrite photolysis

Photolysis of a nitrite ester to a γ-oximino alcohol.

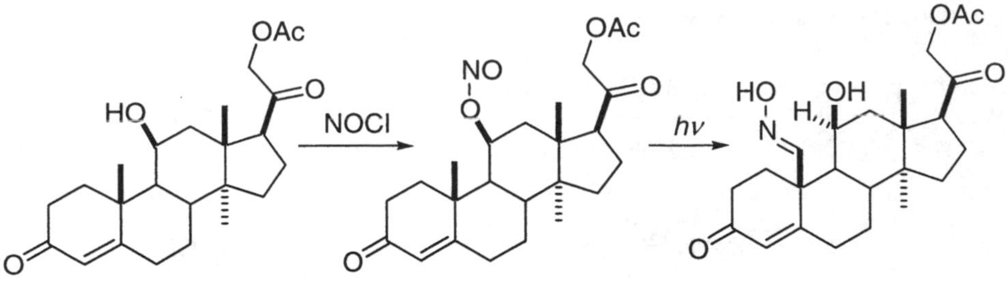

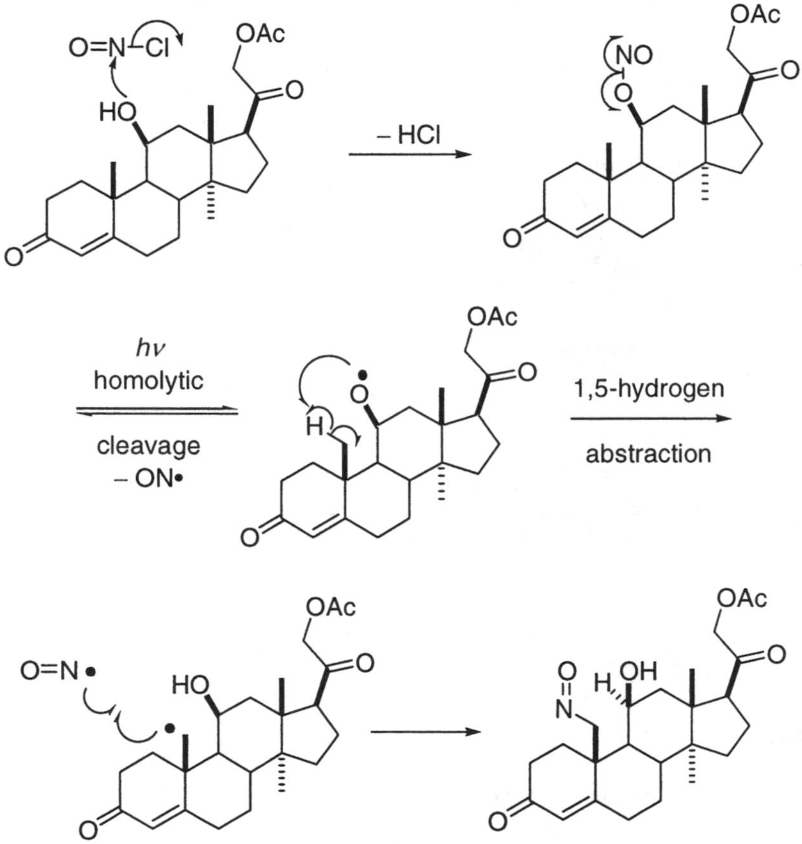

Nitric oxide radical is a stable, and therefore, long-lived radical

nitroso intermediate

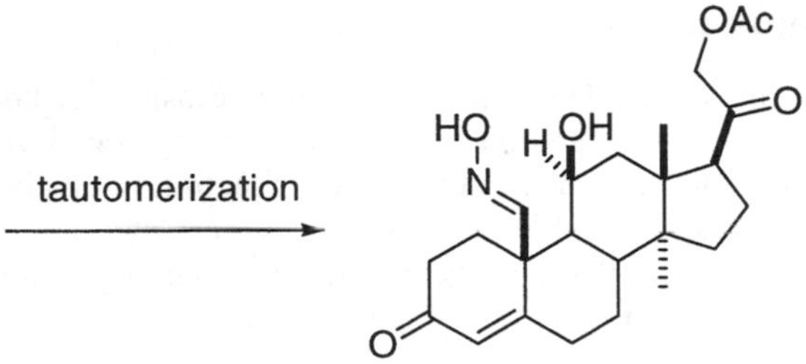

tautomerization

References

1. Barton, D. H. R.; Beaton, J. M.; Geller, L. E.; Pechet, M. M. *J. Am. Chem. Soc.* **1960**, *82*, 2640.
2. Barton, D. H. R.; Beaton, J. M. *J. Am. Chem. Soc.* **1960**, *82*, 2641.
3. Barton, D. H. R.; Beaton, J. M.; Geller, L. E.; Pechet, M. M. *J. Am. Chem. Soc.* **1961**, *83*, 4083.
4. Barton, D. H. R.; Hesse, R. H.; Pechet, M. M.; Smith, L. C. *J. Chem. Soc., Perkin Trans. 1* **1979**, 1159.
5. Barton, D. H. R. *Aldrichimica Acta* **1990**, *23*, 3. (Review).
6. Majetich, G.; Wheless, K. *Tetrahedron* **1995**, *51*, 7095.
7. Herzog, A.; Knobler, C. B.; Hawthorne, M. F. *Angew. Chem., Int. Ed. Engl.* **1998**, *37*, 1552.

Baylis–Hillman reaction

Also known as Morita–Baylis–Hillman reaction, and occasionally known as Rau-hut–Currier reaction. It is a carbon-carbon bond-forming transformation of an electron-poor alkene with a carbon nucleophile. Electron-poor alkenes include acrylic esters, acrylonitriles, vinyl ketones, vinyl sulfones, and acroleins. On the other hand, carbon nucleophiles may be aldehydes, α-alkoxycarbonyl ketones, aldimines, and Michael acceptors.

General scheme:

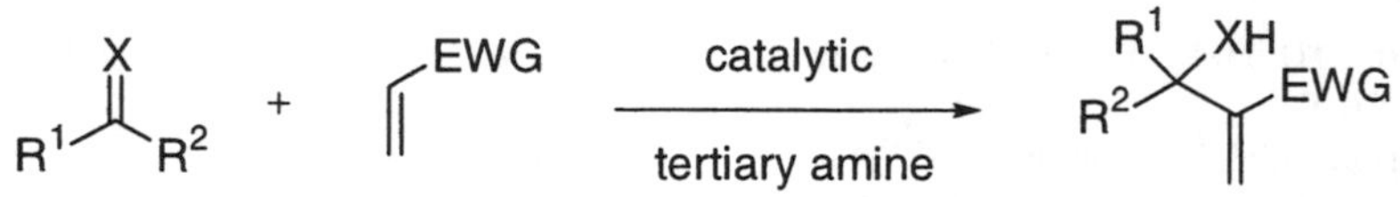

$X = O$, NR_2, $EWR = CO_2R$, COR, CHO, CN, SO_2R, SO_3R, $PO(OEt)_2$, $CONR_2$, $CH_2=CHCO_2Me$

e.g.:

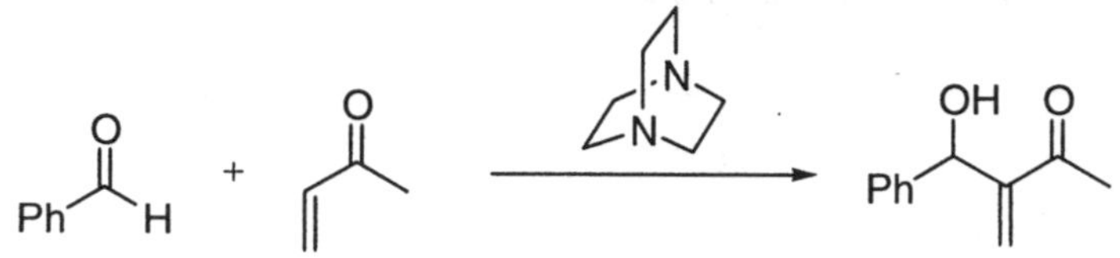

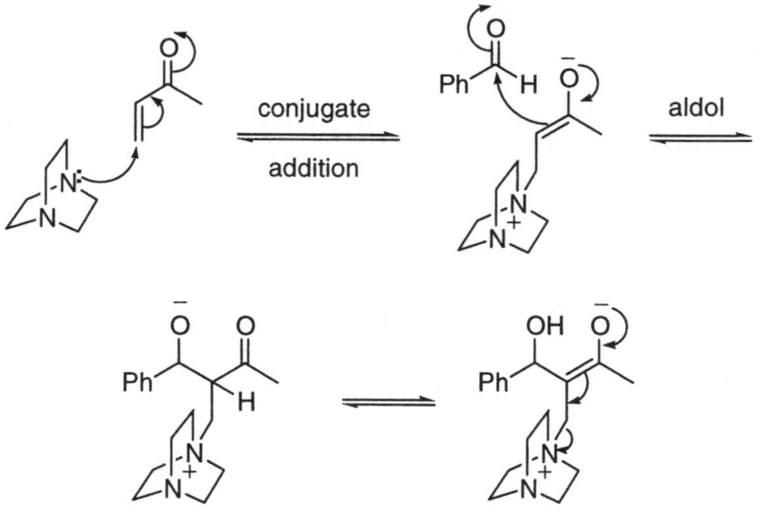

E2 (bimolecular elimination) mechanism is also operative here:

References

1. Baylis, A. B.; Hillman, M. E. D. Ger. Pat. 2,155,113, (**1972**).
2. Drewes, S. E.; Roos, G. H. P. *Tetrahedron* **1988**, *44*, 4653.
3. Basavaiah, D.; Rao, P. D.; Hyma, R. S. *Tetrahedron* **1996**, *52*, 8001.
4. Ciganek, E. *Org. React.* **1997**, *51,* 201. (Review).
5. Shi, M.; Feng, Y.-S. *J. Org. Chem.* **2001**, *66*, 406.
6. Kim, J. N.; Im, Y. J.; Gong, J. H.; Imaeda, K. *Tetrahedron Lett.* **2001**, *42*, 4195.
7. Shi, M.; Li, C.-Q.; Jiang, J.-K. *Helv. Chim. Acta* **2002**, *85*, 1051.
8. Yu, C.; Hu, L. *J. Org. Chem.* **2002**, *67*, 219.
9. Wang L.-C.; Luis A. L.; Agapiou K.; Jang H.-Y.; Krische M. J. *J. Am. Chem. Soc.* **2002**, *124*, 2402.
10. Frank, S. A.; Mergott, D. J.; Roush, W. R. *J. Am. Chem. Soc.* **2002**, *124*, 2404.
11. Shi, M.; Li, C.-Q.; Jiang, J.-K. *Tetrahedron* **2003**, *59*, 1181.

Beckmann rearrangement

Acid-mediated isomerization of oximes to amides.

the substituent *trans* to the leaving group migrates

References

1. Beckmann, E. *Chem. Ber.* **1886**, *89*, 988.
2. Chatterjea, J. N.; Singh, K. R. R. P. *J. Indian Chem. Soc.* **1982**, *59*, 527.
3. Gawley, R. E. *Org. React.* **1988**, *35*, 1. (Review).
4. Catsoulacos, P.; Catsoulacos, D. *J. Heterocycl. Chem.* **1993**, *30*, 1.
5. Anilkumar, R.; Chandrasekhar, S. *Tetrahedron Lett.* **2000**, *41*, 7235.
6. Khodaei, M. M.; Meybodi, F. A.; Rezai, N.; Salehi, P. *Synth. Commun.* **2001**, *31*, 2047.
7. Torisawa, Y.; Nishi, T.; Minamikawa, J.-i. *Bioorg. Med. Chem. Lett.* **2002**, *12*, 387.
8. Sharghi, H.; Hosseini, M. *Synthesis* **2002**, 1057.
9. Chandrasekhar, S.; Copalaiah, K. *Tetrahedron Lett.* **2003**, *44*, 755.

Beirut reaction

Synthesis of quinoxaline-1,4-dioxide from benzofurazan oxide.

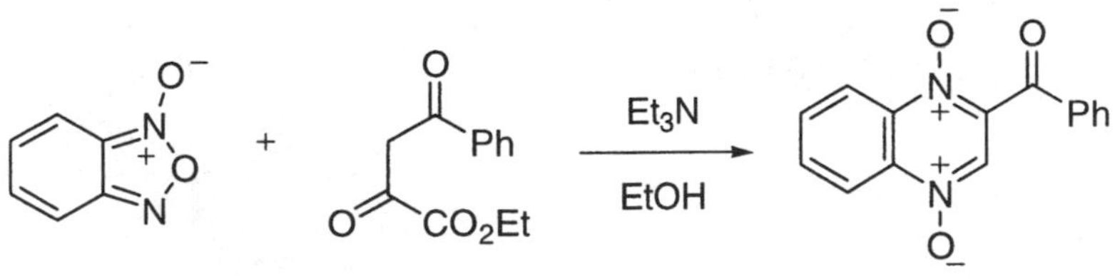

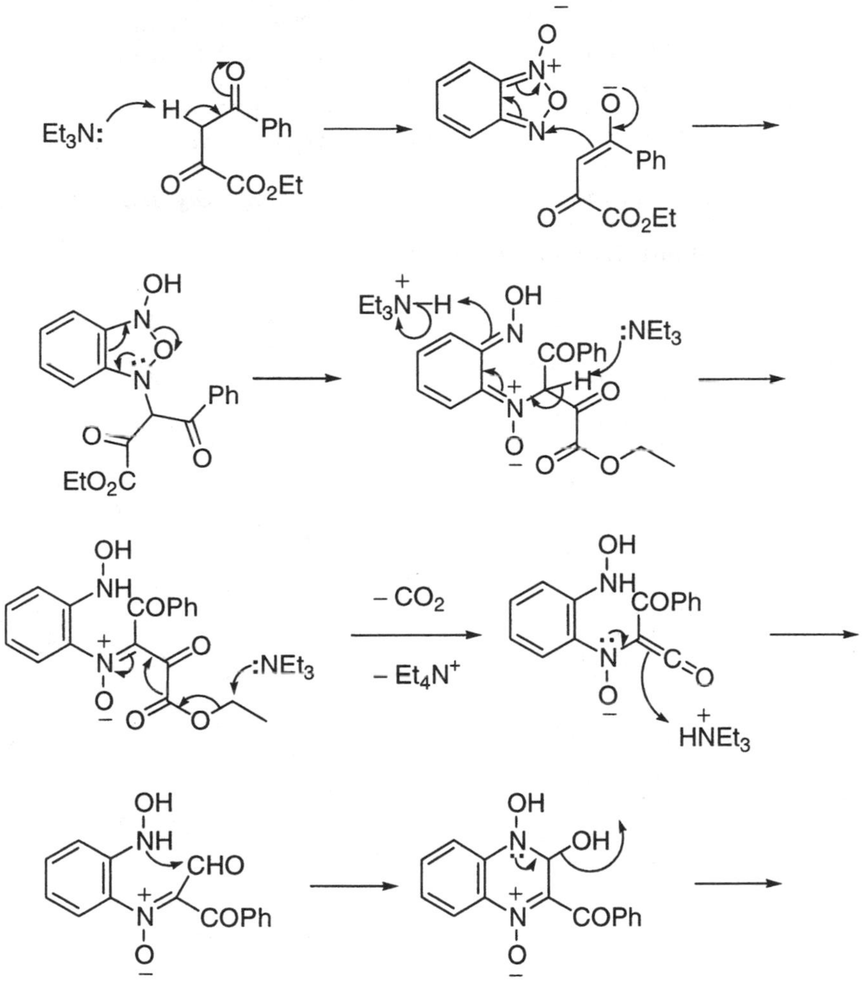

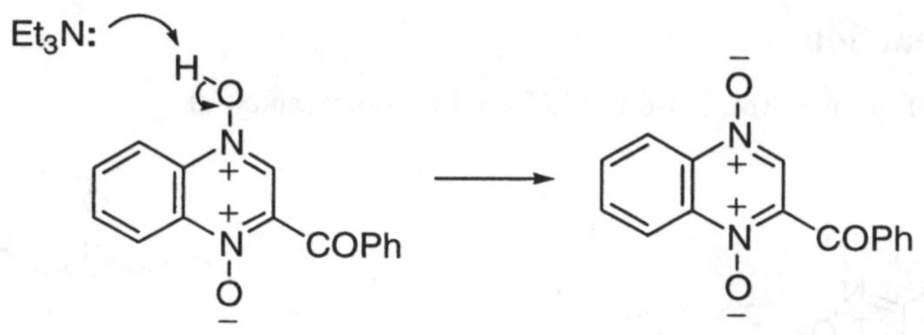

30

References

1. Haddadin, M. J.; Issidorides, C. H. *Heterocycles* **1976**, *4*, 767.
2. Gaso, A.; Boulton, A. J. In *Advances in Heterocycl. Chem.;* Vol. 29, Katritzky, A. R.; Boulton, A. J., eds.; Academic Press Inc.: New York, **1981**, 251. (Review).
3. Atfah, A.; Hill, J. *J. Chem. Soc., Perkin Trans. 1* **1989**, 221.
4. Haddadin, M. J.; Issidorides, C. H. *Heterocycles* **1993**, *35*, 1503.
5. El-Abadelah, M. M.; Nazer, M. Z.; El-Abadla, N. S.; Meier, H. *Heterocycles* **1995**, *41*, 2203.
6. Panasyuk, P. M.; Mel'nikova, S. F.; Tselinskii, I. V. *Russ. J. Org. Chem.* **2001**, *37*, 892.
7. Turker, L.; Dura, E. *Theochem* **2002**, *593*, 143.

Benzilic acid rearrangement

Rearrangement of benzil to benzylic acid *via* aryl migration.

Final deprotonation of the carboxylic acid drives the reaction forward.

References

1. Zinin, N. *Justus Liebigs Ann. Chem.* **1839**, *31*, 329.
2. Rajyaguru, I.; Rzepa, H. S. *J. Chem. Soc., Perkin Trans. 2* **1987**, 1819.
3. Toda, F.; Tanaka, K.; Kagawa, Y.; Sakaino, Y. *Chem. Lett.* **1990**, 373.
4. Robinson, J.; Flynn, E. T.; McMahan, T. L.; Simpson, S. L.; Trisler, J. C.; Conn, K. B. *J. Org. Chem.* **1991**, *56*, 6709.
5. Hatsui, T.; Wang, J.-J.; Ikeda, S.-y.; Takeshita, H. *Synlett* **1995**, 35.
6. Yu, H.-M.; Chen, S.-T.; Tseng, M.-J.; Chen, S.-T.; Wang, K.-T. *J. Chem. Res., (S)* **1999**, 62.
7. Zhang, K.; Corrie, J. E. T.; Munasinghe, V. R. N.; Wan, P. *J. Am. Chem. Soc.* **1999**, *121*, 5625
8. Fohlisch, B.; Radl, A.; Schwetzler-Raschke, R.; Henkel, S. *Eur. J. Org. Chem.* **2001**, 4357.

32

Benzoin condensation

Cyanide-catalyzed condensation of aryl aldehyde to benzoin.

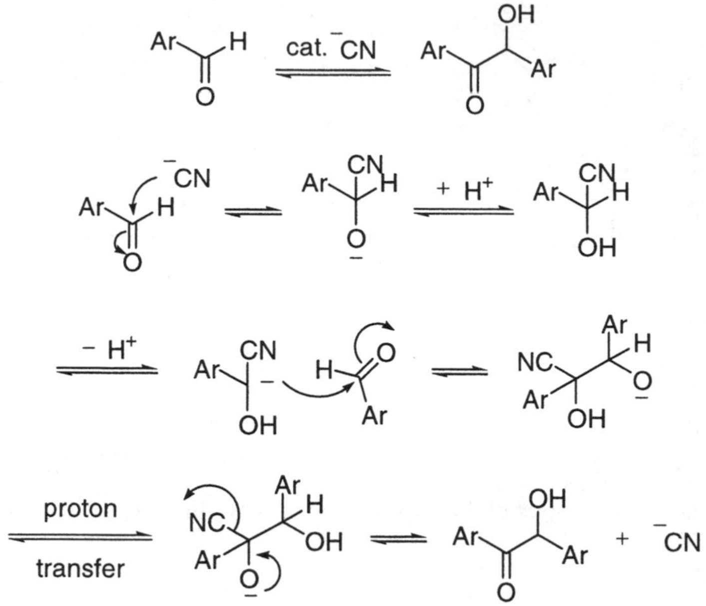

References

1. Lapworth, A. J. *J. Chem. Soc.* **1903**, *83*, 995.
2. Kluger, R. *Pure Appl. Chem.* **1997**, *69*, 1957.
3. Demir, A. S.; Dunnwald, T.; Iding, H.; Pohl, M.; Muller, M. *Tetrahedron: Asymmetry* **1999**, *10*, 4769.
4. Davis, J. H., Jr.; Forrester, K. J. *Tetrahedron Lett.* **1999**, *40*, 1621.
5. White, M. J.; Leeper, F. J. *J. Org. Chem.* **2001**, *66*, 5124.
6. Enders, D.; Kallfass, U. *Angew. Chem., Int. Ed.* **2002**, *41*, 1743.
7. Duenkelmann, P.; Kolter-Jung, D.; Nitsche, A.; Demir, A. S.; Siegert, P.; Lingen, B.; Baumann, M.; Pohl, M.; Mueller, M. *J. Am. Chem. Soc.* **2002**, *124*, 12084.

Bergman cyclization

1,4-Benzenediyl diradical formation from enediyne *via* electrocyclization.

enediyne   1,4-benzenediyl diradical

References

1. Jones, R. R.; Bergman, R. G. *J. Am. Chem. Soc.* **1972**, *94*, 660.
2. Bergman, R. G. *Acc. Chem. Res.* **1973**, *6*, 25. (Review).
3. Evenzahav, A.; Turro, N. J. *J. Am. Chem. Soc.* **1998**, *120*, 1835.
4. McMahon, R. J.; Halter, R. J.; Fimmen, R. L.; Wilson, R. J.; Peebles, S. A.; Kuczkowski, R. L.; Stanton, J. F. *J. Am. Chem. Soc.* **2000**, *122*, 939.
5. Rawat, D. S.; Zaleski, J. M. *Chem. Commun.* **2000**, 2493.
6. Clark, A. E.; Davidson, E. R.; Zaleski, J. M. *J. Am. Chem. Soc.* **2001**, *123*, 2650.
7. Alabugin, I. V.; Manoharan, M.; Kovalenko, S. V. *Org. Lett.* **2002**, *4*, 1119.
8. Stahl, F.; Moran, D.; Schleyer, P. von R.; Prall, M.; Schreiner, P. R. *J. Org. Chem.* **2002**, *67*, 1453.
9. Eshdat, L.; Berger, H.; Hopf, H.; Rabinovitz, M. *J. Am. Chem. Soc.* **2002**, *124*, 3822.
10. Feng, L.; Kumar, D.; Kerwin, S. M. *J. Org. Chem.* **2003**, *68*, 2234.

Biginelli pyrimidone synthesis

One-pot condensation of an aromatic aldehyde, urea, and ethyl acetoacetate in acidic ethanolic solution and expansion of such a condensation thereof.

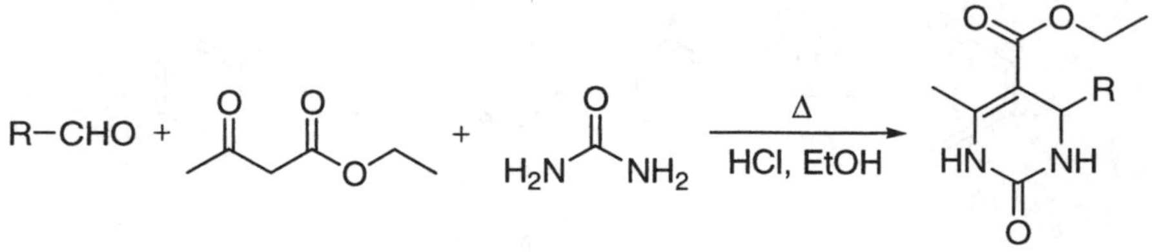

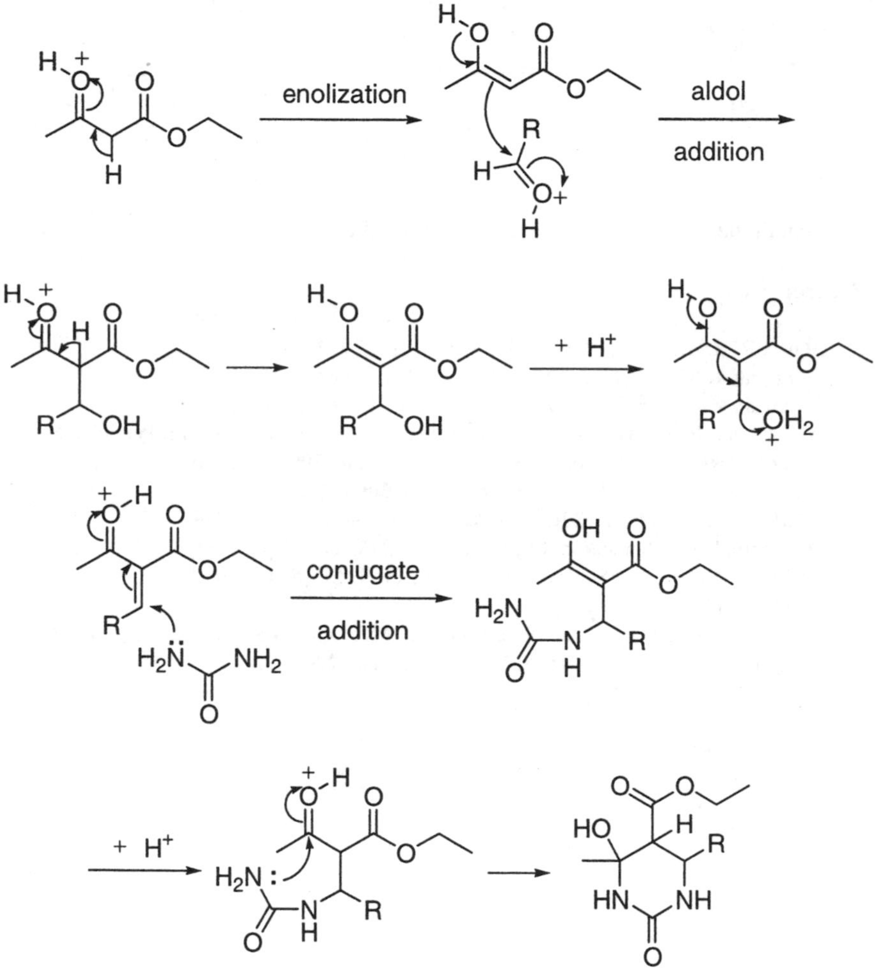

References

1. Biginelli, P. *Ber. Dtsch. Chem. Ges.* **1891**, *24*, 1317.
2. Sweet, F.; Fissekis, J. D. *J. Am. Chem. Soc.* **1973**, *95*, 8741.
3. Kappe, C. O. *Tetrahedron* **1993**, *49*, 6937. (Rev iew).
4. Kappe, C. O. *Acc. Chem. Res.* **2000**, *33*, 879. (Rev iew).
5. Kappe, C. O. *Eur. J. Med. Chem.* **2000**, *35*, 1043. (Rev iew).
6. Lu, J.; Bai, Y.; Wang, Z.; Yang, W.; Ma, H. *Tetrahedron Lett.* **2000**, *41*, 9075.
7. Garcia Valverde, M.; Dallinger, D.; Kappe, C. O. *Synlett* **2001**, 741.
8. Stadler, A.; Kappe, C. O. *J. Comb. Chem.* **2001**, *3*, 624.
9. Rani, V. R.; Srinivas, N.; Kishan, M. R.; Kulkarni, S. J.; Raghavan, K. V. *Green Chem.* **2001**, *3*, 305.
10. Lu, J.; Bai, Y. *Synthesis* **2002**, 466.
11. Perez, R.; Beryozkina, T.; Zbruyev, O. I.; Haas, W.; Kappe, C. O. *J. Comb. Chem.* **2002**, *4*, 501.
12. Varala, R.; Alam, M. M.; Adapa, S. R. *Synlett* **2003**, 67.
13. Martínez, S.; Meseguer, M.; Casas, L.; Rodríguez, E.; Molins, E.; Moreno-Mañas, M.; Roig, A.; Sebastián, R. M.; Valribera, A. *Tetrahedron* **2003**, *59*, 1553.

Birch reduction

Benzene ring bearing an electron-donating substituent:

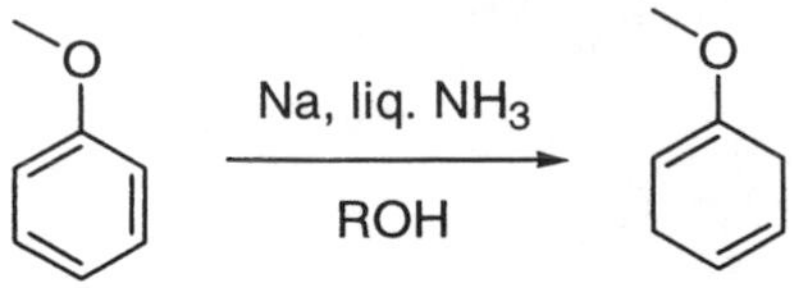

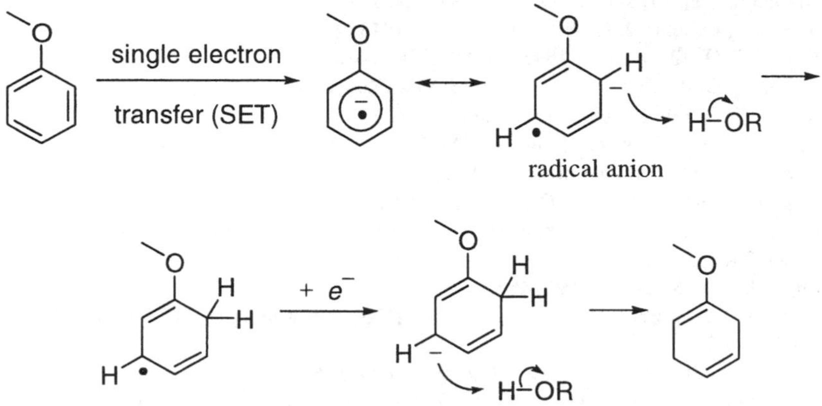

Benzene ring with an electron-withdrawing substituent:

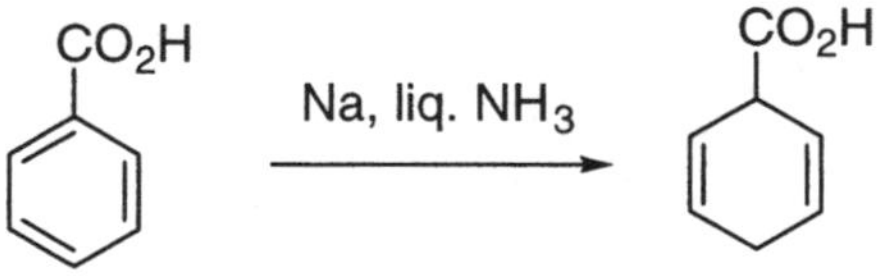

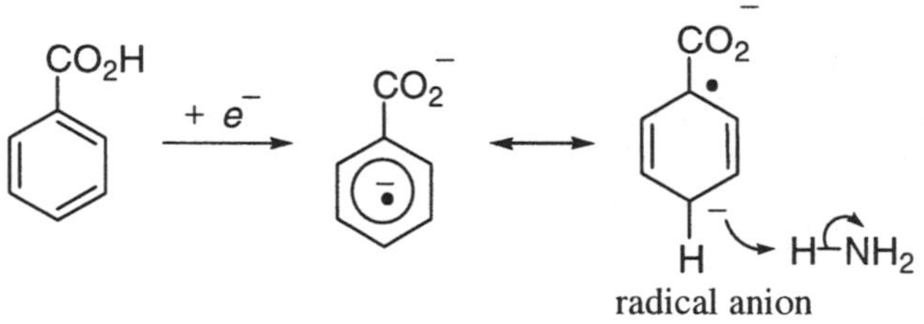

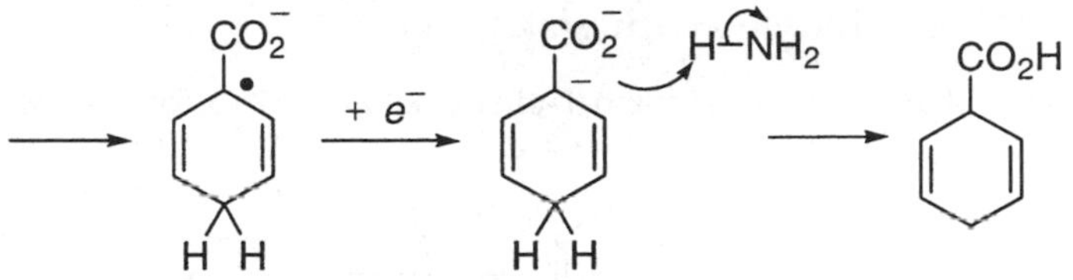

References

1. Birch, A. J. *J. Chem. Soc.* **1944**, 430.
2. Rabideau, P. W.; Marcinow, Z. *Org. React.* **1992**, *42*, 1–334. (Review).
3. Birch, A. J. *Pure Appl. Chem.* **1996**, *68*, 553.
4. Schultz, A. G. *Chem. Commun.* **1999**, 1263.
5. Ohta, Y.; Doe, M.; Morimoto, Y.; Kinoshita, T. *J. Heterocycl. Chem.* **2000**, *37*, 751.
6. Labadie, G. R.; Cravero, R. M.; Gonzalez-Sierra, M. *Synth. Commun.* **2000**, *30*, 4065.
7. Guo, Z.; Schultz, A. G. *J. Org. Chem.* **2001**, *66*, 2154.
8. Yamaguchi, S.; Hamade, E.; Yokoyama, H.; Hirai, Y.; Shiotani, S. *J. Heterocycl. Chem.* **2002**, *39*, 335.
9. Jiang, J.; Lai, Y.-H. *Tetrahedron Lett.* **2003**, *44*, 1271.

38

Bischler–Möhlau indole synthesis

Heating excess of aniline with 2-bromo-1-phenyl-ethanone to afford 2-arylindoles.

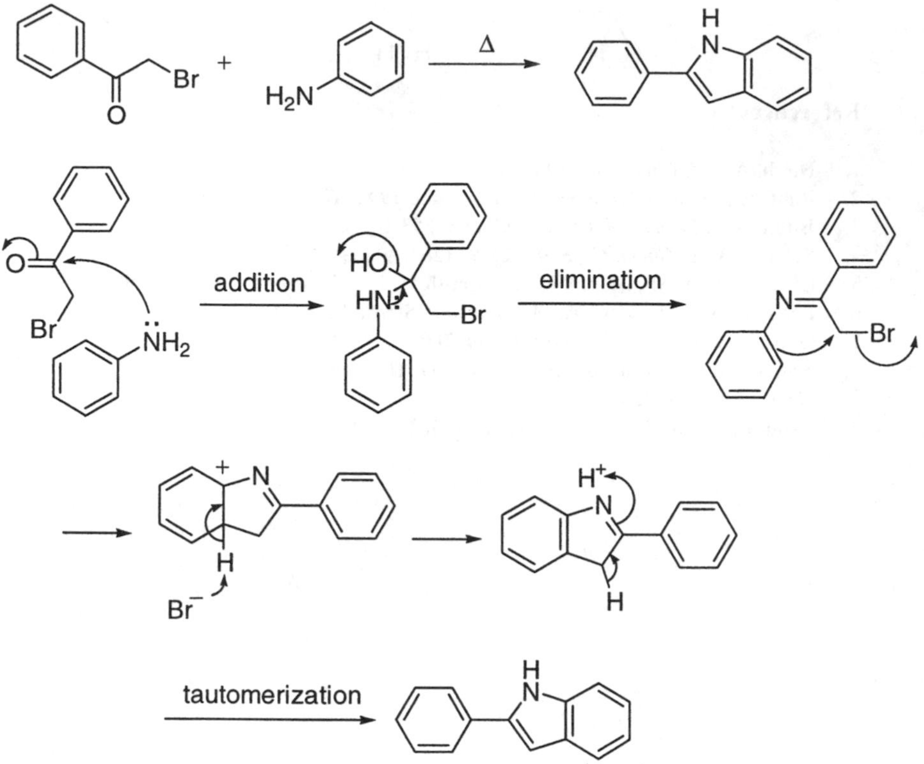

References

1. Möhlau, R. *Ber. Dtsch. Chem. Ges.* **1881**, *14*, 171.
2. Sundberg, R. J. *The Chemistry of Indoles* Academic Press: New York, **1970**, p 164. (Review).
3. Buu-Hoï, N. P.; Saint-Ruf, G.; Deschamps, D.; Bigot, P. *J. Chem. Soc. (C)* **1971**, 2606.
4. *The Chemistry of Heterocycl. Compounds, Indoles (Part 1),* Houlihan, W. J., ed.; Wiley & Sons: New York, 1972. (Review).
5. Bancroft, K. C. C.; Ward, T. J. *J. Chem. Soc., Perkin 1* **1974**, 1852.
6. Coic, J. P.; Saint-Ruf, G. *J. Heterocycl. Chem.* **1978**, *15*, 1367.
7. Henry, J. R.; Dodd, J. H. *Tetrahedron Lett.* **1998**, *39*, 8763.

Bischler–Napieralski reaction

Dihydroisoquinolines from β-phenethylamides using phosphorus oxychloride.

References

1. Bischler, A.; Napieralski, B. *Ber. Dtsch. Chem. Ges.* **1893**, *26*, 1903.
2. Fodor, G.; Nagubandi, S. *Heterocycles* **1981**, *15*, 165.
3. Rozwadowska, M. D. *Heterocycles* **1994**, *39*, 903.
4. Sotomayor, N.; Dominguez, E.; Lete, E. *J. Org. Chem.* **1996**, *61*, 4062.
5. Doi, S.; Shirai, N.; Sato, Y. *J. Chem. Soc., Perkin Trans. 1* **1997**, 2217.
6. Sanchez-Sancho, F.; Mann, E.; Herradon, B. *Synlett* **2000**, 509.
7. Ishikawa, T.; Shimooka, K.; Narioka, T.; Noguchi, S.; Saito, T.; Ishikawa, A.; Yamazaki, E.; Harayama, T.; Seki, H.; Yamaguchi, K. *J. Org. Chem.* **2000**, *65*, 9143.
8. Miyatani, K.; Ohno, M.; Tatsumi, K.; Ohishi, Y.; Kunitomo, J.-I.; Kawasaki, I.; Yamashita, M.; Ohta, S. *Heterocycles* **2001**, *55*, 589.
9. Nicoletti, M.; O'Hagan, D.; Slawin, A. M. Z. *J. Chem. Soc., Perkin Trans. 1* **2002**, 116.

Blaise reaction

β-Ketoesters from nitriles and α-haloesters using Zn.

References

1. Blaise, E. E. *C. R. Hebd. Seances Acad. Sci.* **1901**, *132*, 478.
2. Hannick, S. M.; Kishi, Y. *J. Org. Chem.* **1983**, *48*, 3833.
3. Krepski, L. R.; Lynch, L. E.; Heilmann, S. M.; Rasmussen, J. K. *Tetrahedron Lett.* **1985**, *26*, 981.
4. Beard, R. L.; Meyers, A. I. *J. Org. Chem.* **1991**, *56*, 2091.
5. Syed, J.; Forster, S.; Effenberger, F. *Tetrahedron: Asymmetry* **1998**, *9*, 805.
6. Narkunan, K.; Uang, B.-J. *Synthesis* **1998**, 1713.
7. Erian, A. W. *J. Prakt. Chem.* **1999**, *341*, 147.
8. Deutsch, H. M.; Ye, X.; Shi, Q.; Liu, Z.; Schweri, M. M. *Eur. J. Med. Chem.* **2001**, *36*, 303.
9. Creemers, A. F. L.; Lugtenburg, J. *J. Am. Chem. Soc.* **2002**, *124*, 6324.

Blanc chloromethylation reaction

Lewis acid-promoted chloromethyl group installation onto the aromatics rings with formalin and HCl.

formalin

References

1. Blanc, G. *Bull. Soc. Chim. Fr.* **1923**, *33*, 313.
2. Fuson, R. C.; McKeever, C. H. *Org. React.* **1942**, *1*, 63. (Review).
3. Olah, G.; Tolgyesi, W. S. In *Friedel-Crafts and Related Reactions* vol. II, Part 2, Olah, G., Ed.; Interscience: New York, **1963**, pp 659–784. (Review).
4. Sekine, Y.; Boekelheide, V. *J. Am. Chem. Soc.* **1981**, *103*, 1777.
5. Mallory, F. B.; Rudolph, M. J.; Oh, S. M. *J. Org. Chem.* **1989**, *54*, 4619.
6. Witiak, D. T.; Loper, J. T.; Ananthan, S.; Almerico, A. M.; Verhoef, V. L.; Filppi, J. A. *J. Med. Chem.* **1989**, *32*, 1636.
7. De Mendoza, J.; Nieto, P. M.; Prados, P.; Sanchez, C. *Tetrahedron* **1990**, *46*, 671.
8. Tashiro, M.; Tsuge, A.; Sawada, T.; Makishima, T.; Horie, S.; Arimura, T.; Mataka, S.; Yamato, T. *J. Org. Chem.* **1990**, *55*, 2404.
9. Miller, D. D.; Hamada, A.; Clark, M. T.; Adejare, A.; Patil, P. N.; Shams, G.; Romstedt, K. J.; Kim, S. U.; Intrasuksri, U.; *et al. J. Med. Chem.* **1990**, *33*, 1138.
10. Ito, K.; Ohba, Y.; Shinagawa, E.; Nakayama, S.; Takahashi, S.; Honda, K.; Nagafuji, H.; Suzuki, A.; Sone, T. *J. Heterocycl. Chem.* **2000**, *37*, 1479.
11. Qiao, K.; Deng, Y.-Q. *Huaxue Xuebao* **2003**, *61*, 133.

42

Boekelheide reaction

Treatment of 2-methylpyridine *N*-oxide with trifluoroacetic anhydride gives rise to 2-hydroxymethylpyridine.

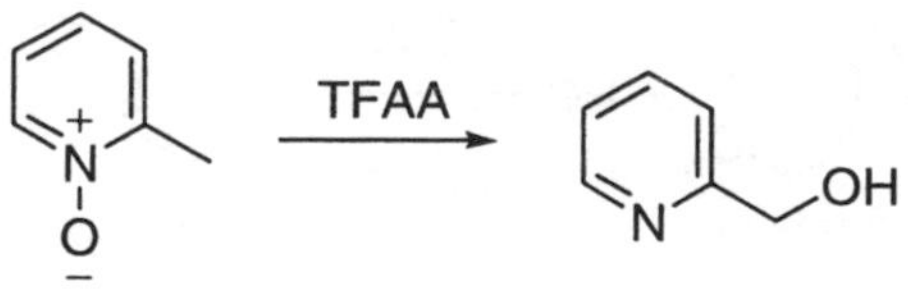

TFAA, trifluoroacetic anhydride

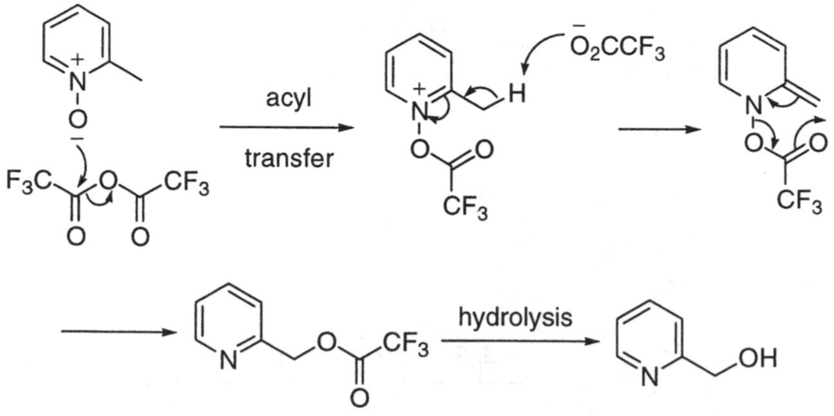

References

1. Bell, T. W.; Firestone, A. *J. Org. Chem.* **1986**, *51*, 764.
2. Newkome, G. R.; Theriot, K. J.; Gupta, V. K.; Fronczek, F. R.; Baker, G. R. *J. Org. Chem.* **1986**, *54*, 1766.
3. Goerlitzer, K.; Schmidt, E. *Arch. Pharm.* **1991**, *324*, 359.
4. Fontenas, C.; Bejan, E.; Haddon, H. A.; Balavoine, G. G. A. *Synth. Commun.* **1995**, *25*, 629.
5. Goerlitzer, K.; Bartke, U. *Pharmazie* **2002**, *57*, 804.
6. Higashibayashi, S.; Mori, T.; Shinko, K.; Hashimoto, K.; Nakata, M. *Heterocycles* **2002**, *57*, 111.

Boger pyridine synthesis

Pyridine synthesis *via* hetero–Diels–Alder reaction of 1,2,4-triazines and dieno-philes (e.g. enamine) followed by extrusion of N_2.

Retro-Diels-Alder
reaction, $- N_2$

Hetero-Diels-Alder
reaction

$- H_2O$

References

1. Boger, D. L.; Panek, J. S.; Meier, M. M. *J. Org. Chem.* **1982**, *47*, 895.
2. Boger, D. L.; Panek, J. S.; Yasuda, M. *Org. Synth.* **1988**, *66*, 142.
3. Boger, D. L. In *Comprehensive Organic Synthesis;* Trost, B. M.; Fleming, I., Eds.; Pergamon, **1991**, *Vol. 5*, 451–512. (Review).
4. Golka, A.; Keyte, P. J.; Paddon-Row, M. N. *Tetrahedron* **1992**, *48*, 7663.
5. Sauer, J.; Heldmann, D. K.; Pabst, G. R. *Eur. J. Org. Chem.* **1999**, 313.
6. Rykowski, A.; Olender, E.; Branowska, D.; Van der Plas, H. C. *Org. Prep. Proced. Int.* **2001**, *33*, 501.
7. Stanforth, S. P.; Tarbit, B.; Watson, M. D. *Tetrahedron Lett.* **2003**, *44*, 693.

44

Boord reaction

Olefin from the treatment of β-halo-alkoxide with zinc.

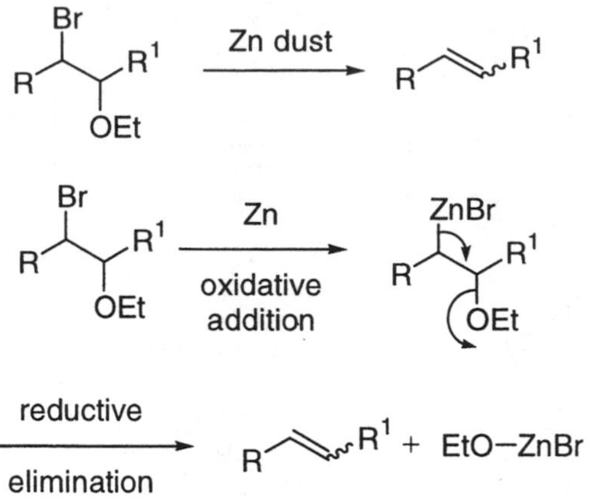

References

1. Swallen, L. C.; Boord, C. E. *J. Am. Chem. Soc.* **1930**, *52*, 651.
2. Schmitt, Claude G.; Boord, Cecil E. *J. Am. Chem. Soc.* **1931**, *53*, 2427.
3. Seifert, H. *Monatsh.* **1948**, *79*, 198.
4. Hatch, C. E., III; Baum, J. S.; Takashima, T.; Kondo, K. *J. Org. Chem.* **1980**, *45*, 3181.
5. Halton, B.; Russell, S. G. G. *J. Org. Chem.* **1991**, *56*, 5553.
6. Yadav, J. S.; Ravishankar, R.; Lakshman, S. *Tetrahedron Lett.* **1994**, *35*, 3617.
7. Yadav, J. S.; Ravishankar, R.; Lakshman, S. *Tetrahedron Lett.* **1994**, *35*, 3621.
8. Beusker, P. H.; Aben, R. W. M.; Seerden, J.-P. G.; Smits, J. M. M.; Scheeren, H. W. *Eur. J. Org. Chem.* **1998**, 2483.

Borsche–Drechsel cyclization

Tetrahydrocarbazole synthesis from cyclohexanone phenylhydrazone.
Cf. Fisher indole synthesis.

References

1. Drechsel, E. *J. Prakt. Chem.* **1858**, *38*, 69.
2. Atkinson, C. M.; Biddle, B. N. *J. Chem. Soc. (C)* **1966**, 2053.
3. Bruck, P. *J. Org. Chem.* **1970**, *35*, 2222.
4. Gazengel, J. M.; Lancelot, J. C.; Rault, S.; Robba, M. *J. Heterocycl. Chem.* **1990**, *27*, 1947.
5. Abramovitch, R. A.; Bulman, A. *Synlett* **1992**, 795.
6. Murakami, Y.; Yokooa, H.; Watanabe, T. *Heterocycles* **1998**, *49*, 127.
7. Lin, G.; Zhang, A. *Tetrahedron* **2000**, *56*, 7163.
8. Ergun, Y.; Bayraktar, N.; Patir, S.; Okay, G. *J. Heterocycl. Chem.* **2000**, *37*, 11.
9. Rebeiro, G. L.; Khadilkar, B. M. *Synthesis* **2001**, 370.
10. Bremner, J B.; Coates, J A.; Keller, P A.; Pyne, S G.; Witchard, H. M. *Synlett* **2002**, 219.

Boulton–Katritzky rearrangement

Rearrangement of one five-membered heterocycle into another under thermolysis.

e.g. [ref. 9]:

References

1. Boulton, A. J.; Katritzky, A. R.; Hamid, A. M. *J. Chem. Soc. (C)* **1967**, 2005.
2. Ruccia, M.; Vivona, N.; Spinelli, D. *Adv. Heterocyl. Chem.* **1981**, *29*, 141. (Review).
3. Butler, R. N.; Fitzgerald, K. J. *J. Chem. Soc., Perkin Trans. 1* **1988**, 1587.
4. Takakis, I. M.; Hadjimihalakis, P. M.; Tsantali, G. G. *Tetrahedron* **1991**, *47*, 7157.
5. Takakis, I. M.; Hadjimihalakis, P. M. *J. Heterocycl. Chem.* **1992**, *29*, 121.
6. Vivona, N.; Buscemi, S.; Frenna, V.; Cusmano, C. *Adv. Heterocyl. Chem.* **1993**, *56*, 49.
7. Katayama, H.; Takatsu, N.; Sakurada, M.; Kawada, Y. *Heterocycles* **1993**, *35*, 453.
8. Rauhut, G. *J. Org. Chem.* **2001**, *66*, 5444.
9. Crampton, M. R.; Pearce, L. M.; Rabbitt, L. C. *J. Chem. Soc., Perkin Trans. 2* **2002**, 257.

Bouveault aldehyde synthesis

Formylation of an alkyl or aryl halide to the homologous aldehyde by transformation to the corresponding organometallic reagent then addition of DMF.

$$R-X \xrightarrow[\text{3. H}^+]{\substack{\text{1. M} \\ \text{2. DMF}}} R-CHO$$

$$R-X \xrightarrow{M} \underset{\text{R-M}}{\overset{}{}} \longrightarrow Me_2N-\overset{O-M}{\underset{R}{<}} \xrightarrow{H^+} R-CHO$$

A modification by Comins [6]:

$$R_2N-CHO \xrightarrow[\text{2. H}^+]{\text{1. R'MgX}} R'-CHO$$

$$\longrightarrow R_2N-\overset{O}{<}_{R'} \longrightarrow R'-CHO + \ ^-NR_2$$

References

1. Bouveault, L. *Bull. Soc. Chim. Fr.* **1904**, *31*, 1306.
2. Maxim, N.; Mavrodineanu, R. *Bull. Soc. Chim. Fr.* **1935**, *2*, 591.
3. Maxim, N.; Mavrodineanu, R. *Bull. Soc. Chim. Fr.* **1936**, *3*, 1084.
4. Smith, L. I.; Bayliss, M. *J. Org. Chem.* **1941**, *6*, 437.
5. Petrier, C.; Gemal, A. L.; Luche, J. L. *Tetrahedron Lett.* **1982**, *23*, 3361.
6. Comins, D. L.; Brown, J. D. *J. Org. Chem.* **1984**, *49*, 1078.
7. Einhorn, J.; Luche, J. L. *Tetrahedron Lett.* **1986**, *27*, 1791.
8. Einhorn, J.; Luche, J. L. *Tetrahedron Lett.* **1986**, *27*, 1793.
9. Denton, S. M.; Wood, A. *Synlett* **1999**, 55.
10. Meier, H.; Aust, H. *J. Prakt. Chem.* **1999**, *341*, 466.

Bouveault–Blanc reduction

Reduction of esters to the corresponding alcohols using sodium in an alcoholic solvent.

ketyl (radical anion)

References

1. Bouveault, L.; Blanc, G. *Compt. Rend.* **1903**, *136*, 1676.
2. Ruehlmann, K.; Seefluth, H.; Kiriakidis, T.; Michael, G.; Jancke, H.; Kriegsmann, H. *J. Organomet. Chem.* **1971**, *27*, 327.
3. Castells, J.; Grandes, D.; Moreno-Manas, M.; Virgili, A. *An. Quim.* **1976**, *72*, 74.
4. Sharda, R.; Krishnamurthy, H. G. *Indian J. Chem., Sect. B* **1980**, *19B*, 405.
5. Banerji, J.; Bose, P.; Chakrabarti, R.; Das, B. *Indian J. Chem., Sect. B* **1993**, *32B*, 709.
6. Seo, B. I.; Wall, L. K.; Lee, H.; Buttrum, J. W.; Lewis, D. E. *Synth. Commun.* **1993**, *23*, 15.
7. Zhang, Y.; Ding, C. *Huaxue Tongbao* **1997**, 36.

Boyland–Sims oxidation

Oxidation of aromatic amines to phenols using alkaline persulfate.

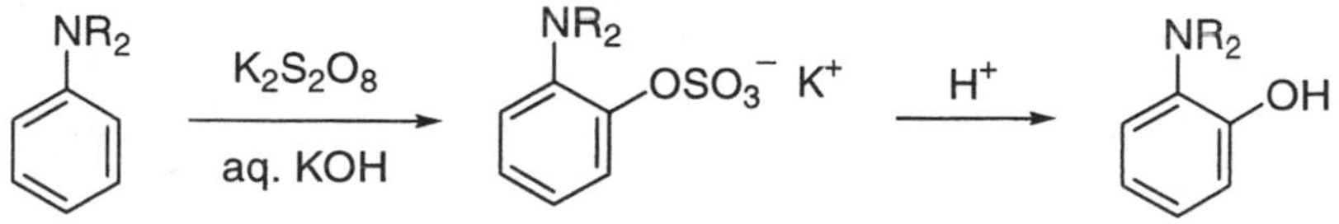

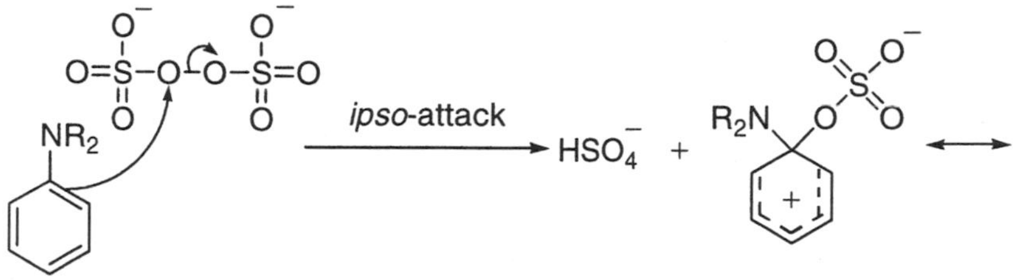

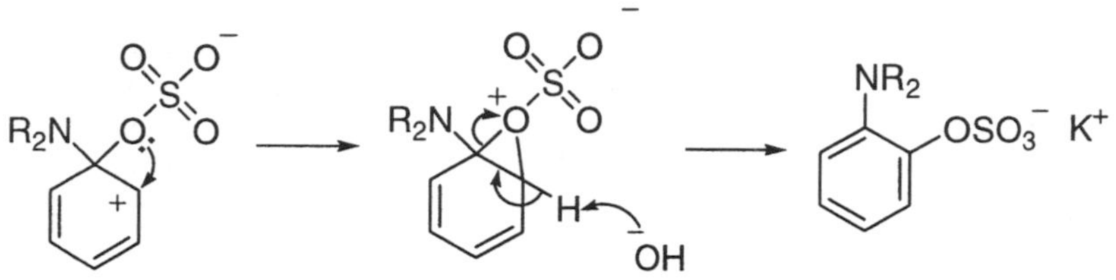

Another pathway is also operative:

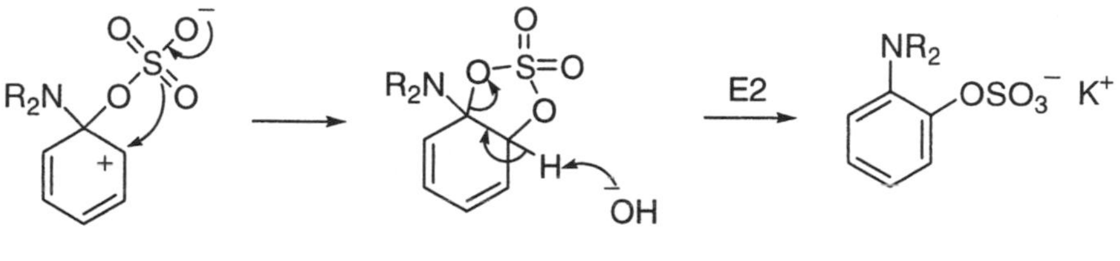

References

1. Boyland, E.; Manson, D.; Sims, P. *J. Chem. Soc.* **1953**, 3623.
2. Boyland, E.; Sims, P. *J. Chem. Soc.* **1954**, 980.
3. Behrman, E. J. *J. Am. Chem. Soc.* **1967**, 89, 2424.
4. Krishnamurthi, T. K.; Venkatasubramanian, N. *Indian J. Chem., Sect. A* **1978**, *16A*, 28.
5. Behrman, E. J.; Behrman, D. M. *J. Org. Chem.* **1978**, 43, 4551.

50

6. Srinivasan, C.; Perumal, S.; Arumugam, N. *J. Chem. Soc., Perkin Trans. 2* **1985**, 1855.
7. Behrman, E. J. *Org. React.* **1988**, *35,* 421–511. (Review).
8. Behrman, E. J. *J. Org. Chem.* **1992**, *57,* 2266.

Bradsher reaction

Anthracenes from *ortho*-acyl diarylmethanes *via* acid-catalyzed cyclodehydration.

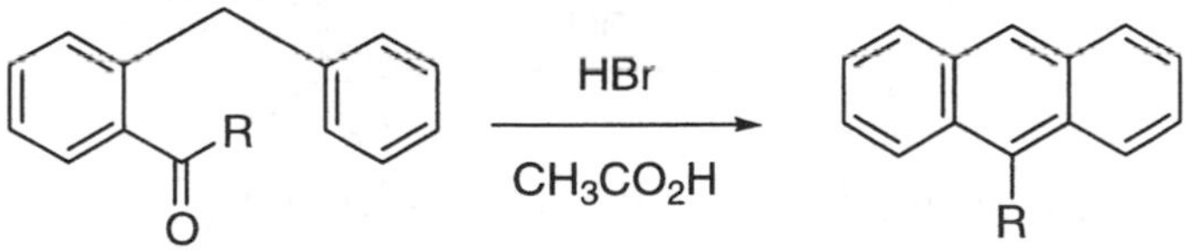

anthracene

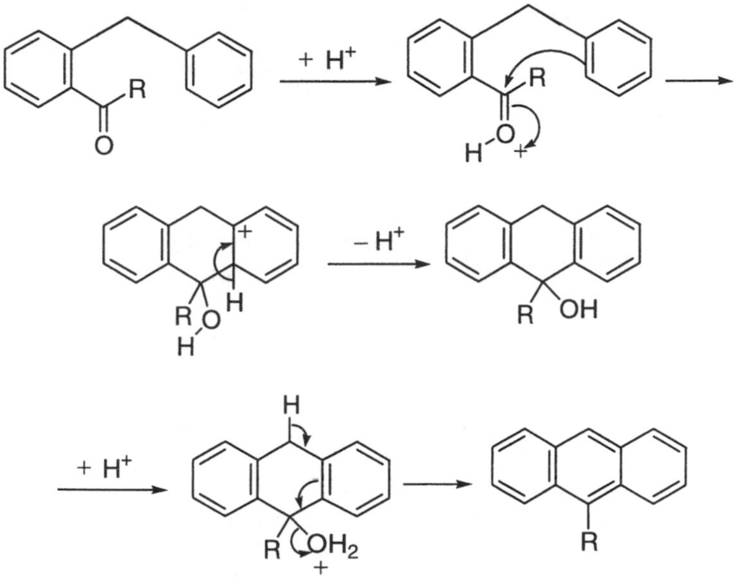

References

1. Bradsher, C. K. *J. Am. Chem. Soc.* **1940**, *62*, 486.
2. Bradshcr, C. K.; Sinclair, E. F. *J. Org. Chem.* **1957**, *22*, 79.
3. Vingiello, F. A.; Spangler, M. O. L.; Bondurant, J. E. *J. Org. Chem.* **1960**, *25*, 2091.
4. Brice, L. K.; Katstra, R. D. *J. Am. Chem. Soc.* **1960**, *82*, 2669.
5. Saraf, S. D.; Vingiello, F. A. *Synthesis* **1970**, 655.
6. Ashby, J.; Ayad, M.; Meth-Cohn, O. *J. Chem. Soc., Perkin Trans. 1* **1974**, 1744.
7. Nicolas, T. E.; Franck, R. W. *J. Org. Chem.* **1995**, *60*, 6904.
8. Magnier, E.; Langlois, Y. *Tetrahedron Lett.* **1998**, *39*, 837.

Brook rearrangement

Base-catalyzed silicon migration from carbon to oxygen.

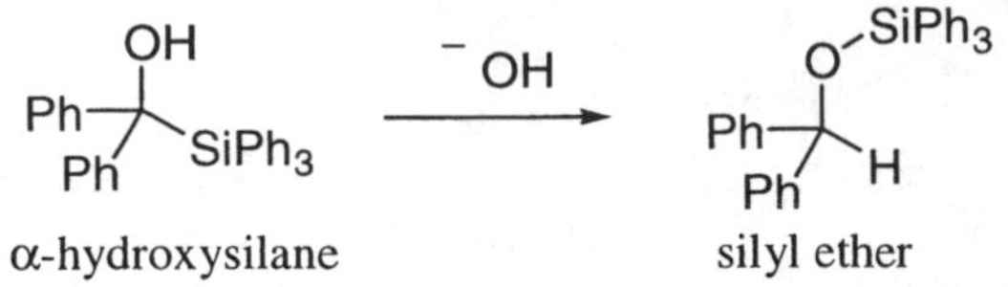

α-hydroxysilane   silyl ether

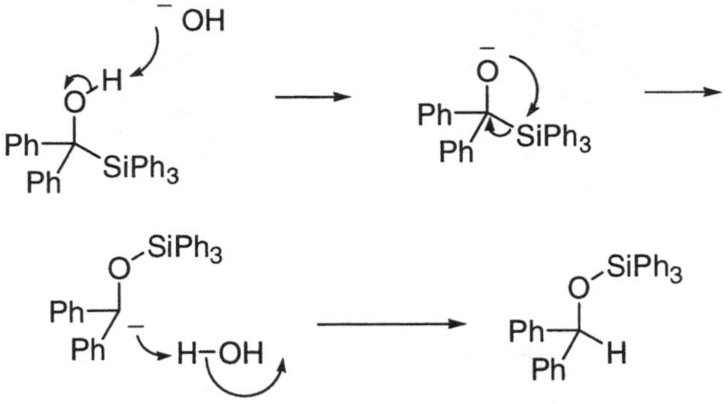

References

1. Brook, A. G. *J. Am. Chem. Soc.* **1958**, *80*, 1886.
2. Brook, A. G. *Acc. Chem. Res.* **1974**, *7*, 77. (Review).
3. Page, P. C. B.; Klair, S. S.; Rosenthal, S. *Chem. Soc. Rev.* **1990**, *19*, 147. (Review).
4. Takeda, K.; Nakatani, J.; Nakamura, H.; Yosgii, E.; Yamaguchi, K. *Synlett* **1993**, 841.
5. Fleming, I.; Ghosh, U. *J. Chem. Soc., Perkin Trans. 1* **1994**, 257.
6. Takeda, K.; Takeda, K.; Ohnishi, Y. *Tetrahedron Lett.* **2000**, *41*, 4169.
7. Sumi, K.; Hagisawa, S. *J. Organomet. Chem.* **2000**, *611*, 449.
8. Moser, W. H. *Tetrahedron* **2001**, *57*, 2065.
9. Takeda, K.; Sawada, Y.; Sumi, K. *Org. Lett.* **2002**, *4*, 1031.

Brown hydroboration reaction

Addition of boranes to olefins, followed by basic oxidation of the organoborane adducts, resulting in alcohols.

References

1. Brown, H. C.; Tierney, P. A. *J. Am. Chem. Soc.* **1958**, *80*, 1552.
2. Nussium, M.; Mazur, Y.; Sondheimer, F. *J. Org. Chem.* **1964**, *29*, 1120.
3. Nussium, M.; Mazur, Y.; Sondheimer, F. *J. Org. Chem.* **1964**, *29*, 1131.
4. Streitwieser, A., Jr.; Verbit, L.; Bittman, R. *J. Org. Chem.* **1967**, *32*, 1530.
5. Herz, J. E.; Marquez, L. A. *J. Chem. Soc. (C)* **1971**, 3504.
6. Pelter, A.; Smith, K.; Brown, H. C. *Borane Reagents* Academic Press: New York, **1972**. (Review).
7. Brewster, J. H.; Negishi, E. *Science* **1980**, *207*, 44. (Review).
8. Brown, H. C.; vara Prasad, J. V. N. *Heterocycles* **1987**, *25*, 641.
9. Fu, G. C.; Evans, D. A.; Muci, A. R. *Advances in Catalytic Processes* **1995**, *1*, 95–121. (Review).
10. Hayashi, T. *Comprehensive Asymmetric Catalysis I-III* **19995**, *1*, 351–364. (Review).
11. Pender, M. J.; Carroll, P. J.; Sneddon, L. G. *J. Am. Chem. Soc.* **2001**, *123*, 12222.
12. Morrill, T. C.; D'Souza, C. A.; Yang, L.; Sampognaro, A. J. *J. Org. Chem.* **2001**, *123*, 2481.
13. Hupe, E.; Calaza, M. I.; Knochel, P. *Tetrahedron Lett.* **2001**, *42*, 8829.

Bucherer carbazole synthesis

Carbazoles from naphthols and aryl hydrazines promoted by sodium bisulfite.

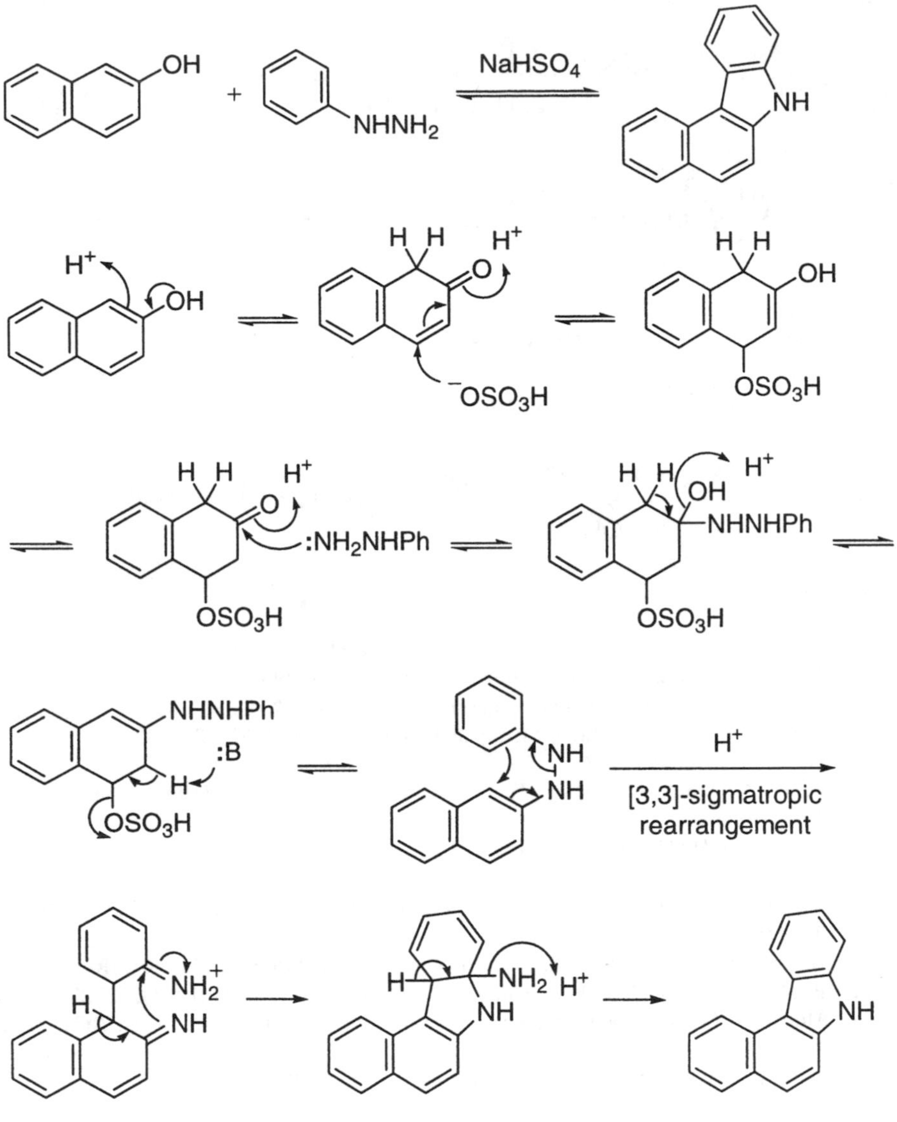

References

1. Bucherer, H. T.; Seyde, F. *J. Prakt. Chem.* **1908**, *77*, 403.
2. Seeboth, H. *Deut. Akad. Wiss. Berlin* **1961**, *3*, 48.

3. Seeboth, H.; Baerwolff, D.; Becker, B. *Justus Liebigs Ann. Chem.* **1965**, *683*, 85.
4. Seeboth, H. *Angew. Chem., Int. Ed. Engl.* **1967**, *6*, 307.

56

Bucherer reaction

Transformation of β-naphthols to β-naphthylamines using ammonium sulfite.

References

1. Bucherer, H. T. *J. Prakt. Chem.* **1904**, *69*, 49.
2. Reiche, A.; Seeboth, H. *Justus Liebigs Ann. Chem.* **1960**, *638*, 66.
3. Gilbert, E. E. *Sulfonation and Related Reactions* Wiley: New York, **1965**, p166. (Review).
4. Seeboth, H. *Angew. Chem., Int. Ed. Engl.* **1967**, *6*, 307.
5. Gruszecka, E.; Shine, H. J. *J. Labeled. Compd. Radiopharm.* **1983**, *20*, 1257.
6. Rebek, J., Jr.; Marshall, L.; Wolak, R.; Parris, K.; Killoran, M.; Askew, B.; Nemeth, D.; Islam, N. *J. Am. Chem. Soc.* **1985**, *107*, 7476.
7. Belica, P. S.; Manchand, P S. *Synthesis* **1990**, 539.
8. Singer, R. A.; Buchwald, S. L. *Tetrahedron Lett.* **1999**, *40*, 1095.
9. Canete, A.; Melendrez, M. X.; Saitz, C.; Zanocco, A. L. *Synth. Commun.* **2001**, *31*, 2143.

Bucherer–Bergs reaction

The formation of hydantoin from carbonyl compound with potassium cyanide (KCN) and ammonium carbonate [$(NH_4)_2CO_3$] or from cyanohydrin and ammonium carbonate. It belongs to the category of multiple component reaction (MCR).

cyanohydrin

isocyanate intermediate

References

1. Bergs, H. Ger. Pat. 566,094, (**1929**).
2. Bucherer, H. T., Fischbeck, H. T. *J. Prakt. Chem.* **1934**, *140*, 69.
3. Bucherer, H. T., Steiner, W. *J. Prakt. Chem.* **1934**, *140*, 291.
4. E. Ware, *Chem. Rev.* **46**, 422 (1950). (Review).
5. Chubb, F. L.; Edward, J. T.; Wong, S. C. *J. Org. Chem.* **1980**, *45*, 2315.
6. Rousset, A.; Laspéras, M.; llades, J.; Commeyras, A. *Tetrahedron* **1980**, *36*, 2649.
7. Bowness, W. G.; Howe, R.; Rao, B. S. *J. Chem. Soc., Perkin Trans. 1* **1983**, 2649.
8. Taillades, J.; Rousset, A.; Laspéras, M.; Commeyras, A. *Bull. Soc. Chim. Fr.* **1986**, 650.
9. Herdeis, C.; Gebhard, R. *Heterocycles* **1986**, *24*, 1019.
10. Haroutounian, S. A.; Georgiadis, M. P.; Polissiou, M. G. *J. Heterocycl. Chem.* **1989**, *26*, 1283.
11. Tanaka, K.-i.; Iwabuchi, H.; Sawanishi, H. *Tetrahedron: Asymmetry* **1995**, *6*, 2271.
12. Micova, J.; Steiner, B.; Koos, M.; Langer, V.; Gyepesova, D. *Synlett* **2002**, 1715.

Buchner–Curtius–Schlotterbeck reaction

Carbonyl compounds react with aliphatic diazo compounds to deliver homologated ketones.

References

1. Buchner, E.; Curtius, T. *Ber.* **1989**, *18*, 2371.
2. Gutsche, C. D. *Org. React.* **1954**, *8*, 364. (Review).
3. Bastus, J. *Tetrahedron Lett.* **1963**, 955.
4. Kirmse, W.; Horn, K. *Tetrahedron Lett.* **1967**, 1827.
5. Moody, C. J.; Miah, S.; Slawin, A. M. Z.; Mansfield, D. J.; Richards, I. C. *J. Chem. Soc., Perkin Trans. 1* **1998**, 4067.
6. Maguire, A. R.; Buckley, N. R.; O'Leary, P.; Ferguson, G. *J. Chem. Soc., Perkin Trans. 1* **1998**, 4077.

Buchner method of ring expansion

Benzene reacts with diazoacetic ester to give cyclohepta-2,4,6-trienecarboxylic acid ester. *Cf.* Pfau–Platter azulene synthesis.

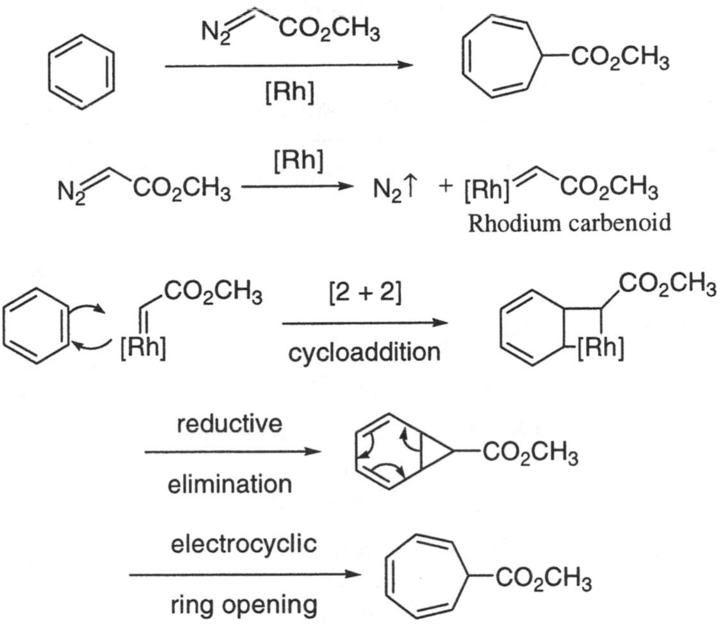

References

1.   Buchner, E. *Ber. Dtsch. Chem. Ges.* **1896**, *29*, 106.
2.   Dev, S. *J. Indian Chem. Soc.* **1955**, *32*, 513.
3.   Von Doering, W.; Knox, L. H. *J. Am. Chem. Soc.* **1957**, *79*, 352.
4.   Marchard, A. P.; Macbrockway, N. *Chem. Rev.* **1974**, *74*, 431. (Review).
5.   Nakamura, A.; Konischi, A.; Tsujitani, R.; Kudo, M.; Otsuka, S. *J. Am. Chem. Soc.* **1978**, *100*, 3449.
6.   Anciaux, A. J.; Noels, A. F.; Hubert, A. J.; Warin, R.; Teyssié, P. *J. Org. Chem.* **1981**, *46*, 873.
7.   Doyle, M. P.; Hu, W.; Timmons, D. J. *Org. Lett.* **2001**, *3*, 933.
8.   Doyle, M. P.; Phillips, I. M. *Tetrahedron Lett.* **2001**, *42*, 3155.

Buchwald–Hartwig C–N bond and C–O bond formation reactions

Direct Pd-catalyzed C–N and C–O bond formation from aryl halides and amines in the presence of stoichiometric amount of base.

The C–O bond formation reaction follows a similar mechanistic pathway [7–9].

References

1. Paul, F.; Patt, J.; Hartwig, J. F. *J. Am. Chem. Soc.* **1994**, *116*, 5969.
2. Guram, A. S.; Buchwald, S. L. *J. Am. Chem. Soc.* **1994**, *116*, 7901.
3. Palucki, M.; Wolfe, J. P.; Buchwald, S. L. *J. Am. Chem. Soc.* **1996**, *118*, 10333.
4. Mann, G.; Hartwig, J. F. *J. Org. Chem.* **1997**, *62*, 5413.
5. Mann, G.; Hartwig, J. F. *Tetrahedron Lett.* **1997**, *38*, 8005.
6. Wolfe, J. P.; Wagaw, S.; Marcoux, J.-F.; Buchwald, S. L. *Acc. Chem. Res.* **1998**, *31*, 805. (Review).
7. Hartwig, J. F. *Acc. Chem. Res.* **1998**, *31*, 852. (Review).
8. Frost, C. G.; Mendonça, P. *J. Chem. Soc., Perkin Trans. 1* **1998**, 2615.
9. Yang, B. H.; Buchwald, S. L. *J. Organomet. Chem.* **1999**, *576*, 125.
10. Browning, R. G.; Mahmud, H.; Badarinarayana, V.; Lovely, C. J. *Tetrahedron Lett.* **2001**, *42*, 7155.
11. Lee, J.-H.; Cho, C.-G. *Tetrahedron Lett.* **2003**, *44*, 65.
12. Ferreira, I. C. F. R.; Queiroz, M.-J. R. P.; Kirsch, G. *Tetrahedron* **2003**, *59*, 975.

Burgess dehydrating reagent

Burgess dehydrating reagent is efficient at generating olefins from secondary and tertiary alcohols where Ei (during the elimination, the two groups leave at about the same time and bond to each other concurrently) mechanism prevails.

References

1. Burgess, E. M. *J. Org. Chem.* **1973**, *38*, 26.
2. Claremon, D. A.; Philips, B. T. *Tetrahedron Lett.* **1988**, *29*, 2155.
3. Creedon, S. M.; Crowley, H. K.; McCarthy, D. G. *J. Chem. Soc., Perkin Trans. 1* **1998**, 1015.
4. Lamberth, C. *J. Prakt. Chem.* **2000**, *342*, 518.
5. Svenja, B. *Synlett.* **2000**, 559.
6. Miller, C. P.; Kaufman, D. H. *Synlett* **2000**, 1169.
7. Nicolaou, K. C.; Huang, X.; Snyder, S. A.; Rao, P. B.; Bella, Reddy, M. V. *Angew. Chem., Int. Ed.* **2002**, *41*, 834.
8. Jose, B.; Unni, M. V. V.; Prathapan, S.; Vadakkan, J. J. *Synth. Commun.* **2002**, *32*, 2495.

62

Cadiot–Chodkiewicz coupling

Bis-acetylene synthesis from alkynyl halides and alkynyl copper reagents.
Cf. Castro–Stephens reaction.

$$R^1\!\!-\!\!\equiv\!\!-X \ + \ Cu\!\!-\!\!\equiv\!\!-R^2 \longrightarrow R^1\!\!-\!\!\equiv\!\!-\!\!\equiv\!\!-R^2$$

$$R^1\!\!-\!\!\equiv\!\!-X \ + \ Cu\!\!-\!\!\equiv\!\!-R^2 \xrightarrow[\text{addition}]{\text{oxidative}} R^1\!\!-\!\!\equiv\!\!-\overset{X}{\underset{|}{Cu}}\!\!-\!\!\equiv\!\!-R$$

Cu(III) intermediate

$$\xrightarrow[\text{elimination}]{\text{reductive}} CuX \ + \ R^1\!\!-\!\!\equiv\!\!-\!\!\equiv\!\!-R^2$$

References

1. Chodkiewicz, W.; Cadiot, P. *C. R. Hebd. Seances Acad. Sci.* **1955**, *241*, 1055.
2. Cadiot, P.; Chodkiewicz, W. In *Chemistry of Acetylenes;* Viehe, H. G., ed.; Dekker: New York, **1969**, pp597–647. (Review).
3. Eastmond, R.; Walton, D. R. M. *Tetrahedron* **1972**, *28*, 4591.
4. Ghose, B. N.; Walton, D. R. M. *Synthesis* **1974**, 890.
5. Hopf, H.; Krause, N. *Tetrahedron Lett.* **1985**, *26*, 3323.
6. Bartik, B.; Dembinski, R.; Bartik, T.; Arif, A. M.; Gladysz, J. A. *New J. Chem.* **1997**, *21*, 739.
7. Montierth, J. M.; DeMario, D. R.; Kurth, M. J.; Schore, N. E. *Tetrahedron* **1998**, *54*, 11741.
8. Negishi, E.-i.; Hata, M.; Xu, C. *Org. Lett.* **2000**, *2*, 3687.
9. Steffen, W.; Laskoski, M.; Collins, G.; Bunz, U. H. F. *J. Organomet. Chem.* **2001**, *630*, 132.
10. Marino, J. P.; Nguyen, H. N. *J. Org. Chem.* **2002**, *67*, 6841.
11. Utesch, N. F.; Diederich, F. *Org. Biomol. Chem.* **2003**, *1*, 237.

Cannizzaro disproportionation reaction

Redox reaction between aromatic aldehydes, formaldehyde or other aliphatic aldehydes without α-hydrogen using base to afford the corresponding alcohols and carboxylic acids.

Pathway a:

Final deprotonation of the carboxylic acid drives the reaction forward.

Pathway b:

References

1. Cannizzaro, S. *Justus Liebigs Ann. Chem.* **1853**, *88*, 129.
2. Pfeil, E. *Chem. Ber.* **1951**, *84*, 229.
3. Hazlet, S. E.; Stauffer, D. A. *J. Org. Chem.* **1962**, *27*, 2021.
4. Hazlet, S. E.; Bosmajian, G., Jr.; Estes, J. H.; Tallyn, E. F. *J. Org. Chem.* **1964**, *29*, 2034.
5. Sen Gupta, A. K. *Tetrahedron Lett.* **1968**, 5205.
6. Griengl, H.; Nowak, P. *Monatsh. Chem.* **1977**, *108*, 407.

64

7.   Swain, C. G.; Powell, A. L.; Sheppard, W. A.; Morgan, C. R. *J. Am. Chem. Soc.* **1979**, *101*, 3576.
8.   Mehta, G.; Padma, S. *J. Org. Chem.* **1991**, *56*, 1298.
9.   Sheldon, J. C.; *et. al. J. Org. Chem.* **1997**, *62*, 3931.
10. Thakuria, J. A.; Baruah, M.; Sandhu, J. S. *Chem. Lett.* **1999**, 995.
11. Russell, A. E.; Miller, S. P.; Morken, J. P. *J. Org. Chem.* **2000**, *65*, 8381.
12. Reddy, B. V. S; Srinvas, R.; Yadav, J. S.; Ramalingam, T. *Synth. Commun.* **2002**, *32*, 1489.

Carroll rearrangement

Base-catalyzed transformation of allylic alcohol and β-ketoester to γ-ketoolefin *via* anion-assisted Claisen reaction.

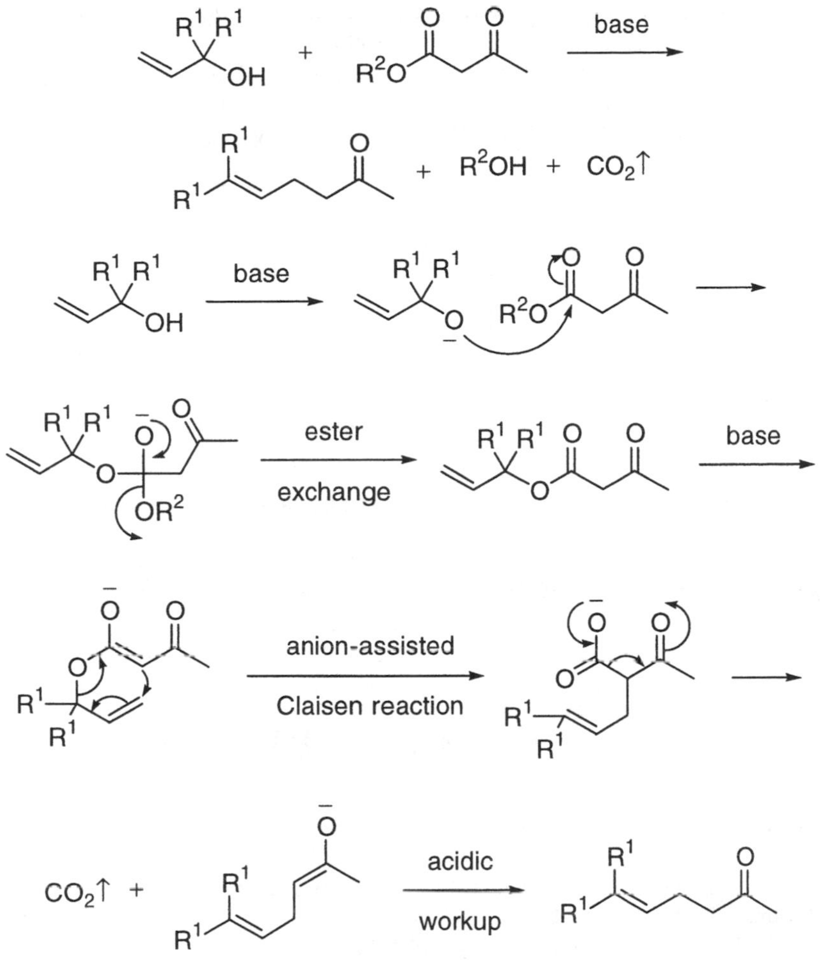

References

1. Carroll, M. F. *J. Chem. Soc.* **1940**, 704.
2. Wilson, S. R.; Price, M. F. *J. Org. Chem.* **1984**, *49*, 722.
3. Gilbert, J. C.; Kelly, T. A. *Tetrahedron* **1988**, *44*, 7587.
4. Enders, D.; Knopp, M.; Runsink, J.; Raabe, G. *Angew. Chem., Int. Ed. Engl.* **1995**, *34*, 2278.
5. Enders, D.; Knopp, M.; Runsink, J.; Raabe, G. *Liebigs Ann.* **1996**, 1095.
6. Hatcher, M. A.; Posner, G. H. *Tetrahedron Lett.* **2002**, *43*, 5009.

Castro–Stephens coupling

Aryl-acetylene synthesis, *Cf.* Cadiot–Chodkiewicz coupling.

$$Ar-X \ + \ Cu-\!\!\equiv\!\!-R \ \xrightarrow[\text{reflux}]{\text{pyridine}} \ Ar-\!\!\equiv\!\!-R$$

$$Ar-X + L_3Cu-\!\!\equiv\!\!-R \ \longrightarrow \ ArX-\overset{L}{\underset{L}{Cu}}-\!\!\equiv\!\!-R$$

$$\longrightarrow \ \left[\ \overset{X}{\underset{R}{Ar\!\!\diagdown\!\!\diagup Cu}} \ \right] \ \longrightarrow \ CuX \ + \ Ar-\!\!\equiv\!\!-R$$

An alternative mechanism similar to that of the Cadiot–Chodkiewicz coupling:

$$Ar-\!\!\equiv\!\!-X + Cu-\!\!\equiv\!\!-R \ \xrightarrow[\text{addition}]{\text{oxidative}} \ Ar-\!\!\equiv\!\!-\overset{X}{Cu}-\!\!\equiv\!\!-R$$

Cu(III) intermediate

$$\xrightarrow[\text{elimination}]{\text{reductive}} \ CuX \ + \ Ar-\!\!\equiv\!\!-\!\!\equiv\!\!-R$$

References

1. Castro, C. E.; Stephens, R. D. *J. Org. Chem.* **1963**, *28*, 2163.
2. Castro, C. E.; Stephens, R. D. *J. Org. Chem.* **1963**, *28*, 3313.
3. Staab, H. A.; Neunhoeffer, K. *Synthesis* **1974**, 424.
4. Kabbara, J.; Hoffmann, C.; Schinzer, D. *Synthesis* **1995**, 299.
5. von der Ohe, F.; Bruckner, R. *New J. Chem.* **2000**, *24*, 659.
6. White, J. D.; Carter, R. G.; Sundermann, K. F.; Wartmann, M. *J. Am. Chem. Soc.* **2001**, *123*, 5407.
7. Rawat, D. S.; Zaleski, J. M. *Synth. Commun.* **2002**, *32*, 1489.

Chapman rearrangement

Thermal aryl rearrangement of *O*-aryliminoethers to amides.

oxazete intermediate

References

1. Chapman, A. W. *J. Chem. Soc.* **1925**, *127*, 1992.
2. Wheeler, O. H.; Roman, F.; Rosado, O. *J. Org. Chem.* **1969**, *34*, 966.
3. Kimura, M. *J. Chem. Soc., Perkin Trans. 2* **1987**, 205.
4. Kimura, M.; Okabayashi, I.; Isogai, K. *J. Heterocycl. Chem.* **1988**, *25*. 315.
5. Dessolin, M.; Eisenstein, O.; Golfier, M.; Prange, T.; Sautet, P. *J. Chem. Soc., Chem. Commun.* **1992**, 132.
6. Wang, X.; Cai, Y.; Xu, Z. *Zhongguo Yaoxue Zazhi* **1997**, *32*, 774.
7. Shohda, K.-I.; Wada, T.; Sekine, M. *Nucleosides Nucleotides* **1998**, *17*, 2199.

Chichibabin amination reaction

Direct amination of pyridines, quinolines, and other *N*-heterocycles using sodium amide in liquid ammonia.

References

1. Chichibabin, A. E.; Zeide, O. A. *J. Russ. Phys. Chem. Soc.* **1914**, *46*, 1216.
2. Knize, M. G.; Felton, J. S. *Heterocycles* **1986**, *24*, 1815.
3. Chambron, J. Cl.; Sauvage, J. P. *Tetrahedron* **1987**, *43*, 895.
4. McGill, C. K.; Rappa, A. *Adv. Heterocycl. Chem.* **1988**, *44*, 1.
5. Kelly, T. R.; Bridger, G. J.; Zhao, C. *J. Am. Chem. Soc.* **1990**, *112*, 8024.
6. Tanga, M. J.; Bupp, J. E.; Tochimoto, T. K. *J. Heterocycl. Chem.* **1994**, *31*, 1641.
7. Kiselyov, A. S.; Strekowski, L. *Synth. Commun.* **1994**, *24*, 2387.
8. Seko, S.; Miyake, K. *Chem. Commun.* **1998**, 1519.
9. Katritzky, A. R.; Qiu, G.; Long, Q.-H.; He, H.-Y.; Steel, P. J. *J. Org. Chem.* **2000**, *65*, 9201.
10. Palucki, M.; Hughes, D. L.; Yasuda, N.; Yang, C.; Reider, P. J. *Tetrahedron Lett.* **2001**, *42*, 6811.

Chichibabin pyridine synthesis

Condensation of aldehydes with ammonia to afford pyridines.

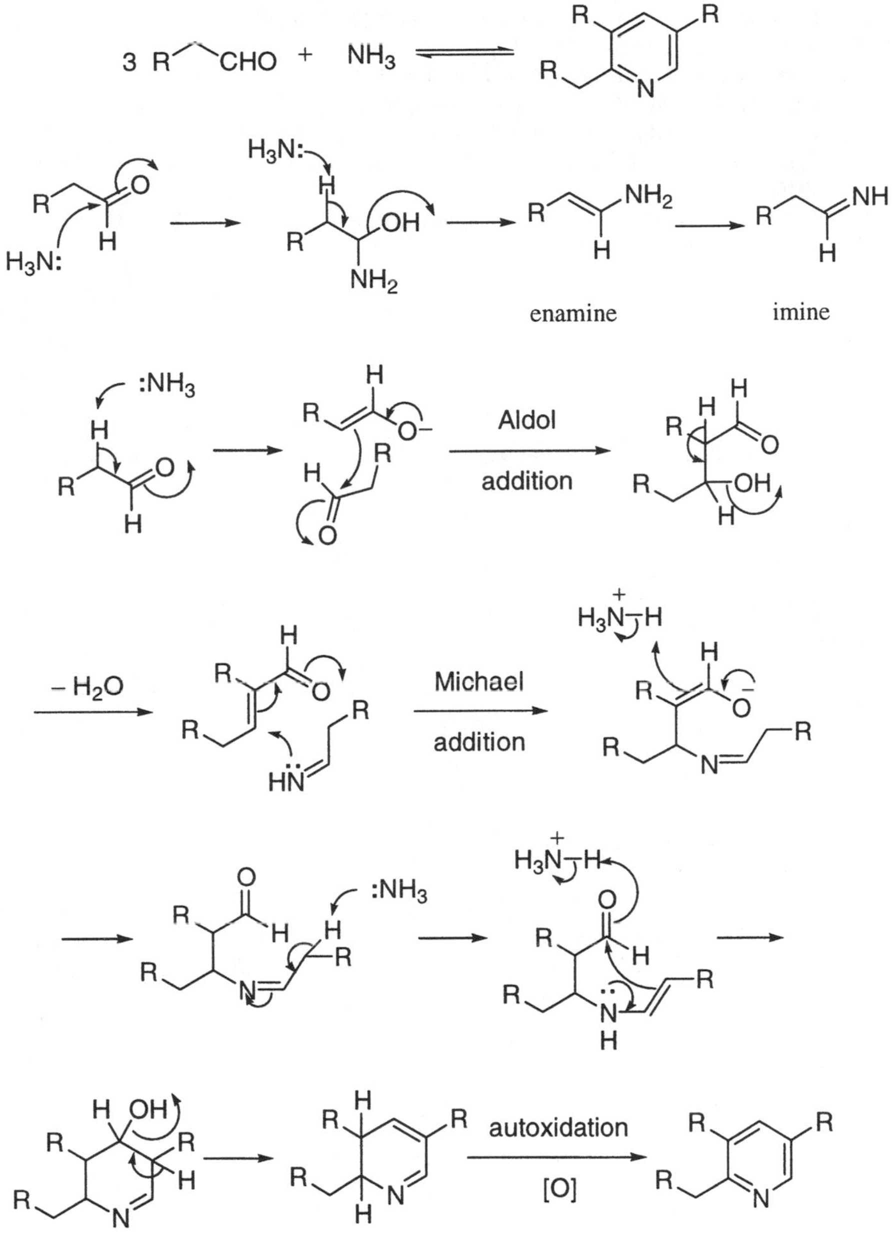

enamine   imine

Aldol
addition

– H₂O   Michael
addition

:NH₃

autoxidation
[O]

70

References

1. Chichibabin, A. E. *J. Russ. Phys. Chem. Soc.* **1906**, *37*, 1229.
2. Frank, R. L.; Seven, R. P. *J. Am. Chem. Soc.* **1949**, *71*, 2629.
3. Frank, R. L.; Riener, E. F. *J. Am. Chem. Soc.* **1950**, *72*, 4182.
4. Weiss, M. *J. Am. Chem. Soc.* **1952**, *74*, 200.
5. Herzenberg, J.; Boccato, G. *Chem. Ind.* **1958**, 248.
6. Levitt, L. S.; Levitt, B. W. *Chem. Ind.* **1963**, 1621.
7. Kessar, S. V.; Nadir, U. K.; Singh, M. *Indian J. Chem.* **1973**, *11*, 825.
8. Sagitullin, R. S.; Shkil, G. P.; Nosonova, I. I.; Ferber, A. A. *Khim. Geterotsikl. Soedin.* **1996**, 147.
9. Shimizu, S.; Abe, N.; Iguchi, A.; Dohba, M.; Sato, H.; Hirose, K.-I. *Microporous Mesoporous Materials* **1998**, *21*, 447.

Chugaev elimination

Thermal elimination of xanthates to olefins.

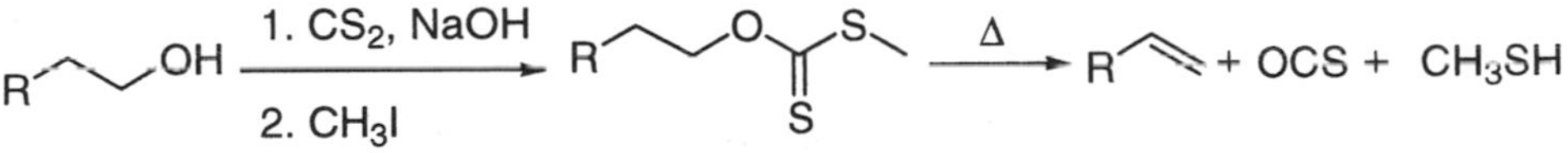

xanthate

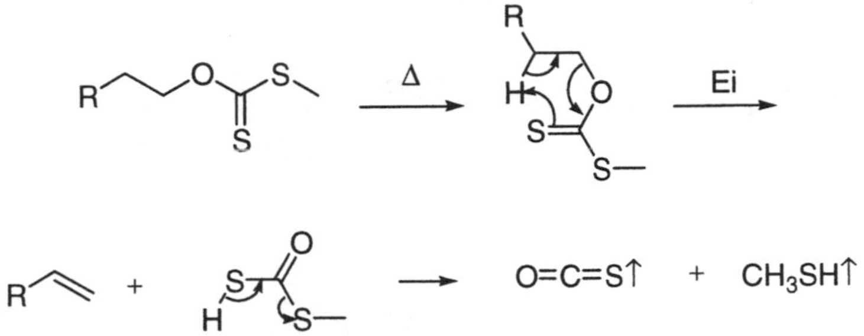

References

1. Chugaev, L. *Ber. Dtsch. Chem. Ges.* **1899**, *32*, 3332.
2. Chande, M. S.; Pranjpe, S. D. *Indian J. Chem.* **1973**, *11*, 1206.
3. Kawata, T.; Harano, K.; Taguchi, T. *Chem. Pharm. Bull.* **1973**, *21*, 604.
4. Harano, K.; Taguchi, T. *Chem. Pharm. Bull.* **1975**, *23*, 467.
5. Ho, T. L.; Liu, S. H. *J. Chem. Soc., Perkin Trans. 1* **1984**, 615.
6. Meulemans, T. M.; Stork, G. Λ.; Macaev, F. Z.; Jansen, B. J. M.; de Groot, Λ. *J. Org. Chem.* **1999**, *64*, 9178.
7. Nakagawa, H.; Sugahara, T.; Ogasawara, K. *Org. Lett.* **2000**, *2*, 3181.
8. Nakagawa, H.; Sugahara, T.; Ogasawara, K. *Tetrahedron Lett.* **2001**, *42*, 4523.

Ciamician–Dennsted rearrangement

Cyclopropanation of a pyrrole with dichlorocarbene generated from $CHCl_3$ and NaOH. Subsequent rearrangement takes place to give 3-chloropyridine.

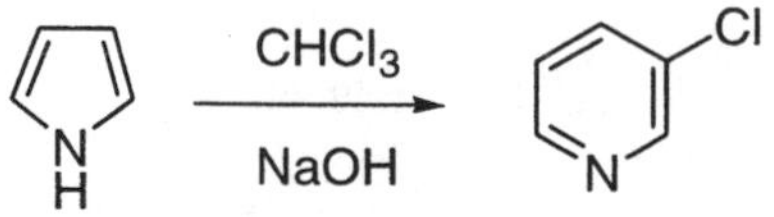

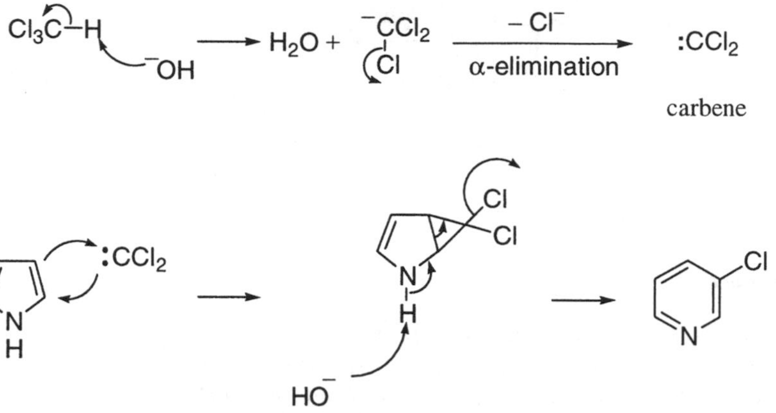

References

1. Ciamician, G. L.; Dennsted, M. *Ber. Dtsch. Chem. Ges.* **1881**, *14*, 1153.
2. Skell, P. S.; Sandler, R. S. *J. Am. Chem. Soc.* **1958**, *80*, 970.
3. Krauch, H.; Kunz, W. *Chemiker-Ztg.* **1959**, *83*, 815.
4. Vogel, E. *Angew. Chem.* **1960**, *72*, 8.
5. Josey, A. D.; Tuite, R. J.; Snyder, H. R. *J. Am. Chem. Soc.* **1963**, *82*, 1597.

Claisen condensation

Base-catalyzed condensation of esters to afford β-keto esters.

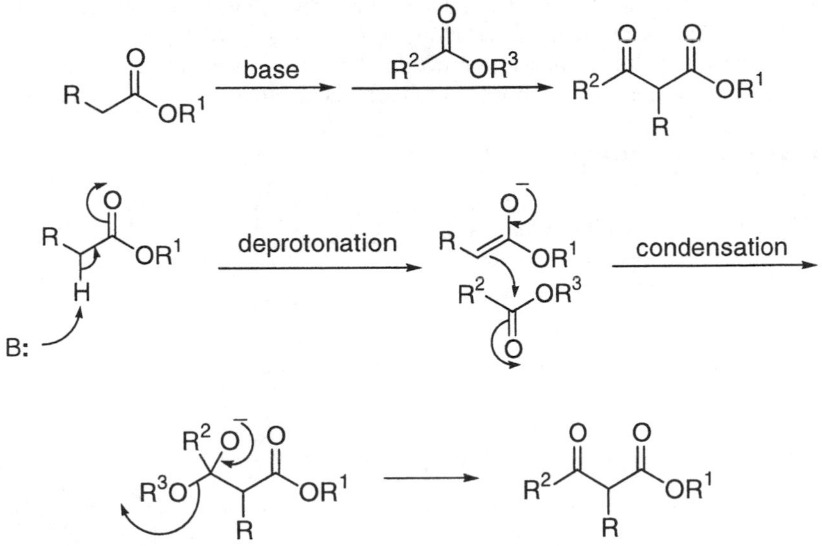

References

1   Claisen, L.; Lowman, O. *Ber. Dtsch. Chem. Ges.* **1887**, *20*, 651.
2   Hauser, C. R.; Hudson, B. E. *Org. React.* **1942**, *1*, 266–302. (Review).
3   Thaker, K. A.; Pathak, U. S. *Indian J. Chem.* **1965**, *3*, 416.
4   Schäfer, J. P.; Bloomfield, J. J. *Org. React.* **1967**, *15*, 1–203. (Review).
5   Tanabe, Y. *Bull. Chem. Soc. Jpn.* **1989**, *62*, 1917.
6   Kashima, C.; Takahashi, K.; Fukusaka, K. *J. Heterocycl. Chem.* **1995**, *32*, 1775.
7   Tanabe, Y.; Hamasaki, R.; Funakoshi, S. *Chem. Commun.* **2001**, 1674.
8   Mogilaiah, K.; Kankaiah, G. *Indian J. Chem., Sect. B* **2002**, *41B*, 2194.
9   Mogilaiah, K.; Reddy, N. V. *Synth. Commun.* **2003**, *33,* 73.

Claisen, Eschenmoser–Claisen, Johnson–Claisen, and Ireland–Claisen rearrangements

The Claisen, Johnson–Claisen, Ireland–Claisen, para-Claisen rearrangements, along with the Carroll rearrangement belong to the category of *[3,3]-sigmatropic rearrangements*, which is a concerted process. The arrow-pushing here is merely illustrative. For the abnormal Claisen rearrangement, see page 1.

Claisen rearrangement

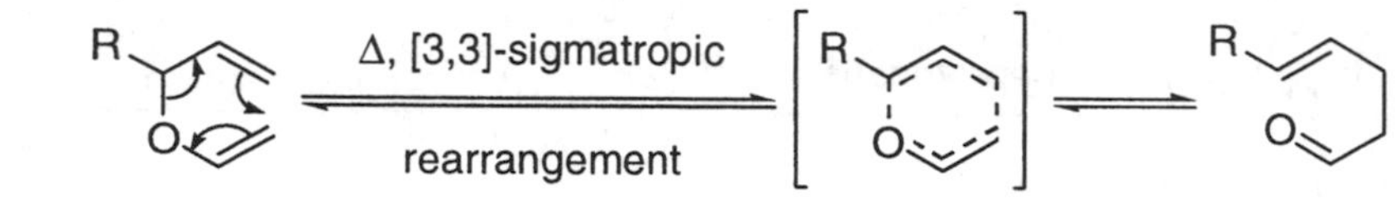

Eschenmoser–Claisen (amide acetal) rearrangement

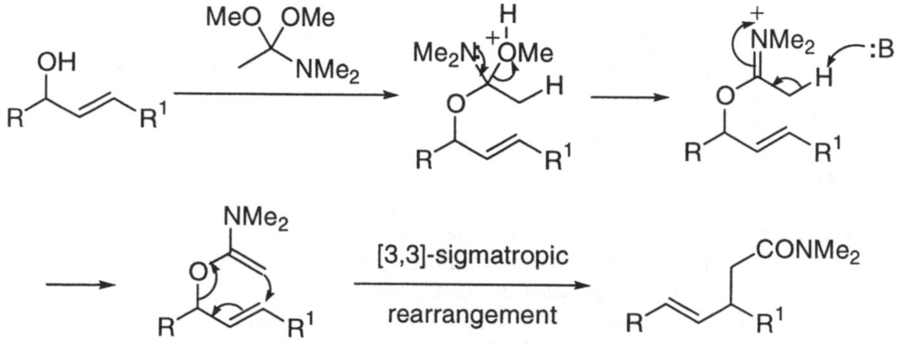

Johnson–Claisen (orthoester) rearrangement

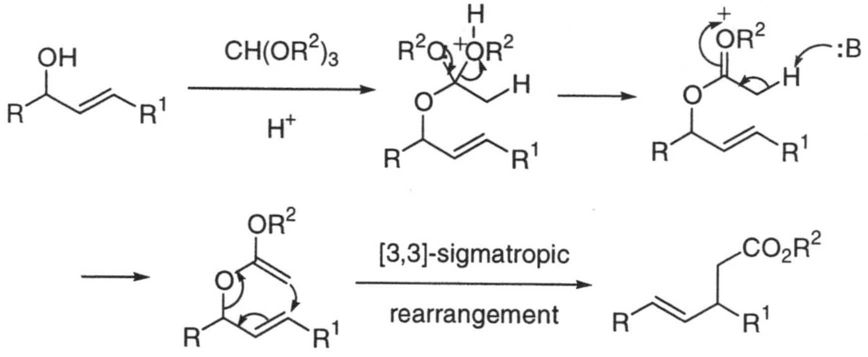

Ireland–Claisen (silyl ester) rearrangement

References

1. Claisen, L. *Ber. Dtsch. Chem. Ges.* **1912**, *45*, 3157.
2. Wick, A. E.; Felix, D.; Steen, K.; Eschenmoser, A. *Helv. Chim. Acta* **1964**, *47*, 2425.
3. Johnson, W. S.; Werthemann, L.; Bartlett, W. R.; Brocksom, T. J.; Li, T.-T.; Faulkner, D. J.; Peterson, M. R. *J. Am. Chem. Soc.* **1970**, *92*, 741.
4. Ireland, R. E.; Mueller, R. H. *J. Am. Chem. Soc.* **1972**, *94*, 5897.
5. Wipf, P. In *Comprehensive Organic Synthesis;* Trost, B. M.; Fleming, I., Eds.; Pergamon, **1991**, *Vol. 5*, 827–873. (Review).
6. Pereira, S.; Srebnik, M. *Aldrichimica Acta* **1993**, *26*, 17. (Review).
7. Ganem, B. *Angew. Chem., Int. Ed. Engl.* **1996**, *35*, 936.
8. Ito, H.; Taguchi, T. *Chem. Soc. Rev.* **1999**, *28*, 43. (Review).
9. Mohamed, M.; Brook, M. A. *Tetrahedron Lett.* **2001**, *42*, 191.
10. Loh, T.-P.; Hu, Q.-Y. *Org. Lett.* **2001**, *3*, 279.
11. Hong, S.-p.; Lindsay, H. A.; Yaramasu, T.; Zhang, X.; McIntosh, M. C. *J. Org. Chem.* **2002**, *67*, 2042.
12. Chai, Y.; Hong, S.-p.; Lindsay, H. A.; McFarland, C.; McIntosh, M. C. *Tetrahedron* **2002**, *58*, 2905–2928. (Review).
13. Khaledy, M. M.; Salani, M. Y. S.; Khuong, K. S.; Houk, K. N.; Aviyente, V.; Neier, R.; Soldermann, N.; Velker, J. *J. Org. Chem.* **2003**, *68*, 572.

Clarke–Eschweiler reductive alkylation of amines

Reductive methylation of primary or secondary amines using formaldehyde and formic acid.

$$R-NH_2 \ + \ CH_2O \ + \ HCO_2H \ \longrightarrow \ R-N\big\langle$$

formic acid is the hydrogen source as a reducing agent

References

1. Eschweiler, W. *Chem. Ber.* **1905**, *38*, 880.
2. Clarke, H. T. *J. Am. Chem. Soc.* **1933**, *55*, 4571.
3. Moore, M. L. *Org. React.* **1949**, *5*, 301. (Review).
4. Pine, S. H.; Sanchez, B. L. *J. Org. Chem.* **1971**, *36*, 829.
5. Bobowski, G. *J. Org. Chem.* **1985**, *50*, 929.
6. Alder, R. W.; Colclough, D.; Mowlam, R. W. *Tetrahedron Lett.* **1991**, *32*, 7755.
7. Fache, F.; Jacquot, L.; Lemaire, M. *Tetrahedron Lett.* **1994**, *35*, 3313.
8. Bulman P., Philip C.; Heaney, H.; Rassias, G. A.; Reignier, S.; Sampler, E. P.; Talib, S. *Synlett* **2000**, 104.
9. Torchy, S.; Barbry, D. *J. Chem. Soc. (C)* **2001**, 292.
10. Rosenau, T.; Potthast, A.; Rohrling, J.; Hofinger, A.; Sixta, H.; Kosma, P. *Synth. Commun.* **2002**, *32*, 457.

Clemmensen reduction

Reduction of aldehydes and ketones to the corresponding methylene compounds using amalgamated zinc and hydrogen chloride.

$$R \overset{O}{\underset{}{\|}} R^1 \quad \xrightarrow[\text{HCl}]{\text{Zn(Hg)}} \quad R \overset{H\ H}{\underset{}{}} R^1$$

e.g.:

$$Ph \overset{O}{\underset{}{\|}} CH_3 \quad \xrightarrow[\text{HCl}]{\text{Zn(Hg)}} \quad Ph \overset{H\ H}{\underset{}{}} CH_3$$

$$Ph \overset{O}{\underset{}{\|}} CH_3 \quad \xrightarrow[\text{SET}]{\text{Zn(Hg), HCl}} \quad \underset{\text{radical anion}}{Ph \overset{OZnCl}{\underset{}{\bullet}} CH_3} \quad \xrightarrow{{}^{\ominus}H^+} \quad Ph \overset{OZnCl}{\underset{H}{}} CH_3$$

$$\xrightarrow{H^+} \quad Ph \overset{H_{\overset{+}{}}OZnCl}{\underset{H}{}} CH_3 \quad \xrightarrow{S_N2} \quad Ph \overset{H}{\underset{Cl}{}} CH_3$$
$$Cl^-$$

$$\xrightarrow{e} \quad Ph \overset{H}{\underset{\bullet}{}} CH_3 \quad \xrightarrow{{}^{\ominus}H^+} \quad Ph \overset{H\ H}{\underset{}{}} CH_3$$

References

1. Clemmensen, E. *Ber. Dtsch. Chem. Ges.* **1913**, *46*, 1837.
2. Martin, E. L. *Org. React.* **1942**, *1*, 155–209. (Review).
3. Brewster, J. H. *J. Am. Chem. Soc.* **1954**, *76*, 6360.
4. Vedejs, E. *Org. React.* **1975**, *22*, 401–422. (Review).
5. Elphimoff-Felkin, I.; Sarda, P. *Tetrahedron Lett.* **1983**, *24*, 4425.
6. Talpatra, S. K.; Chakrabati, S.; Mallik, A.; Talpatra, B. *Tetrahedron* **1990**, *46*, 6047.
7. Luchetti, L.; Rosnati, V. *J. Org. Chem.* **1991**, *56*, 6836.

8. Cheng, L.; Ma, J. *Org. Prep. Proced. Int.* **1995**, *27*, 224.
9. Kappe, T.; Aigner, R.; Roschger, P.; Schnell, B.; Stadbauer, W. *Tetrahedron* **1995**, *51*, 12923.
10. Kadam, A. J.; Baraskar, U. K.; Mane, R. B. *Indian J. Chem., Sect. B* **2000**, *39B*, 822.
11. Kohara, T.; Tanaka, H.; Kimura, K.; Fujimoto, T.; Yamamoto, I.; Arita, M. *Synthesis* **2002**, 355.

Combes quinoline synthesis

Acid-catalyzed condensation of anilines and β-diketones to assemble quinolines.

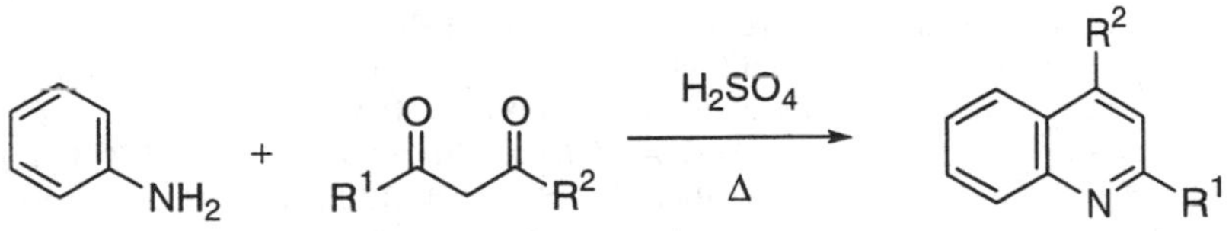
H2SO4
Δ

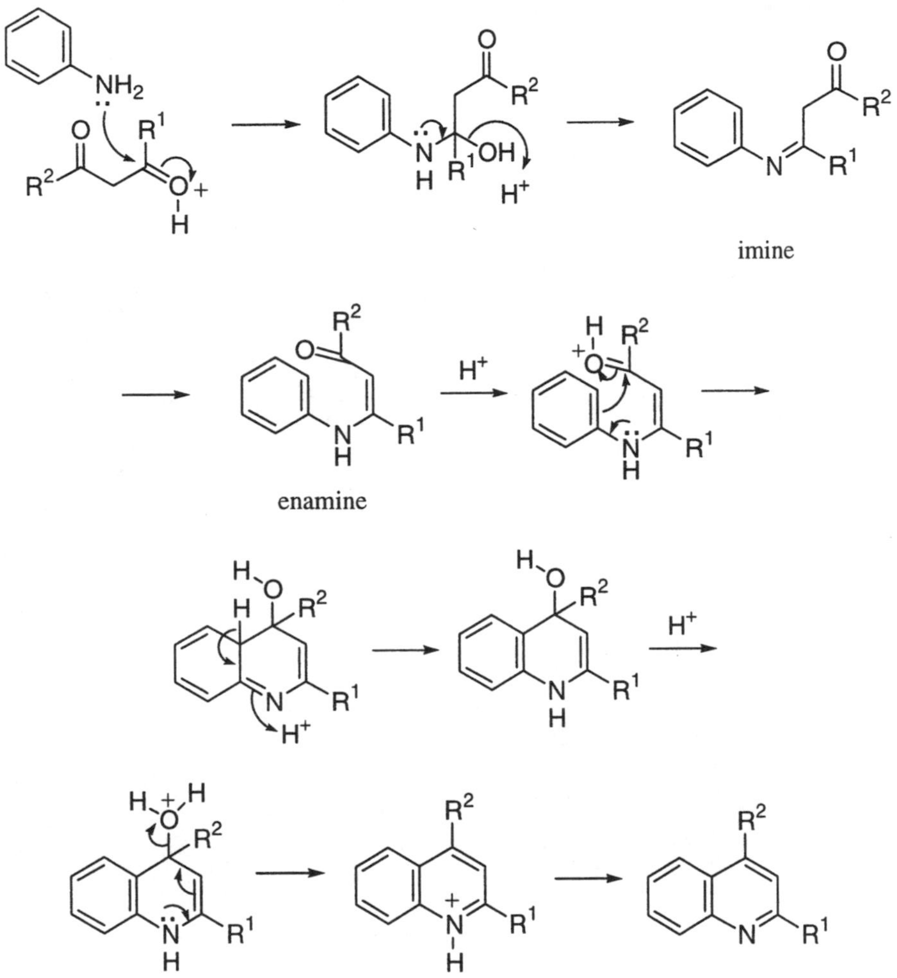
imine
enamine

80

References

1. Combes, A. *Bull. Soc. Chim. Fr.* **1888**, *49*, 89.
2. Coscia, A. T.; Dickerman, S. C. *J. Am. Chem. Soc.* **1959**, *81*, 3098.
3. Claret, P. A.; Osborne, A. G. *Org. Prep. Proced. Int.* **1970**, *2*, 305.
4. Born, J. L. *J. Org. Chem.* **1972**, *37*, 3952.
5. Ruhland, B.; Leclerc, G. *J. Heterocycl. Chem.* **1989**, *26*, 469.
6. Yamashkin, S. A.; Yudin, L. G.; Kost, A. N. *Khim. Geterotsikl. Soedin.* **1992**, 1011.
7. Davioud-Charvet, E.; Delarue, S.; Biot, C.; Schwoebel, B.; Boehme, C. C.; Muessig-brodt, A.; Maes, L.; Sergheraert, C.; Grellier, P.; Schirmer, R. H; Becker, K. *J. Med. Chem.* **2001**, *44*, 4268.

Conrad–Limpach reaction

Thermal or acid-catalyzed condensation of anilines with β-ketoesters leads to quinolin-4-ones.

Schiff base

6π electron

electrocyclization

References

1. Conrad, M.; Limpach, L. *Ber. Dtsch. Chem. Ges.* **1891**, *20*, 944.
2. Heindel, N. D.; Brodof, T. A.; Kogelschatz, J. E. *J. Heterocycl. Chem.* **1966**, *3*, 222.
3. Heindel, N. D.; Bechara, I. S.; Kennewell, P. D.; *et al. J. Med. Chem.* **1968**, *11*, 1218.
4. Perche, J. C.; Saint-Ruf, G. *J. Heterocycl. Chem.* **1974**, *11*, 93.
5. Barker, J. M.; Huddleston, P. R.; Jones, A. W.; Edwards, M. *J. Chem. Res., (S)* **1980**, 4.
6. Guay, V.; Brassard, P. *J. Heterocycl. Chem.* **1987**, *24*, 1649.
7. Deady, L. W.; Werden, D. M. *J. Org. Chem.* **1987**, *52*, 3930.
8. Hormi, O. E. O.; Peltonen, C.; Heikkila, L. *J. Org. Chem.* **1990**, *55*, 2513.
9. Jaroszewski, J. W. *J. Heterocycl. Chem.* **1990**, *27*, 1227.
10. Billah, M.; Buckley, G. M.; Cooper, N; *et al. Bioorg. Med. Chem.* **2002**, *12*, 1617.

Cook–Heilbron thiazole synthesis

5-Amino-thiazole synthesis from α-aminonitriles and carbon disulfide, or carbon oxysulfide, isothiocyanate, dithioacids.

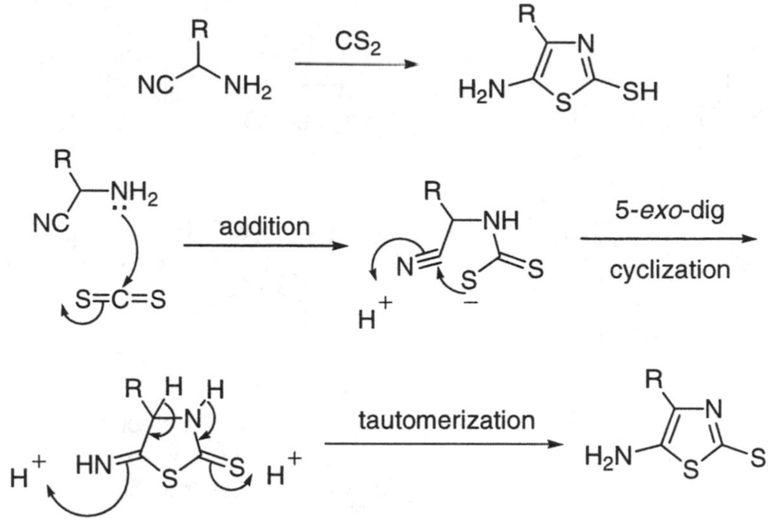

References

1. Cook, A. H.; Heilbron, I.; Macdonald, S. F.; Mahadevan, A. P. *J. Chem. Soc.* **1949**, 1064.
2. Davis, A. C.; Levy, A. L. *J. Chem. Soc.* **1951**, 2419.
3. L'Abbe, G.; Meutermans, W.; Bruynseels, M. *Bull. Soc. Chim. Belg.* **1986**, *95*, 1129.
4. El-Bayouki, Khairy A. M.; Basyouni, W. M. *Bull. Soc. Chim. Jpn.* **1988**, *61*, 3794.
5. Balquist, J. M.; Goetz, F. J. *J. Heterocycl. Chem.* **1972**, *9*, 937.

Cope elimination reaction

Thermal elimination of *N*-oxides to olefins.

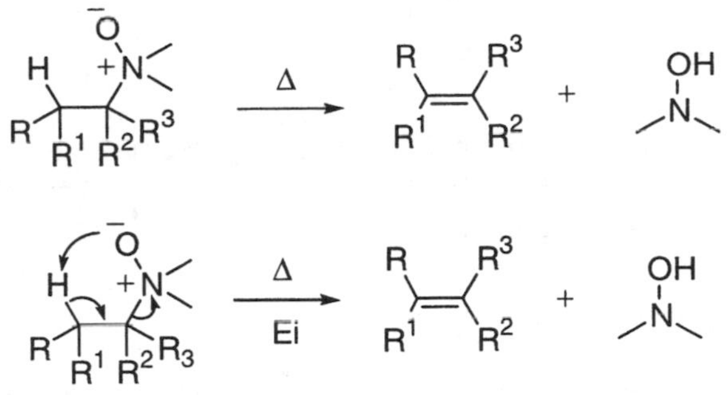

References

1. Cope, A. C.; Foster, T. T.; Towle, P. H. *J. Am. Chem. Soc.* **1949**, *71*, 3929.
2. Frey, H. M.; Walsh, R. *Chem. Rev.* **1969**, *69*, 103. (Review).
3. Gallagher, B. M.; Pearson, W. H. *Chemtracts: Org. Chem.* **1996**, *9*, 126.
4. Vidal, T.; Magnier, E.; Langlois, Y. *Tetrahedron* **1998**, *54*, 5959.
5. Gravestock, M. B.; Knight, D. W.; Malik, K. M. A.; Thornton, S. R. *Perkin 1* **2000**, 3292.
6. Bagley, M. C.; Tovey, J. *Tetrahedron Lett.* **2001**, *42*, 351.
7. Remen, L.; Vasella, A. *Helv. Chim. Acta* **2002**, *85*, 1118.
8. Garcia Martinez, A.; Teso Vilar, E.; Garcia Fraile, A.; de la Moya Cerero, S.; Lora Maroto, B. *Tetrahedron: Asymmetry* **2002**, *13*, 17.

Cope, oxy-Cope, and anionic oxy-Cope rearrangements

The Cope, oxy-Cope, and anionic oxy-Cope rearrangements belong to the category of *[3,3]-sigmatropic rearrangement*, which is a concerted process. The arrow-pushing here is only illustrative.

Cope rearrangement

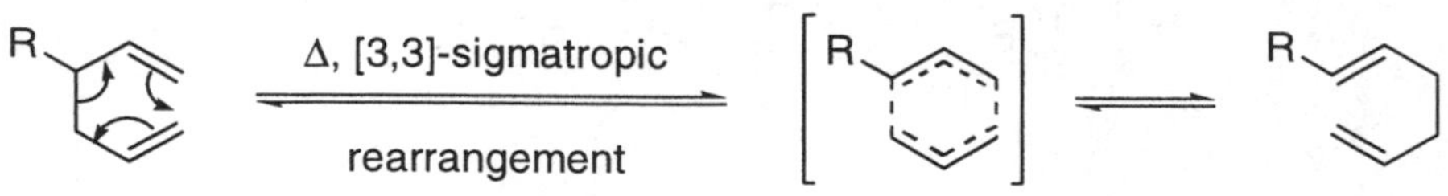

oxy-Cope rearrangement

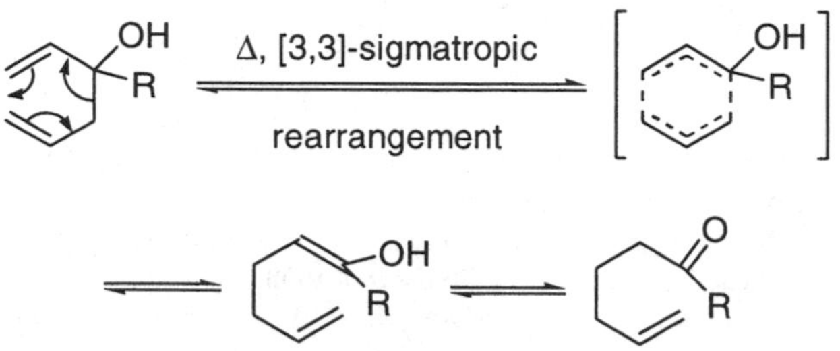

anionic oxy-Cope rearrangement

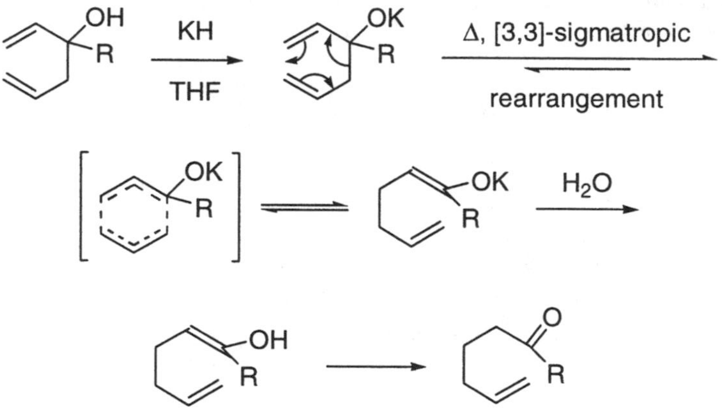

References

1. Cope, A. C.; Hardy, E. M. *J. Am. Chem. Soc.* **1940**, *62*, 441.
2. Evans, D. A.; Golob, A. M. *J. Am. Chem. Soc.* **1975**, *97*, 4765.
3. Paquette, L. A. *Angew. Chem.* **1990**, *102*, 642.

4. Hill, R. K. In *Comprehensive Organic Synthesis* Trost, B. M.; Fleming, I., Eds., Pergamon, **1991**, *Vol. 5*, 785–826. (Review).
5. Davies, H. M. L. *Tetrahedron* **1993**, *49*, 5203. (Review).
6. Paquette, L. A. *Tetrahedron* **1997**, *53*, 13971. (Review).
7. Miyashi, T.; Ikeda, H.; Takahashi, Y. *Acc. Chem. Res.* **1999**, *32*, 815. (Review).
8. Nakamura, H.; Yamamoto, Y. *Handbook of Organopalladium Chemistry for Organic Synthesis* **2002**, *2*, 2919–2934. (Review).
9. Ogawa, Y.; Ueno, T.; Karikomi, M.; Seki, K.; Haga, K.; Uyehara, T. *Tetrahedron Lett.* **2002**, *43*, 7827.
10. Clive, D. L. J.; Ou, L. *Tetrahedron Lett.* **2002**, *43*, 4559.
11. Ogawa, Y.; Toyama, M.; Karikomi, M.; Seki, K.; Haga, K.; Uyehara, T. *Tetrahedron Lett.* **2003**, *44*, 2167.

Corey–Bakshi–Shibata (CBS) reduction

Enantioselective borane reduction of ketones catalyzed by chiral oxaborolidines.

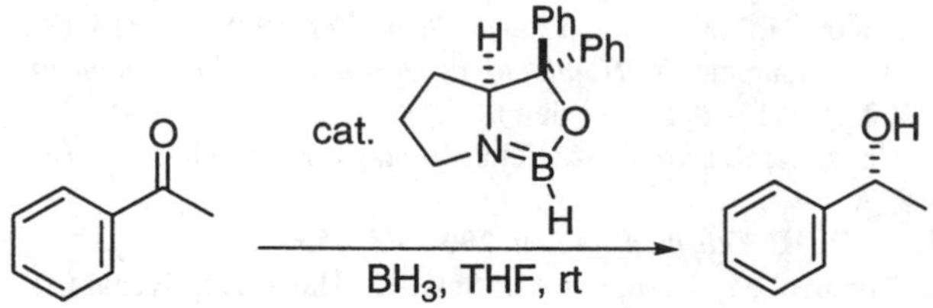

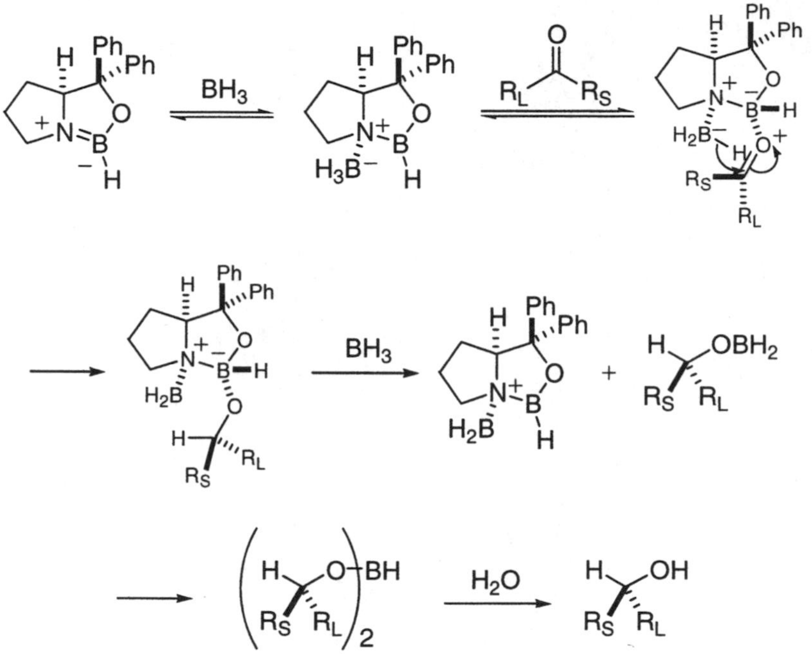

The catalytic cycle:

References

1. Corey, E. J.; Bakshi, R. K.; Shibata, S. *J. Am. Chem. Soc.* **1987**, *109*, 5551.
2. Corey, E. J.; Bakshi, R. K.; Shibata, S.; Chen, C. P.; Singh, V. K. *J. Am. Chem. Soc.* **1987**, *109*, 7925.
3. Corey, E. J.; Shibata, S.; Bakshi, R. K. *J. Org. Chem.* **1988**, *53*, 2861.
4. Cho, B. T.; Chun, Y. S. *Tetrahedron: Asymmetry* **1992**, *3*, 1583.
5. Corey, E. J.; Helal, C. J. *Tetrahedron Lett.* **1996**, *37*, 4837.
6. Clark, W. M.; Tickner-Eldridge, A. M.; Huang, G. K.; Pridgen, L. N.; Olsen, M. A.; Mills, R. J.; Lantos, I.; Baine, N. H. *J. Am. Chem. Soc.* **1998**, *120*, 4550.
7. Itsuno, S. *Org. React.* **1998**, *52*, 395–576. (Review).
8. de Koning, C. B.; Giles, R. G. F.; Green, I. R.; Jahed, N. M. *Tetrahedron Lett.* **2002**, *43*, 4199.
9. Price, M. D.; Sui, J. K.; Kurth, M. J.; Schore, N. E. *J. Org. Chem.* **2002**, *67*, 8086.
10. Fu, X.; McAllister, T. L.; Thiruvengadam, T. K.; Tann, C.-H.; Su, D. *Tetrahedron Lett.* **2003**, *44*, 801.

Corey–Chaykovsky reaction

The Corey–Chaykovsky reaction entails the reaction of a sulfur ylide, either dimethylsulfoxonium methylide **1**, Corey's ylide, sometimes known as DMSY) or dimethylsulfonium methylide **2**, with electrophile **3** such as carbonyl, olefin, imine, or thiocarbonyl, to offer **4** as the corresponding epoxide, cyclopropane, aziridine, or thiirane.

$$X = O, CH_2, NR^2, S, CHCOR^3,$$
$$CHCO_2R^3, CHCONR_2, CHCN$$

e.g.

References

1. Corey, E. J.; Chaykovsky, M. *J. Am. Chem. Soc.* **1962**, *84*, 867.
2. Corey, E. J.; Chaykovsky, M. *Tetrahedron Lett.* **1963**, 169.
3. Corey, E. J.; Chaykovsky, M. *J. Am. Chem. Soc.* **1964**, *86*, 1640.
4. Corey, E. J.; Chaykovsky, M. *J. Am. Chem. Soc.* **1965**, *87*, 1353.
5. Trost, B. M.; Melvin, L. S., Jr. *Sulfur Ylides* Academic Press: New York, **1975**. (Review).
6. Block, E. *Reactions of Organosulfur Compounds* Academic Press: New York, **1978**. (Review).
7. Gololobov, Y. G.; Nesmeyanov, A. N. *Tetrahedron* **1987**, *43*, 2609. (Review).
8. Aubé, J. In *Comprehensive Organic Synthesis;* Trost, B. M.; Fleming, I., Ed.; Pergamon: Oxford, **1991**, *vol. 1*, pp 820–825. (Review).

9. Okazaki, R.; Tokitoh, N. In *Encyclopedia of Reagents in Organic Synthesis;* Paquette, L. A., Ed.; Wiley: New York, **1995**, pp 2139–41. (Review).

10. Ng, J. S.; Liu, C. In *Encyclopedia of Reagents in Organic Synthesis;* Paquette, L. A., Ed.; Wiley: New York, **1995**, pp 2159–65. (Review).

11. Li, A.-H.; Dai, L.-X.; Aggarwal, V. K. *Chem. Rev.* **1997**, *97*, 2341. (Review).

12. Shea, K. J. *Chem. Eur. J.* **2000**, *6*, 1113. (Review).

13. Saito, T.; Akiba, D.; Sakairi, M. *Tetrahedron Lett.* **2001**, *42*, 5451.

14. Mae, M.; Matsuura, M.; Amii, H.; Uneyama, K. *Tetrahedron Lett.* **2002**, *43*, 2069.

15. Chandrasekhar, S.; Narasihmulu, Ch.; Jagadeshwar, V.; Venkatram, Reddy, K. *Tetrahedron Lett.* **2003**, *44*, 3629.

Corey–Fuchs reaction

One-carbon homologation of aldehyde to dibromoolefin, which is then treated with *n*-BuLi to produce a terminal alkyne.

R—CHO $\xrightarrow[\text{Zn}]{\text{CBr}_4,\ \text{PPh}_3}$ (R)(H)C=C(Br)(Br) $\xrightarrow{n\text{-BuLi}}$ R≡H

$\text{Br}_3\text{C}-\text{Br} \quad :\text{PPh}_3 \xrightarrow{S_N2} {}^-\text{CBr}_3 + \overset{+}{\text{Br}}-\text{PPh}_3$

$\text{Br}-\overset{+}{\text{PPh}}_3 \quad {}^-\text{CBr}_3 \xrightarrow{S_N2} \text{Br}-\underset{\text{Br}}{\overset{\text{Br}}{\text{C}}}\overset{+}{-\text{PPh}_3} \xrightarrow{S_N2}$

$\text{Br}_2 + \underset{\text{Br}}{\overset{\text{Br}}{>}}\overset{-}{\underset{}{}}\overset{+}{-\text{PPh}_3} \quad \xrightarrow{\ R-CHO\ } \quad (R)(H)C=C(Br)(Br) + \text{O}=\text{PPh}_3$

Wittig reaction (see page 396 for the mechanism)

$\text{Br}_2 + \text{Zn} \longrightarrow \text{ZnBr}_2$

(R)(H)C=C(Br)(Br), Bu⁻ $\longrightarrow$ R≡C—Br, Bu

$\longrightarrow$ R≡C⁻ $\xrightarrow[\text{workup}]{\text{acidic}}$ R≡—H

References

1. Corey, E. J.; Fuchs, P. L. *Tetrahedron Lett.* **1972**, 3769.
2. For the synthesis of 1-bromalkynes, see, Grandjean, D.; Pale, P.; Chuche, J. *Tetrahedron Lett.* **1994**, *35*, 3529.
3. Gilbert, A. M.; Miller, R.; Wulff, W. D. *Tetrahedron* **1999**, *55*, 1607.
4. Muller, T. J. J. *Tetrahedron Lett.* **1999**, *40*, 6563.

5. Serrat, X.; Cabarrocas, G.; Rafel, S.; Ventura, M.; Linden, A.; Villalgordo, J. M. *Tetrahedron: Asymmetry* **1999**, *10*, 3417.
6. Okamura, W. H.; Zhu, G.-D.; Hill, D. K.; Thomas, R. J.; Ringe, K.; Borchardt, D. B.; Norman, A. W.; Mueller, L. J. *J. Org. Chem.* **2002**, *67*, 1637.
7. Falomir, E.; Murga, J.; Carda, M.; Marco, J. A. *Tetrahedron Lett.* **2003**, *44*, 539.

92

Corey–Kim oxidation

Oxidation of alcohol to the corresponding aldehyde or ketone using NCS/DMF, followed by treatment with a base.

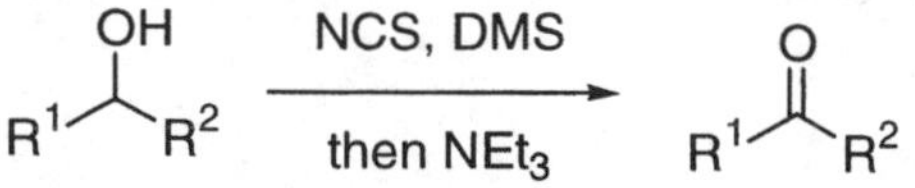

NCS = *N*-Chlorosuccinamide; DMS = **D**imethylsulfide.

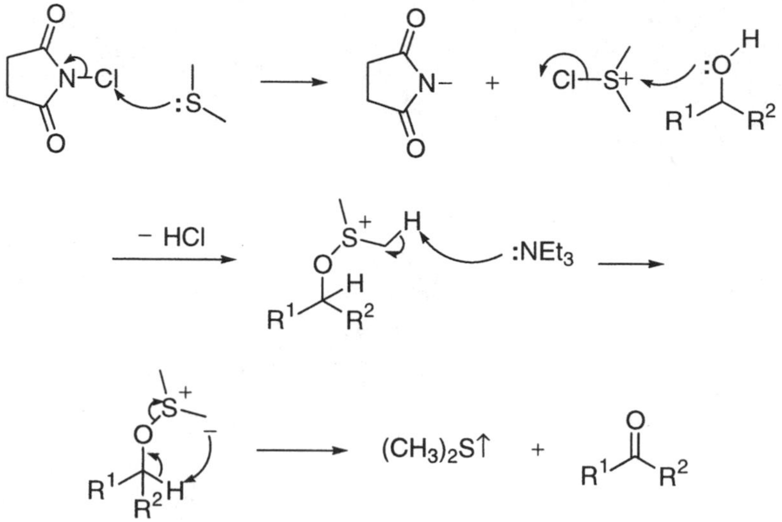

Alternatively:

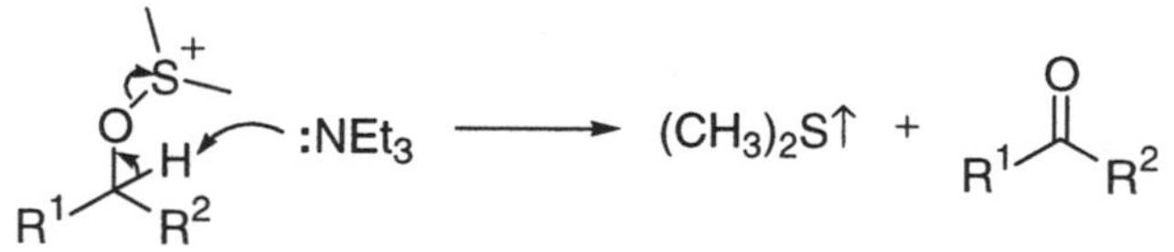

References

1. Corey, E. J.; Kim, C. U. *J. Am. Chem. Soc.* **1972**, *94*, 7586.
2. Katayama, S.; Fukuda, K.; Watanabe, T.; Yamauchi, M. *Synthesis* **1988**, 178.
3. Shapiro, G.; Lavi, Y. *Heterocycles* **1990**, *31*, 2099.
4. Pulkkinen, J. T.; Vepsalainen, J. J. *J. Org. Chem.* **1996**, *61*, 8604.
5. Crich, D.; Neelamkavil, S. *Tetrahedron Lett.* **2002**, *58*, 3865.
6. Nishide, K.; Ohsugi, S.-i.; Fudesaka, M.; Kodama, S.; Node, M. *Tetrahedron Lett.* **2002**, *58*, 5177.

Corey–Winter olefin synthesis

Transformation of diols to the corresponding olefins by sequential treatment with 1,1'-thiocarbonyldiimidazole and trimethylphosphite.

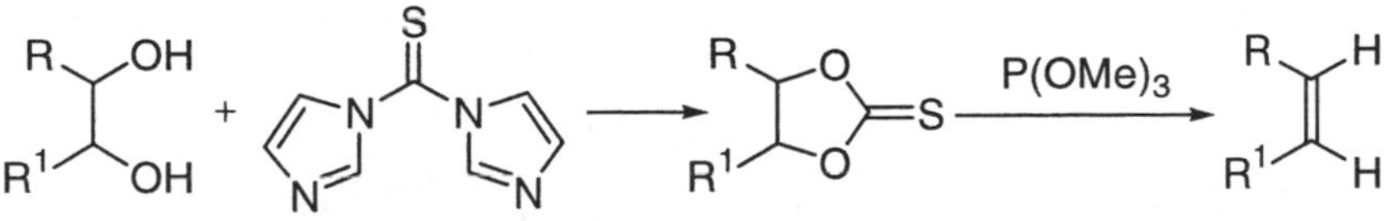

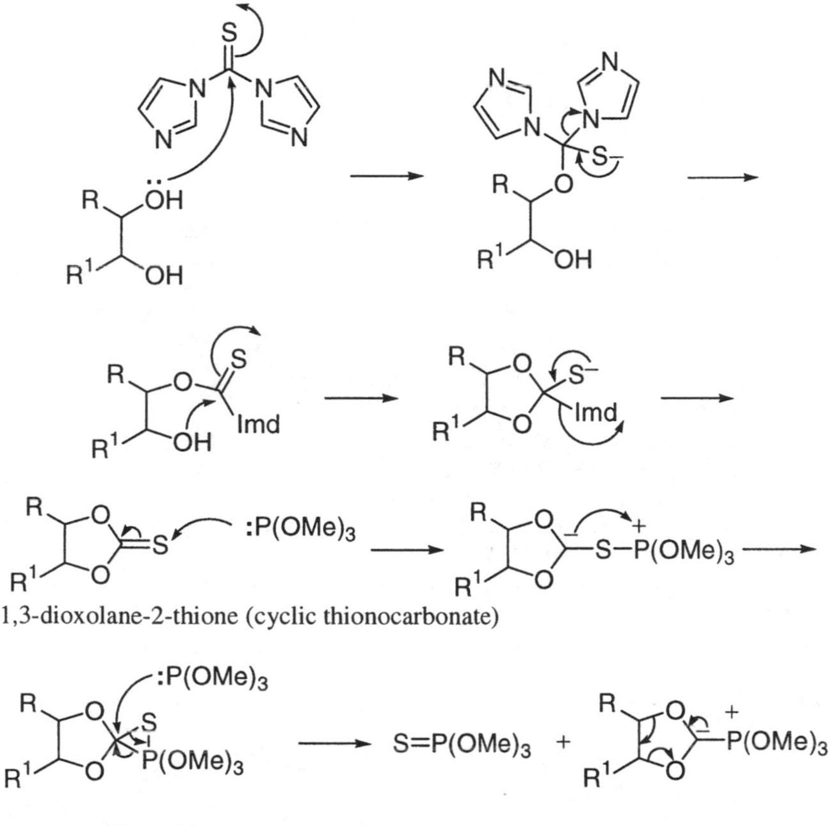

1,3-dioxolane-2-thione (cyclic thionocarbonate)

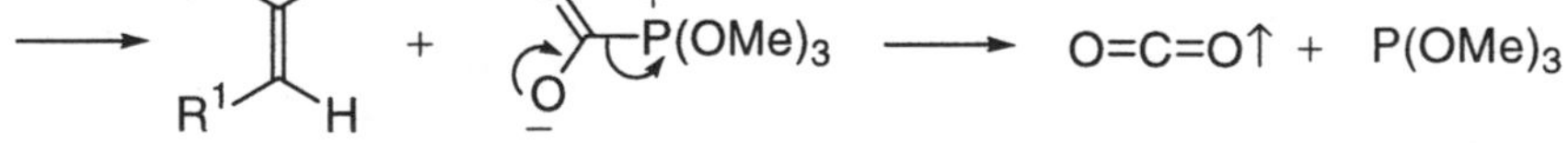

A mechanism involving a carbene intermediate is also viable as it is supported by pyrolysis studies:

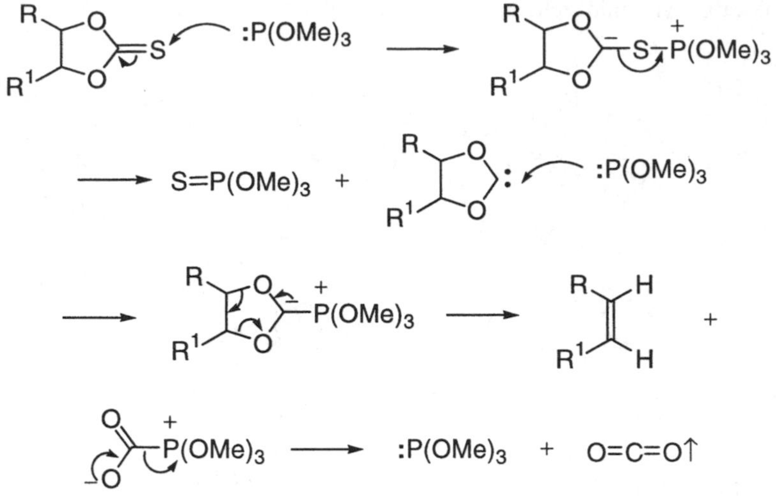

References

1. Corey, E. J.; Winter, E. *J. Am. Chem. Soc.* **1963**, *85*, 2677.
2. Horton, D.; Tindall, C. G., Jr. *J. Org. Chem.* **1970**, *35*, 3558.
3. Hartmann, W.; Fischler, H. M.; Heine, H. G. *Tetrahedron Lett.* **1972**, 853.
4. Block, E. *Org. Recat.* **1984**, *30*, 457. (Review).
5. Dudycz, L. W. *Nucleosides Nucleotides* **1989**, *8*, 35.
6. Carr, R. L. K.; Donovan, T. A., Jr.; Sharma, M. N.; Vizine, C. D.; Wovkulich, M. J. *Org. Prep. Proced. Int.* **1990**, *22*, 245.
7. Crich, D.; Pavlovic, A. B.; Wink, D. J. *Synth. Commun.* **1999**, *29*, 359.
8. Palomo, C.; Oiarbide, M.; Landa, A.; Esnal, A.; Linden, A. *J. Org. Chem.* **2001**, *66*, 4180.

Cornforth rearrangement

Thermal rearrangement of keto-oxazoles.

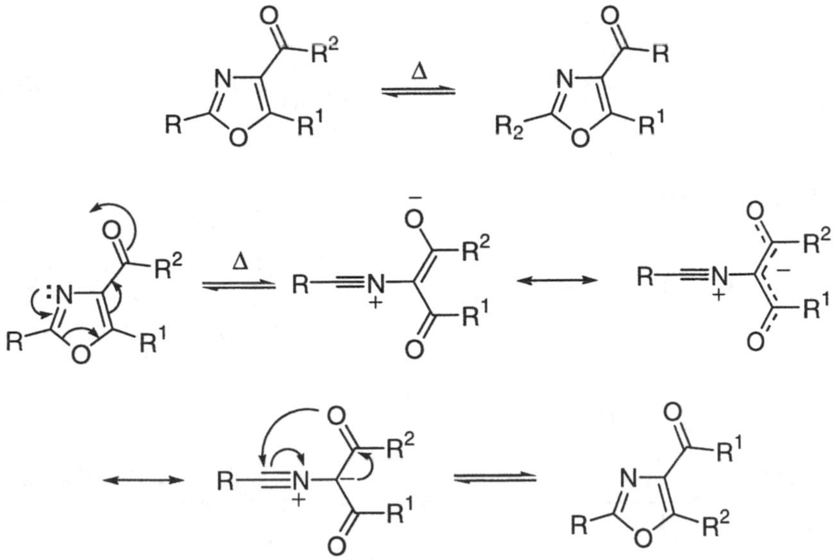

dicarbonyl nitrile ylide intermediate

References

1. Cornforth, J. W. In *The Chemistry of Penicillin* Princeton University Press: New Jersey, **1949**, 700.
2. Dewar, M. J. S.; Spanninger, P. A.; Turchi, I. J. *J. Chem. Soc., Chem. Commun.* **1973**, 925.
3. Dewar, M. J. S. *J. Am. Chem. Soc.* **1974**, *96*, 6148.
4. Dewar, M. J. S.; Turchi, I. J. *J. Org. Chem.* **1975**, *40*, 1521.
5. Williams, D. R.; McClymont, E. L. *Tetrahedron Lett.* **1993**, *34*, 7705.

Criegee glycol cleavage

Vicinal diol is oxidized to the two corresponding carbonyl compounds using $Pb(OAc)_4$.

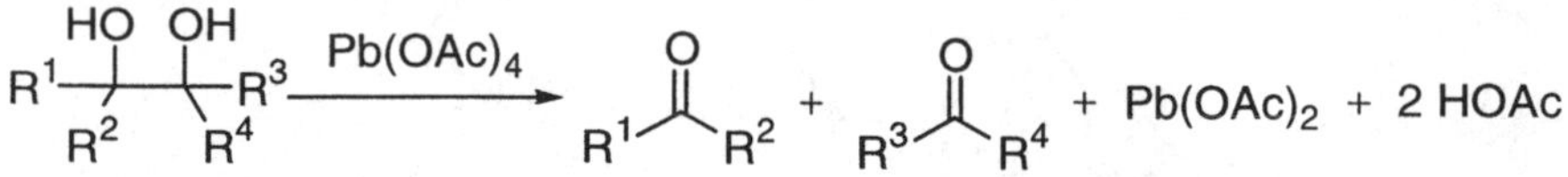

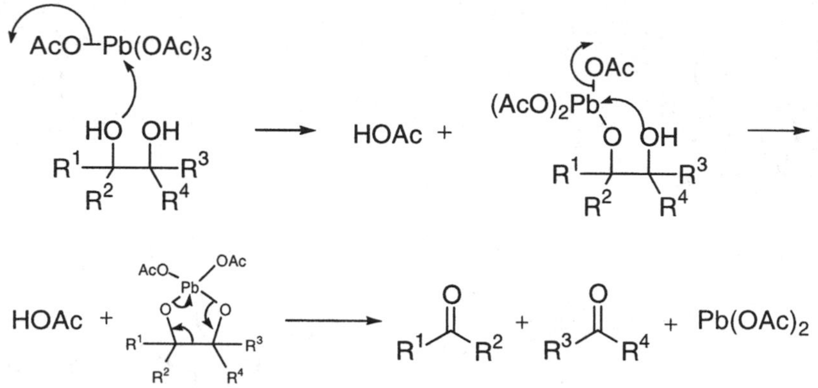

References

1. Criegee, R. *Ber. Dtsch. Chem. Ges.* **1931**, *64*, 260.
2. Michailovici, M. L. *Synthesis* **1970**, 209.
3. Hatakeyama, S.; Numata, H.; Osanai, K.; Takano, S. *J. Org. Chem.* **1989**, *54*, 3515.

Criegee mechanism of ozonolysis

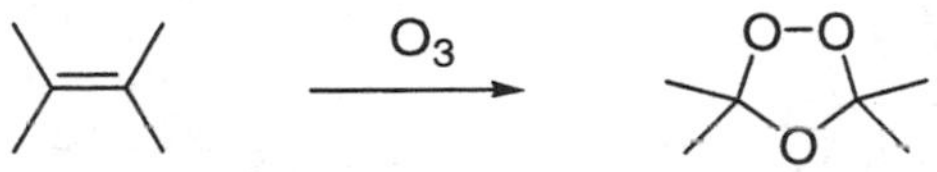

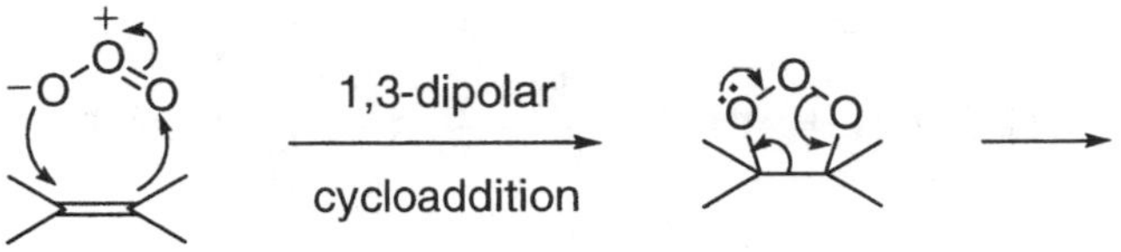

primary ozonide (1,2,3-trioxolane)

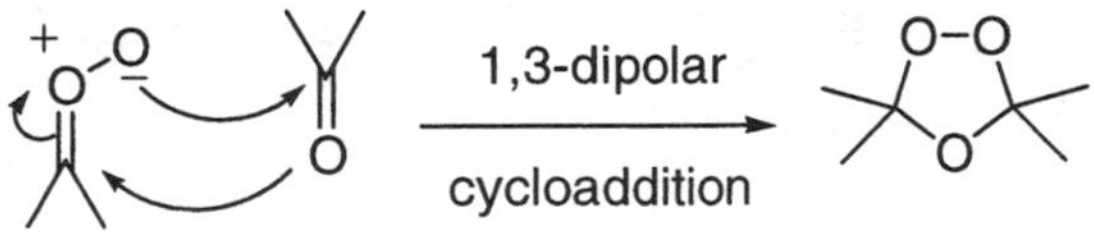

zwitterionic peroxide
(Criegee zwitterion)

secondary ozonide (1,2,4-trioxolane)

References

1. Criegee, R.; Werner, G. *Justus Liebigs Ann. Chem.* **1949**, *9*, 564.
2. Criegee, R. *Rec. Chem. Proc.* **1957**, *18*, 111.
3. Criegee, R. *Angew. Chem.* **1975**, *87*, 765.
4. Kuczkowski, R. L. *Chem. Soc. Rev.* **1992**, *21*, 79. (Review).
5. Ponec, R.; Yuzhakov, G.; Haas, Y.; Samuni, U. *J. Org. Chem.* **1997**, *62*, 2757.
6. Anglada, J. M.; Crehuet, R.; Maria Bofill, J. *Chem.--Eur. J.* **1999**, *5*, 1809.
7. Dussault, P. H.; Raible, J. M. *Org. Lett.* **2000**, *2*, 3377.

98

Curtius rearrangement

Thermal decomposition of acyl azides into amines *via* isocyanate intermediates.

$$R-C(=O)-Cl \xrightarrow{NaN_3} R-C(=O)-N_3 \xrightarrow{\Delta}$$

$$N_2\uparrow + R-N=C=O \xrightarrow{H_2O} R-NH_2 + CO_2\uparrow$$

isocyanate intermediate

$$\xrightarrow{} R-N(H)-C(=O)-O-H \xrightarrow{} R-NH_2 + CO_2\uparrow$$

References

1. Curtius, T. *Ber. Dtsch. Chem. Ges.* **1890**, *23*, 3023.
2. Chen, J. J.; Hinkley, J. M.; Wise, D. S.; Townsend, L. B. *Synth. Commun.* **1996**, *26*, 617.
3. Am Ende, D. J.; DeVries, K. M.; Clifford, P. J.; Brenek, S. J. *Org. Process Res. Dev.* **1998**, *2*, 382.
4. Braibante, M. E. F.; Braibante, H. S.; Costenaro, E. R. *Synthesis* **1999**, 943.
5. Migawa, M. T.; Swayze, E. E. *Org. Lett.* **2000**, *2*, 3309.
6. Haddad, M. E.; Soukri, M.; Lazar, S.; Bennamara, A.; Guillaumet, G.; Akssira, M. *J. Heterocycl. Chem.* **2000**, *37*, 1247.
7. Mamouni, R.; Aadil, M.; Akssira, M.; Lasri, J.; Sepulveda-Arques, J. *Tetrahedron Lett.* **2003**, *44*, 2745.

Dakin oxidation

Cf. Baeyer–Villiger oxidation.

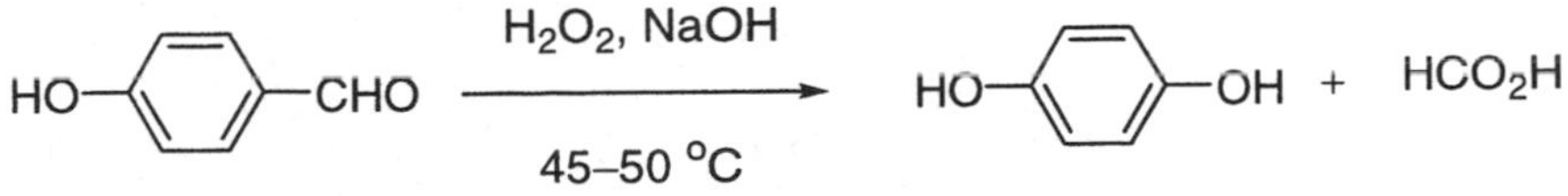

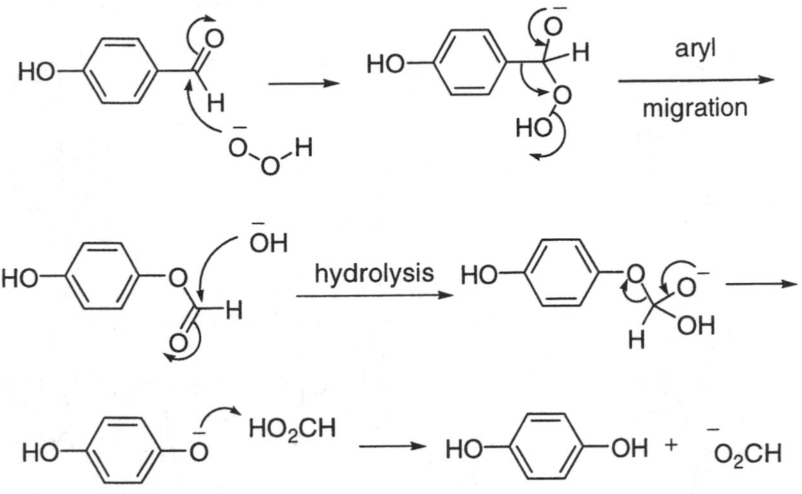

References:

1. Dakin, H. D. *J. Am. Chem. Soc.* **1909**, *42*, 477.
2. Matsumoto, M.; Kobayashi, H.; Hotta, Y. *J. Org. Chem.* **1984**, *49*, 4740.
3. Zhu, J.; Beugelmans, R.; Bigot, A.; Singh, G. P.; Bois-Choussy, M. *Tetrahedron Lett.* **1993**, *34*, 7401.
4. Guzman, J. A.; Mendoza, V.; Garcia, E.; Garibay, C. F.; Olivares, L. Z.; Maldonado, L. A. *Synth. Commun.* **1995**, *25*, 2121.
5. Jung, M. E.; Lazarova, T. I. *J. Org. Chem.* **1997**, *62*, 1553.
6. Varma, R. S.; Naicker, K. P. *Org. Lett.* **1999**, *1*, 189.
7. Roy, A.; Reddy, K. R.; Mohanta, P. K.; Ila, H.; Junjappa, H. *Synth. Commun.* **1999**, *29*, 3781.
8. Lawrence, N. J.; Rennison, D.; Woo, M.; McGown, A. T.; Hadfield, J. A. *Bioorg. Med. Chem. Lett.* **2001**, *11*, 51.

Dakin–West reaction

Acylation of α-amino acids leading to α-acetamido ketones *via* azalactone intermediates.

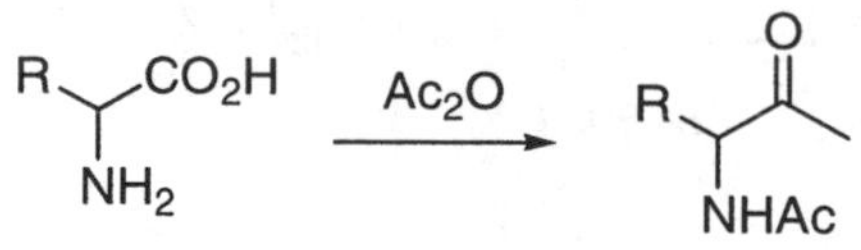

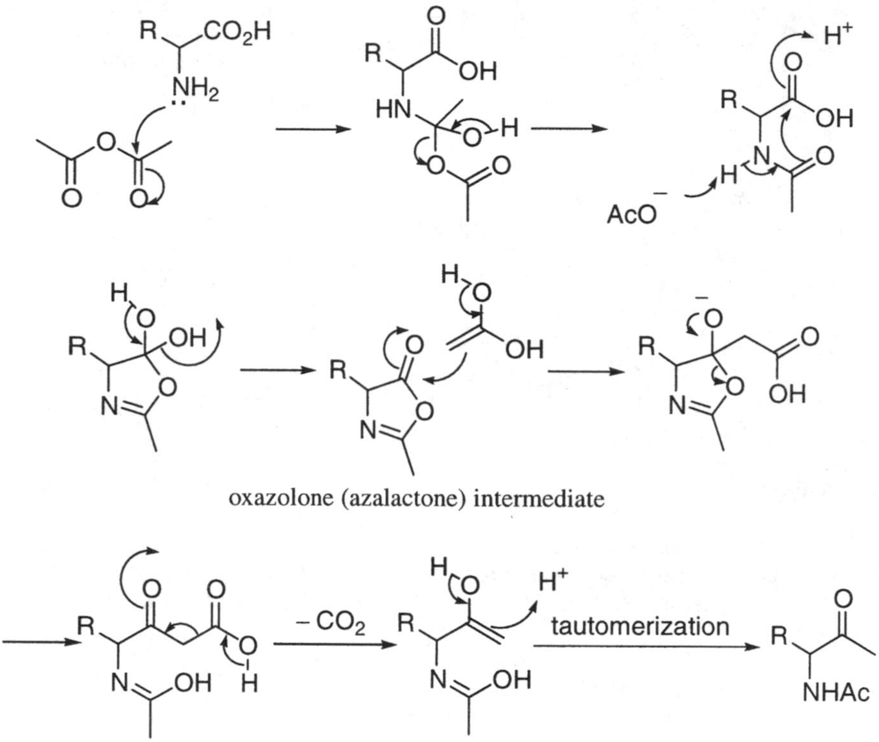

oxazolone (azalactone) intermediate

References:

1. Dakin, H. D.; West, R. *J. Biol. Chem.* **1928**, *91*, 745.
2. Buchanan, G. L. *Chem. Soc. Rev.* **1988**, *17*, 91. (Review).
3. Jung, M. E.; Lazarova, T. I. *J. Org. Chem.* **1997**, *62*, 1553.
4. Kawase, M.; Hirabayashi, M.; Koiwai, H.; Yamamoto, K.; Miyamae, H. *Chem. Commun.* **1998**, 641.
5. Kawase, M.; Okada, Y.; Miyamae, H. *Heterocycles* **1998**, *48*, 285.
6. Kawase, M.; Hirabayashi, M.; Kumakura, H.; Saito, S.; Yamamoto, K. *Chem. Pharm. Bull.* **2000**, *48*, 114.

7. Kawase, M.; Hirabayashi, M.; Saito, S. *Recent Res. Dev. Org. Chem.* **2001**, *4*, 283–293. (Review).

8. Fischer, R. W.; Misun, M. *Org. Proc. Res. Dev.* **2001**, *5*, 581.

9. Orain, D.; Canova, R.; Dattilo, M.; Kloppner, E.; Denay, R.; Koch, G.; Giger, R. *Synlett* **2002**, 1443.

10. Godfrey, A. G.; Brooks, D. A.; Hay, L. A.; Peters, M.; McCarthy, J. R.; Mitchell, D. *J. Org. Chem.* **2003**, *68*, 2623.

Danheiser annulation

Trimethylsilylcyclopentene annulation from an α,β-unsaturated ketone and trimethylsilylallene in the presence of a Lewis acid.

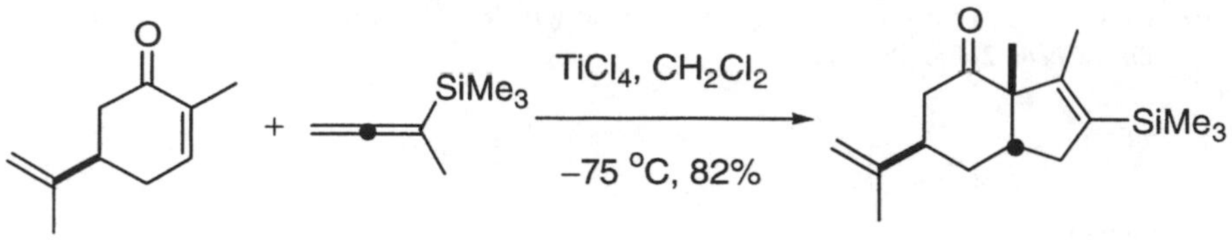

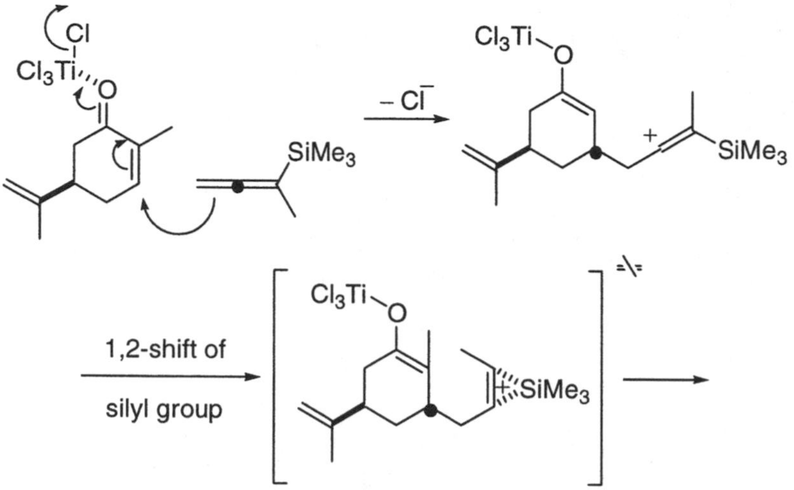

1,2-shift of silyl group

Transition State

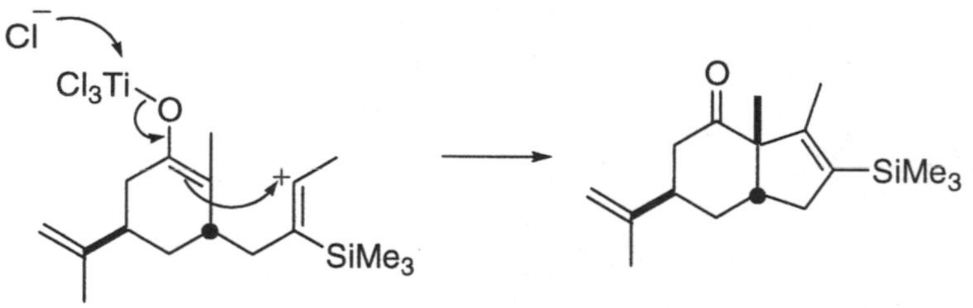

References

1. Danheiser, R. L; Carini, D. J.; Basak, A. *J. Am. Chem. Soc.* **1981**, *103*, 1604.
2. Danheiser, R. L; Carini, D. J.; Fink, D. M.; Basak, A. *Tetrahedron* **1983**, *39*, 935.
3. Danheiser, R. L; Fink, D. M.; Tsai, Y.-M. *Org. Synth.* **1988**, *66*, 8.
4. Engler, T. A.; Agrios, K.; Reddy, J. P.; Iyengar, R. *Tetrahedron Lett.* **1996**, *37*, 327.
5. Friese, J. C.; Krause, S.; Schafer, H. J. *Tetrahedron Lett.* **2002**, *43*, 2683.

Darzens glycidic ester condensation

α,β-Epoxy esters (glycidic esters) from base-catalyzed condensation of α-haloesters with carbonyl compounds.

References

1. Darzens, G. *Compt. Rend.* **1904**, *139*, 1214.
2. Bauman, J. G.; Hawley, R. C.; Rapoport, H. *J. Org. Chem.* **1984**, *49*, 3791.
3. Takahashi, T.; Muraoki, M.; Capo, M.; Koga, K. *Chem. Pharm. Bull.* **1995**, *43*, 1821.
4. Ohkata, K.; Kimura, J.; Shinohara, Y.; Takagi, R.; Hiraga, Y. *Chem. Commun.* **1996**, 2411.
5. Takagi, R.i; Kimura, J.; Shinohara, Y.; Ohba, Y.; Takezono, K.; Hiraga, Y.; Kojima, S.; Ohkata, K. *J. Chem. Soc., Perkin Trans. 1* **1998**, 689.
6. Hirashita, T.; Kinoshita, K.; Yamamura, H.; Kawai, M.; Araki, S. *J. Chem. Soc., Perkin Trans. 1* **2000**, 825.
7. Shinohara, Y.; Ohba, Y.; Takagi, R.; Kojima, S.; Ohkata, K. *Heterocycles* **2001**, *55*, 9.
8. Arai, A.; Suzuki, Y.; Tokumaru, K.; Shioiri, T. *Tetrahedron Lett.* **2002**, *43*, 833.
9. Davis, F. A.; Wu, Y.; Yan, H.; McCoull, W.; Prasad, K. R. *J. Org. Chem.* **2003**, *68*, 2410.

Davis chiral oxaziridine reagents

Chiral *N*-sulfonyloxaziridines employed for asymmetric hydroxylation *etc.*

References

1. Davis, F. A.; Vishwakarma, L. C.; Billmers, J. M.; Finn, J. *J. Org. Chem.* **1984**, *49*, 3241.
2. Davis, F. A.; Billmers, J. M.; Gosciniak, D. J.; Towson, J. C.; Bach, R. D. *J. Org. Chem.* **1986**, *51*, 4240.
3. Davis, F. A.; Chen, B.-C. *Chem. Rev.* **1992**, *92*, 919. (Review).
4. Davis, F. A.; ThimmaReddy, R.; Weismiller, M. C. *J. Am. Chem. Soc.* **1989**, *111*, 5964.
5. Davis, F. A.; Kumar, A.; Chen, B. C. *J. Org. Chem.* **1991**, *56*, 1143.
6. Davis, F. A.; Reddy, R. T.; Han, W.; Carroll, P. J. *J. Am. Chem. Soc.* **1992**, *114*, 1428.
7. Tagami, K.; Nakazawa, N.; Sano, S.; Nagao, Y. *Heterocycles* **2000**, *53*, 771.
8. Takeda, K.; Sawada, Y.; Sumi, K. *Org. Lett.* **2002**, *4*, 1031.

de Mayo reaction

[2 + 2] Photoaddition of enols from 1,3-diketones with olefins is followed by a retro-aldol reaction to give 1,5-diketones.

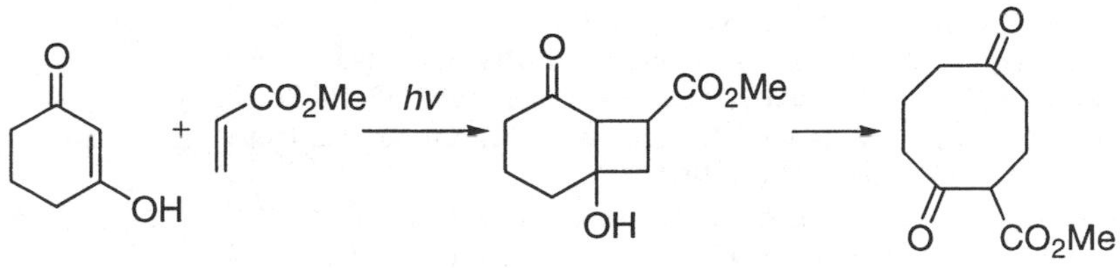

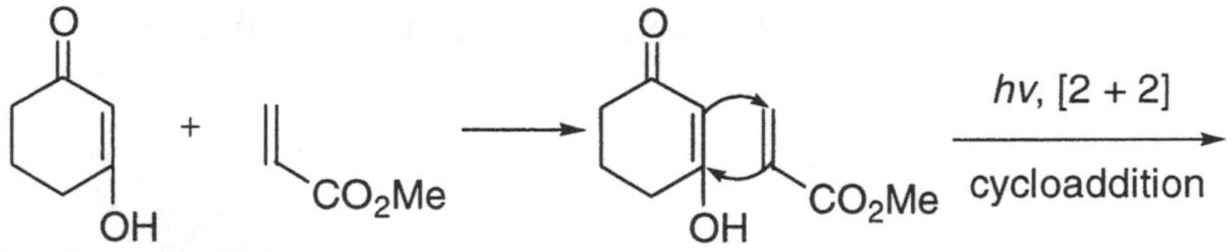

head-to-tail alignment

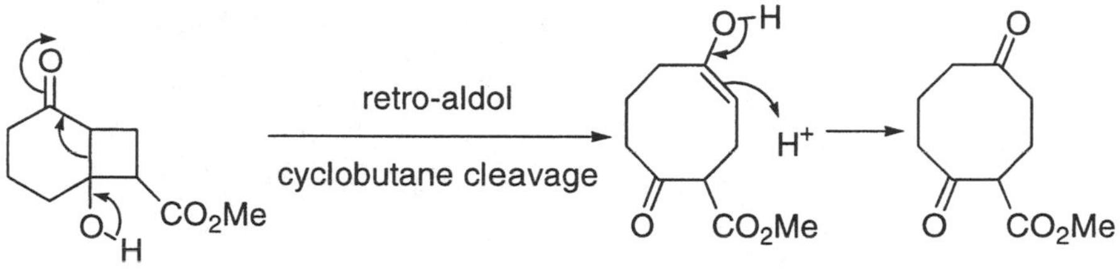

A minor regioisomer:

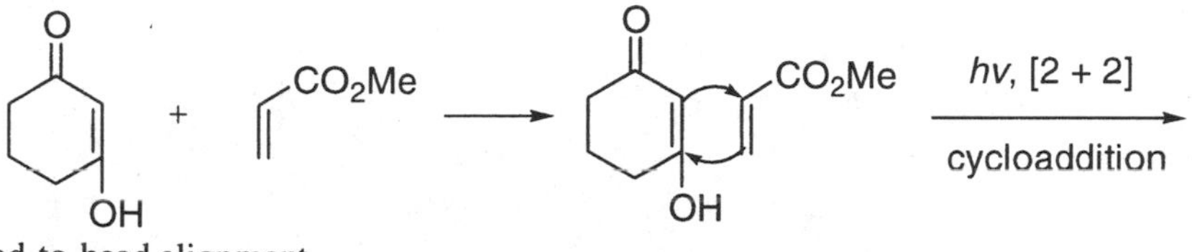

head-to-head alignment

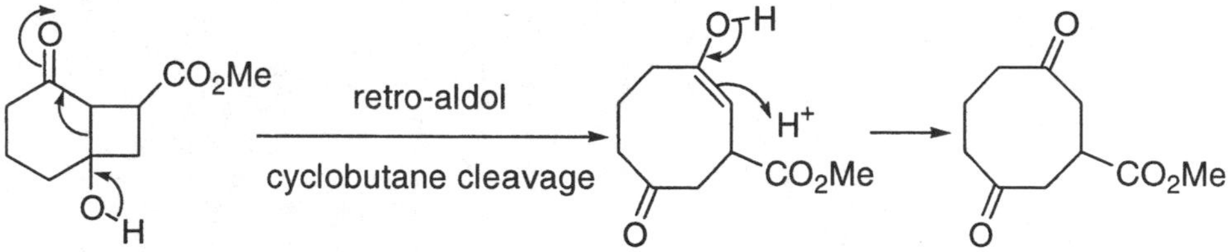

References

1. de Mayo, P.; Takeshita, H.; Sattar, A. B. M. A. *Proc. Chem. Soc., London* **1962**, 119.
2. de Mayo, P. *Acc. Chem. Res.* **1971**, *4*, 49. (Review).
3. Oppolzer, W. *Pure Appl. Chem.* **1981**, *53*, 1189. (Review).
4. Pearlman, B. A. *J. Am. Chem. Soc.* **1979**, *101*, 6398.
5. Kaczmarek, R.; Blechert, S. *Tetrahedron Lett.* **1986**, *27*, 2845.
6. Disanayaka, B. W.; Weedon, A. C. *J. Org. Chem.* **1987**, *52*, 2905.
7. Sato, M.; Abe, Y.; Takayama, K.; Sekiguchi, K.; Kaneko, C.; Inoue, N.; Furuya, T.; Inukai, N. *J. Heterocycl. Chem.* **1991**, *28*, 241.
8. Sato, M.; Sunami, S.; Kogawa, T.; Kaneko, C. *Chem. Lett.* **1994**, 2191.
9. Quevillon, T. M.; Weedon, A. C. *Tetrahedron Lett.* **1996**, *37*, 3939.
10. Blaauw, R. H.; Briere, J.-F.; de Jong, R.; Benningshof, J. C. J.; van Ginkel, A. E.; Fraanje, J.; Goubitz, K.; Schenk, H.; Rutjes, F. P. J. T.; Hiemstra, H. *J. Org. Chem.* **2001**, *66*, 233.

Demjanov rearrangement

Carbocation rearrangement of primary amines *via* diazotization to give alcohols.

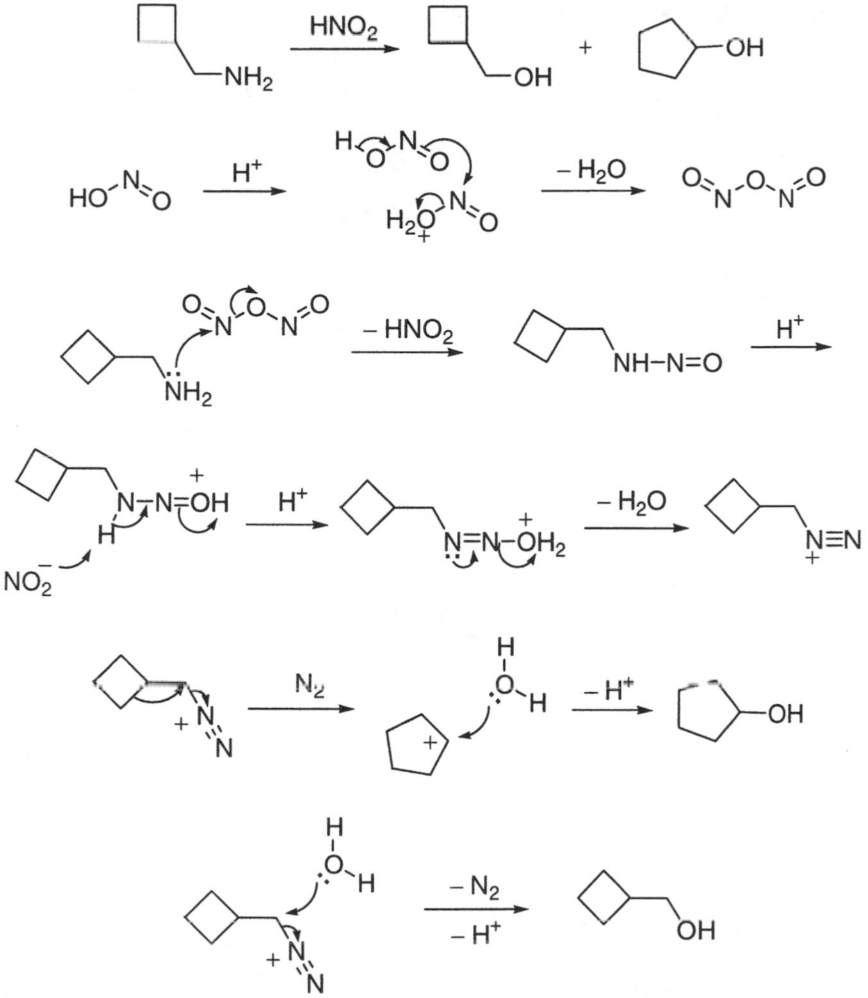

References

1. Demjanov, N. J.; Lushnikov, M. *J. Russ. Phys. Chem. Soc.* **1903**, *35*, 26.
2. Kotani, R. *J. Org. Chem.* **1965**, *30*, 350.
3. Diamond, J.; Bruce, W. F.; Tyson, F. T. *J. Org. Chem.* **1965**, *30*, 1840.
4. Alam, S. N.; MacLean, D. B. *Can. J. Chem.* **1965**, *43*, 3433.
5. Cooper, C. N.; Jenner, P. J.; Perry, N. B.; Russell-King, J.; Storesund, H. J.; Whiting, M. C. *J. Chem. Soc., Perkin Trans. 2* **1982**, 605.

6. Nakazaki, M.; Naemura, K.; Hashimoto, M. *J. Org. Chem.* **1983**, *48*, 2289.
7. Uyehara, T.; Kabasawa, Y.; Furuta, Toshiaki; K., T. *Bull. Chem. Soc. Jpn.* **1986**, *59*, 539.
8. Fattori, D.; Henry, S.; Vogel, P. *Tetrahedron* **1993**, *49*, 1649.
9. Boeckman, R. K. *Org. Synth.* **1999**, *77*, 141.

Dess–Martin periodinane oxidation

Oxidation of alcohols to the corresponding carbonyl compounds using triacetoxyperiodinane.

References

1. Dess, P. B.; Martin, J. C. *J. Am. Chem. Soc.* **1978**, *100*, 300.
2. Dess, P. B.; Martin, J. C. *J. Am. Chem. Soc.* **1979**, *101*, 5294.
3. Dess, P. B.; Martin, J. C. *J. Am. Chem. Soc.* **1991**, *113*, 7277.
4. Ireland, R. E.; Liu, L. *J. Org. Chem.* **1993**, *58*, 2899.
5. Speicher, A.; Bomm, V.; Eicher, T. *J. Prakt. Chem.* **1996**, *338*, 588.
6. Chaudhari, S. S.; Akamanchi, K. G. *Synthesis* **1999**, 760.
7. Nicolaou, K. C.; Zhong, Y.-L.; Baran, P. S. *Angew. Chem., Int. Ed.* **2000**, *39*, 622.
8. Jenkins, N. E.; Ware, R. W., Jr.; Atkinson, R. N.; King, S. B. *Synth. Commun.* **2000**, *30*, 947.
9. Promarak, V.; Burn, P. L. *Perkin 1* **2001**, 14.
10. Wavrin, L.; Viala, J. *Synthesis* **2002**, 326.
11. Wellner, E.; Sandin, H.; Paakkonen, L. *Synthesis* **2003**, 223.
12. Langille, N. F.; Dakin, L. A.; Panek, J. S. *Org. Lett.* **2003**, *5*, 575.

110

Dieckmann condensation

The Dieckmann condensation is the intramolecular version of the Claisen condensation.

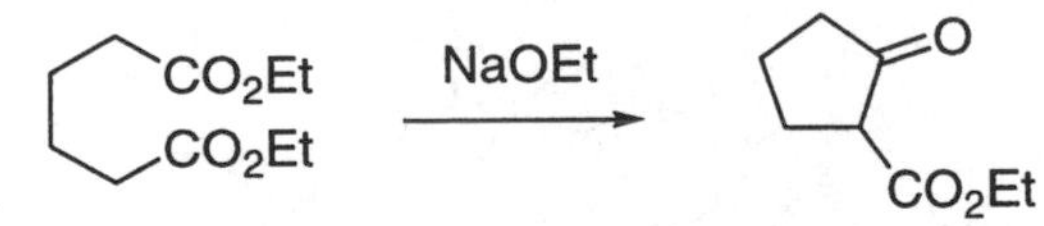

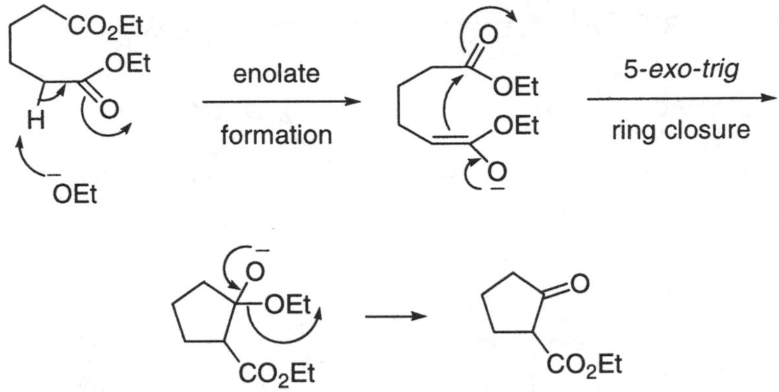

References

1. Dieckmann, W. *Ber. Dtsch. Chem. Ges.* **1894**, *27*, 102.
2. Davis, B. R.; Garrett, P. J. *Comp. Org. Synth.* **1991**, *2*, pp 806–829. (Review).
3. Toda, F.; Suzuki, T.; Higa, S. *J. Chem. Soc., Perkin Trans. 1* **1998**, 3521.
4. Shindo, M.; Sato, Y.; Shishido, K. *J. Am. Chem. Soc.* **1999**, *121*, 6507.
5. Balo, C.; Fernandez, F.; Garcia-Mera, X.; Lopez, C. *Org. Prep. Proced. Int.* **2001**, *32*, 563.
6. Deville, J. P.; Behar, V. *Org. Lett.* **2002**, *4*, 1403.
7. Ho, J. Z.; Mohareb, R. M.; Ahn, J. H.; Sim, T. B.; Rapoport, H. *J. Org. Chem.* **2003**, *68*, 109.

Diels–Alder reaction

The Diels–Alder reaction, reverse electronic demand Diels–Alder reaction, as well as the hetero-Diels–Alder reaction, belong to the category of *[4+2]-cycloaddition reactions*, which are concerted processes. The arrow-pushing here is merely illustrative.

Normal Diels–Alder reaction

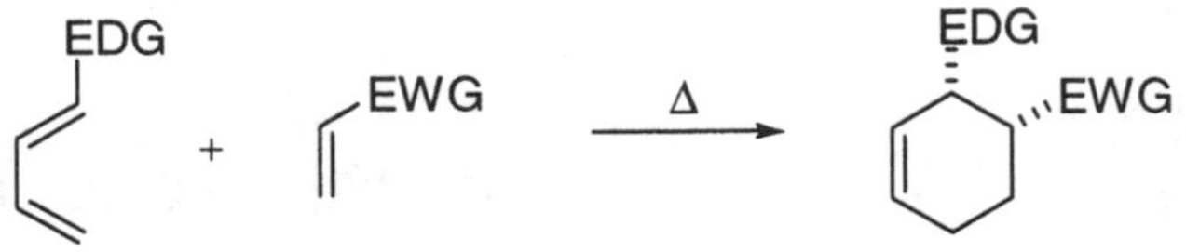

diene   dienophile   adduct

EDG = electron-donating group; EWG = electron-withdrawing group

e.g.

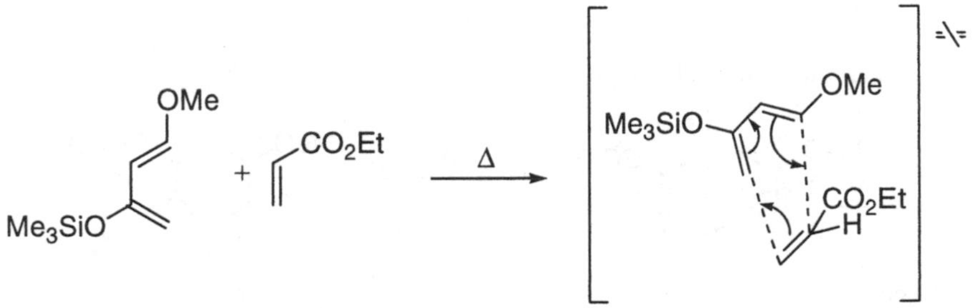

Danishefsky diene   Alder's *endo* rule

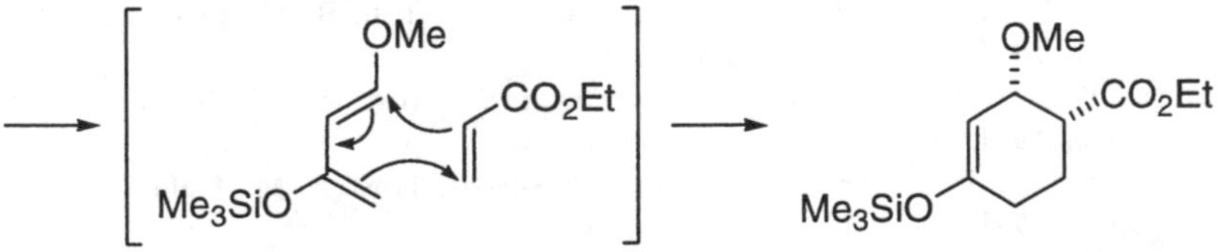

Inverse electronic demand Diels–Alder reaction

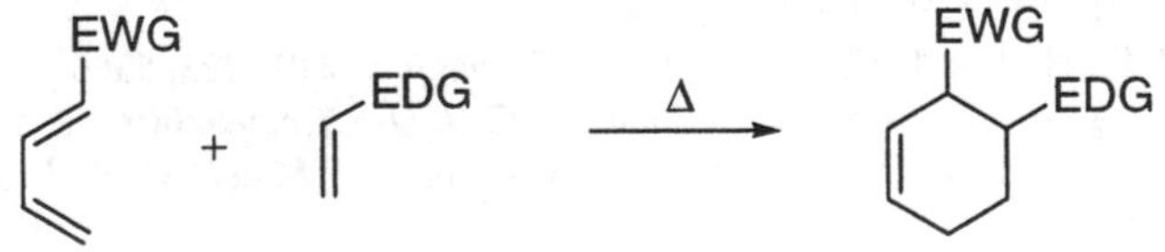

diene   dienophile   adduct

112

e.g.

Hetero-Diels–Alder reaction

a. Heterodiene addition to dienophile

b. Heterodienophile addition to diene

References

1. Diels, O.; Alder, K. *Justus Liebigs Ann. Chem.* **1928**, *460*, 98.
2. Oppolzer, W. In *Comprehensive Organic Synthesis;* Trost, B. M.; Fleming, I., Eds.; Pergamon, **1991**, *Vol. 5*, 315–399. (Review).
3. Boger, D. L. In *Comprehensive Organic Synthesis;* Trost, B. M.; Fleming, I., Eds.; Pergamon, **1991**, *Vol. 5*, 451–512. (Review).
4. Weinreb, S. M. In *Comprehensive Organic Synthesis;* Trost, B. M.; Fleming, I., Eds.; Pergamon, **1991**, *Vol. 5*, 401–499. (Review).
5. Mehta, G.; Uma, R. *Acc. Chem. Res.* **2000**, *33*, 278. (Review).
6. Jorgensen, K. A. *Angew. Chem., Int. Ed.* **2000**, *39*, 3558.
7. Evans, D. A.; Johnson, J. S.; Olhava, E. J. *J. Am. Chem. Soc.* **2000**, *122*, 1635.
8. Huang, Y.; Rawal, V. H. *Org. Lett.* **2000**, *2*, 3321.
9. Doyle, M. P.; Phillips, I. M.; Hu, W. *J. Am. Chem. Soc.* **2001**, *123*, 5366.
10. Placios, F.; Alonso, C.; Amezua, P.; Rubiales, G. *J. Org. Chem.* **2002**, *67*, 1941.
11. Richter, F.; Bauer, M.; Perez, C.; Maichle-Mössmer, C.; Maier, M. E. *J. Org. Chem.* **2002**, *67*, 2474.
12. Gainelli, G.; Galletti, P.; Giacomini, D.; Quintavalla, A. *Tetrahedron Lett.* **2003**, *44*, 93.

Dienone–phenol rearrangement

Acid-promoted rearrangement of 4,4-disubstituted cyclohexadienones to 3,4-disubstituted phenols.

References

1. Shine, H. J. In *Aromatic Rearrangement;* Elsevier: New York, **1967**, pp 55–68. (Review).
2. Schultz, A. G.; Hardinger, S. A. *J. Org. Chem.* **1991**, *56*, 1105.
3. Schultz, A. G.; Green, N. J. *J. Am. Chem. Soc.* **1991**, *114*, 1824.
4. Hart, D. J.; Kim, A.; Krishnamurthy, R.; Merriman, G. H.; Waltos, A. M. *Tetrahedron* **1992**, *48*, 8179.
5. Frimer, A. A.; Marks, V.; Sprecher, M.; Gilinsky-Sharon, P. *J. Org. Chem.* **1994**, *59*, 1831.
6. Oshima, T.; Nakajima, Y.-i.; Nagai, T. *Heterocycles* **1996**, *43*, 619.
7. Draper, R. W.; Puar, M. S.; Vater, E. J.; Mcphail, A. T. *Steroids* **1998**, *63*, 135.
8. Banerjee, A. K.; Castillo-Melendez, J. A.; Vera, W.; Azocar, J. A.; Laya, M. S. *J. Chem. Res., (S)* **2000**, 324.
9. Zimmerman, H. E.; Cirkva, V. *J. Org. Chem.* **2001**, *66*, 1839.

Di-π-methane rearrangement

1,4-Dienes to vinylcyclopropanes under photolysis.

1,4-diene   vinylcyclopropane

diradical

diradical

References

1. Zimmerman, H. E.; Grunewald, G. L. *J. Am. Chem. Soc.* **1966**, *88*, 183.
2. Janz, K. M.; Scheffer, J. R. *Tetrahedron Lett.* **1999**, *40*, 8725.
3. Zimmerman, H. E.; Cirkva, V. *Org. Lett.* **2000**, *2*, 2365.
4. Tu, Y. Q.; Fan, C. A.; Ren, S. K.; Chan, A. S. C. *Perkin 1* **2000**, 3791.
5. Jimenez, M. C.; Miranda, M. A.; Tormos, R. *Chem. Commun.* **2000**, 2341.
6. Ihmels, H.; Mohrschladt, C. J.; Grimme, J. W.; Quast, H. *Synthesis* **2001**, 1175.
7. Altundas, R.; Dastan, A.; Unaldi, N. S.; Guven, K.; Uzun, O.; Balci, M. *Eur. J. Org. Chem.* **2002**, 526.
8. Zimmerman, H. E.; Chen, W. *Org. Lett.* **2002**, *4*, 1155.
9. Tanifuji, N.; Huang, H.; Shinagawa, Y.; Kobayashi, K. *Tetrahedron Lett.* **2003**, *44*, 751.

Doebner reaction

Three-component reaction yielding isoquinolines.

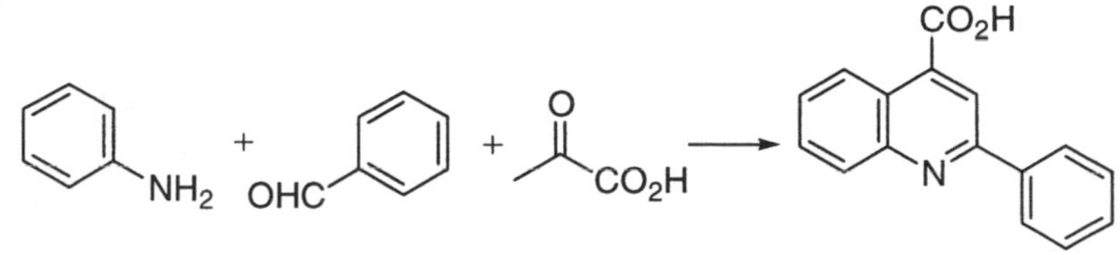

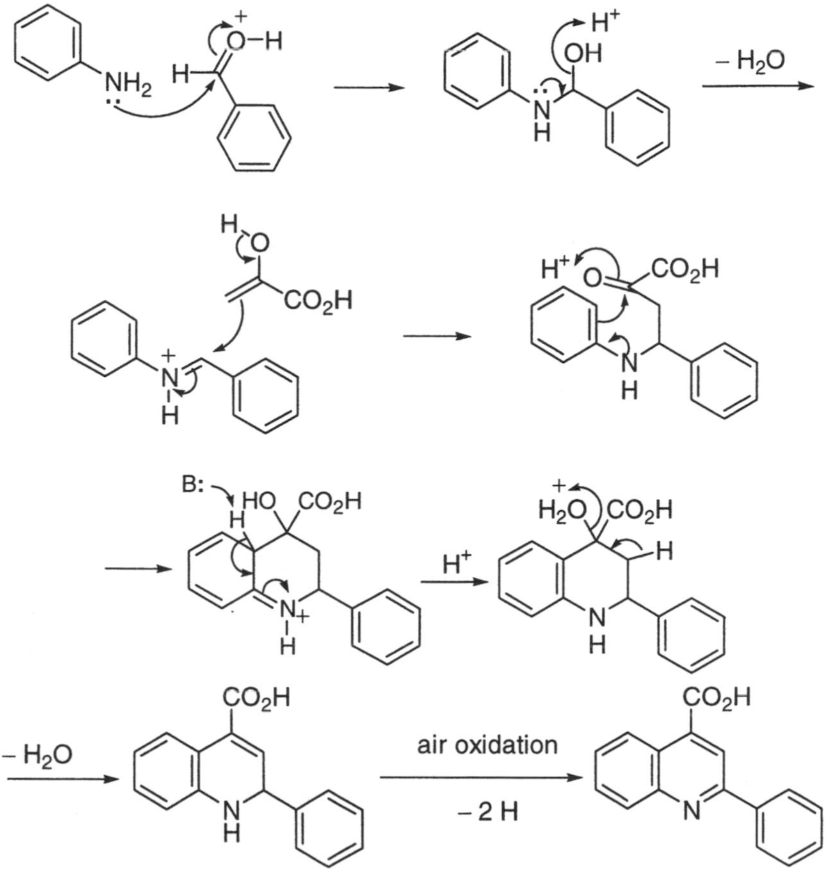

References

1. Doebner, O. *Justus Liebigs Ann. Chem.* **1887**, *242*, 256.
2. Allen, C. F. H.; Spangler, F. W.; Webster, E. R. *J. Org. Chem.* **1951**, *16*, 17.
3. Nitidandhaprabhas, O. *Nature* **1966**, *212*, 5061.

116

4. Herbert, R. B.; Kattah, A. E.; Knagg, E. *Tetrahedron* **1990**, *46*, 7119.
5. Mitra, A. K.; De, A.; Karchaudhuri, N. *Synth. Commun.* **1999**, *29*, 573.

Doebner–von Miller reaction

Doebner–von Miller reaction is a variant of the Skraup quinoline synthesis (page 378). Therefore, the mechanism for the Skraup reaction is also operative for the Doebner–von Miller reaction. An alternative mechanism shown below is based on the fact that the preformed imine (Schiff base) also gives 2-methylquinoline:

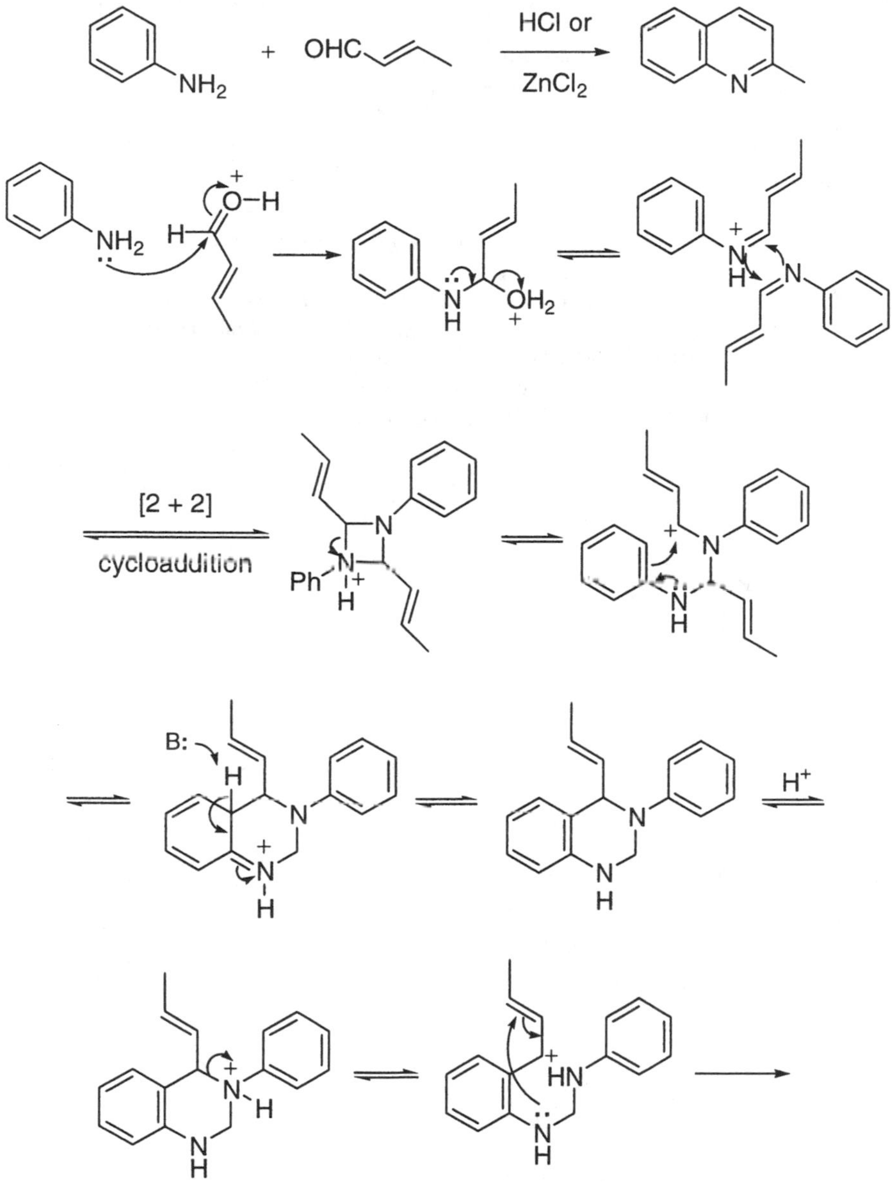

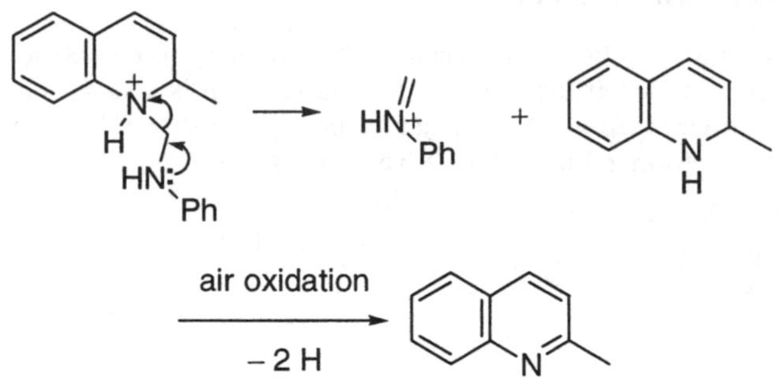

118

References

1. Doebner, O.; von Miller, W. *Ber. Dtsch. Chem. Ges.* **1883**, *16*, 2464.
2. Eisch, J. J.; Dluzniewski, T. *J. Org. Chem.* **1989**, *54*, 1269.
3. Zhang, Z. P.; Tillekeratne, L. M. V.; Hudson, R. A. *Tetrahedron Lett.* **1998**, *39*, 5133.
4. Matsugi, M.; Tabusa, F.; Minamikawa, J.-i. *Tetrahedron Lett.* **2000**, *41*, 8523.
5. Fürstner, A.; Thiel, O. R.; Blanda, G. *Org. Lett.* **2000**, *2*, 3731.
6. Kavitha, J.; Vanisree, M.; Subbaraju, G. V. *Indian J. Chem., Sect. B* **2001**, *40B*, 522.
7. Li, X.-G.; Cheng, X.; Zhou, Q.-L. *Synth. Commun.* **2002**, *32*, 2477.

Doering–LaFlamme allene synthesis

Allenes from gem-dibromocyclopropanes which are installed by treatment of olefins with bromoform and alkoxide.

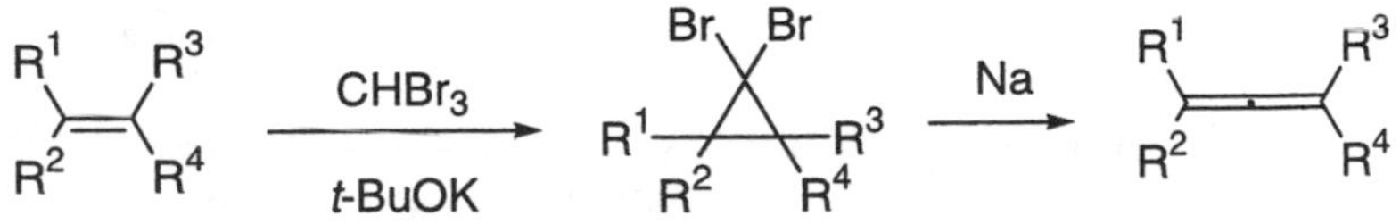

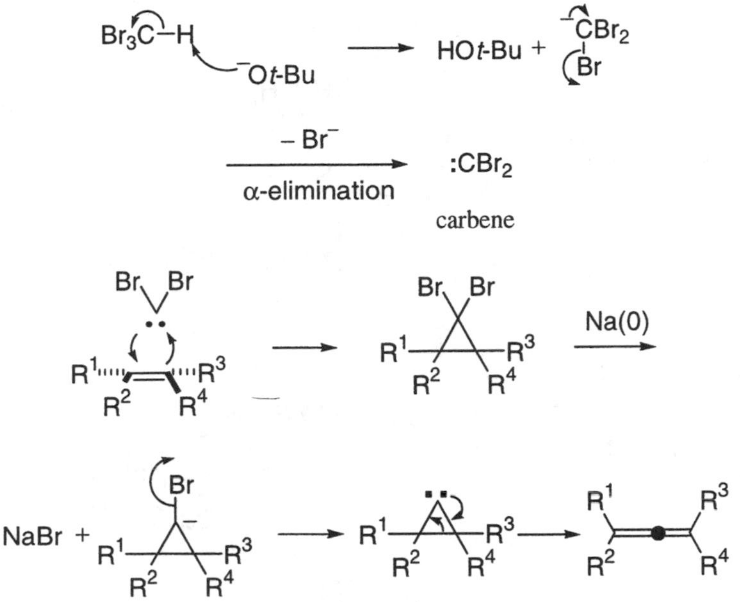

References

1. Doering, W. von E.; LaFlamme, P. M. *Tetrahedron* **1958**, *2*, 75.
2. Skattebol, L. *Tetrahedron Lett.* **1961**, 167.
3. Christl, M.; Braun, M.; Wolz, E.; Wagner, W. *Ber.* **1994**, *127*, 1137.
4. Magid, R. M.; Jones, M., Jr. *Tetrahedron* **1997**, *53*, xiii-xvi (Preface).

Dornow–Wiehler isoxazole synthesis

Isoxazoles from condensation of aryl aldehydes and α-nitro-esters. Interestingly, the nitrogen atom at the isoxazole ring comes from the nitro group.

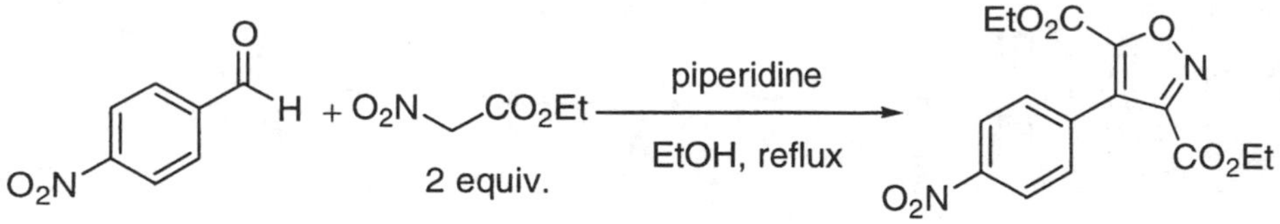

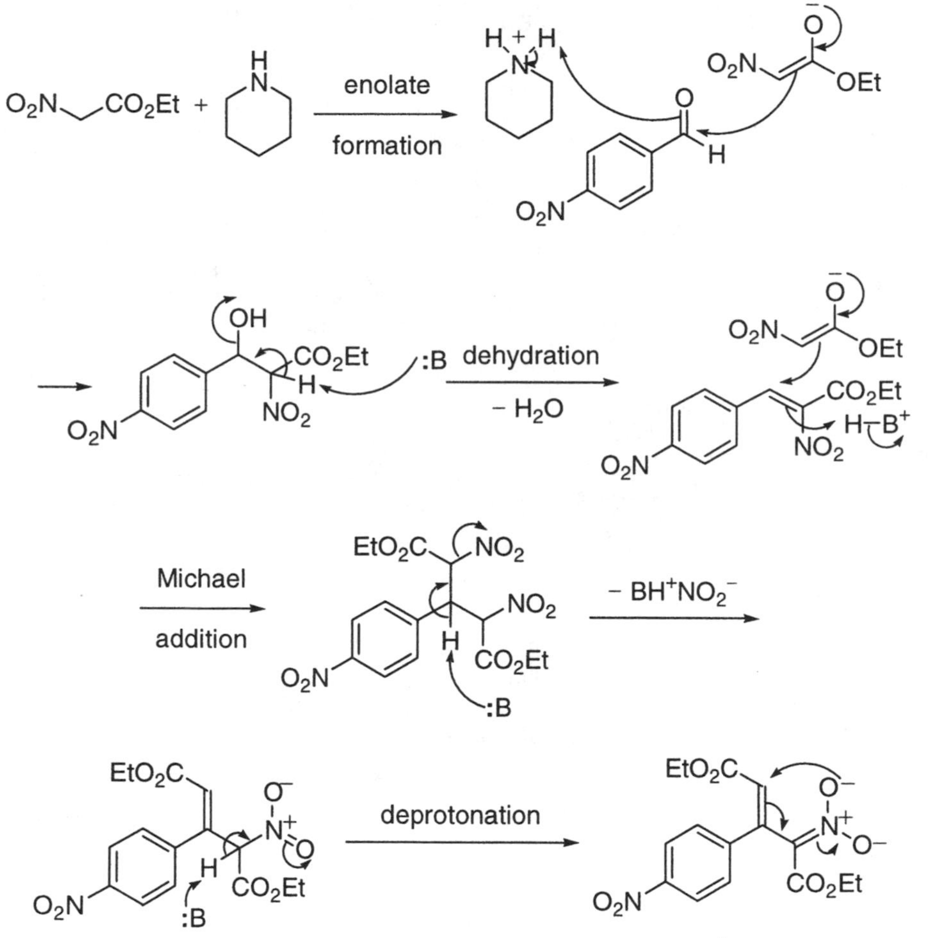

nitronate

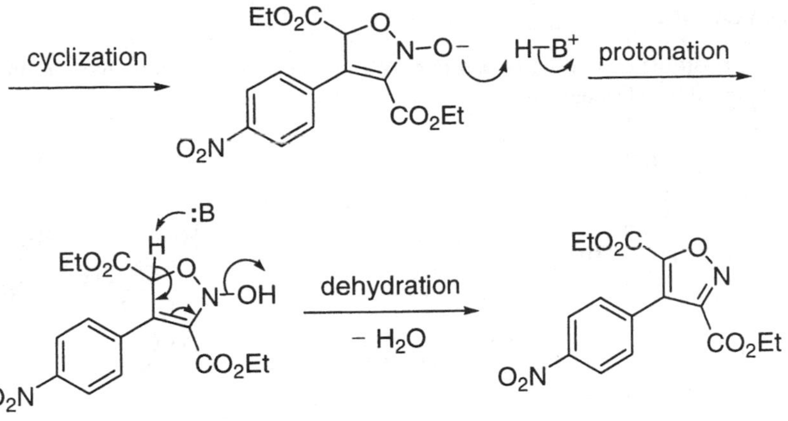

References

1. Dornow, A.; Wiehler, G. *Justus Liebigs Ann. Chem.* **1952**, *578*, 113.
2. Dornow, A.; Wiehler, G. *Justus Liebigs Ann. Chem.* **1952**, *578*, 122.
3. Umezawa, S.; Zen, S. *Bull. Chem. Soc. Jpn.* **1963**, *36*, 1150.

Dötz reaction

Cr(CO)$_3$-coordinated hydroquinone from vinylic alkoxy pentacarbonyl chromium carbene complex (Fischer carbene) and alkynes.

References

1. Dötz, K. H. *Angew. Chem., Int. Ed. Engl.* **1975**, *14*, 644.
2. Wulff, W. D.; Tang, P.; Mccallum, J. S. *J. Am. Chem. Soc.* **1981**, *103*, 7677.
3. Wulff, W. D. In *Advances in Metal-Organic Chemistry*; Liebeskind, L. S., Ed.; JAI Press, Greenwich, CT; 1989; Vol. 1. (Review).
4. Wulff, W. D. In *Comprehensive Organometallic Chemistry II*; Abel, E. W., Stone, F. G. A., Wilkinson, G., Eds.; Pergamon Press: Oxford, UK, 1995; Vol. 12. (Review).
5. Torrent, M. *Chem. Commun.* **1998**, 999.
6. Torrent, M.; Sola, M.; Frenking, G. *Chem. Rev.* **2000**, *100*, 439. (Review).
7. Jackson, T. J.; Herndon, J. W. *Tetrahedron* **2001**, *57*, 3859.
8. Sola, M.; Duran, M.; Torrent, M. *Catalysis Metal Complexes* **2002**, *25*, 269.

Dowd ring expansion

Radical-mediated ring expansion of 2-halomethyl cycloalkanones.

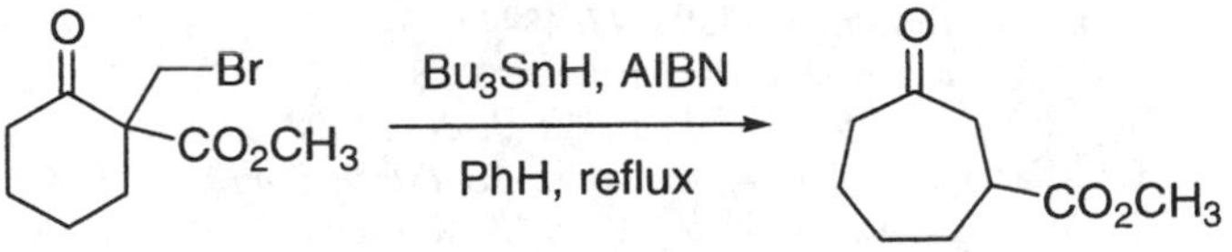

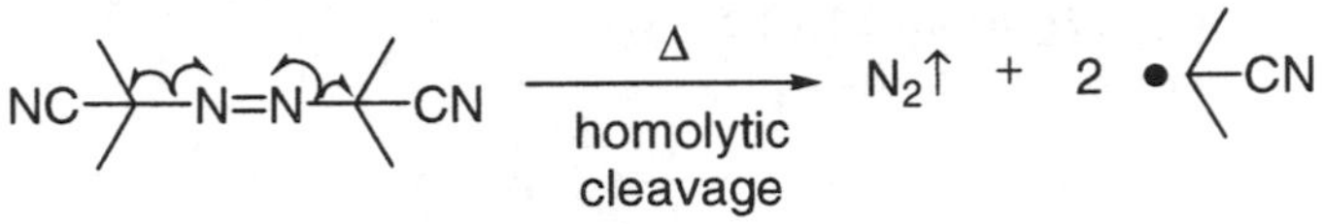

2,2'-azobisisobutyronitrile (AIBN)

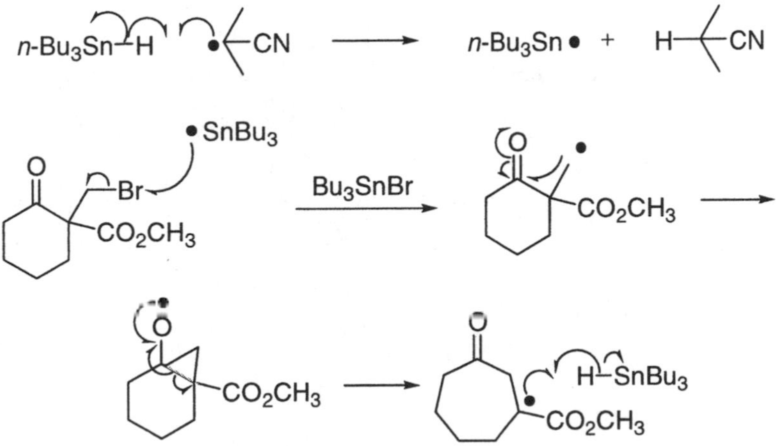

The cyclopropyloxy radical intermediate fragments in this fashion because the resulting tertiary radical is stabilized via the resonance effect excerted by the adja-
cent carbonyl group of the ester.

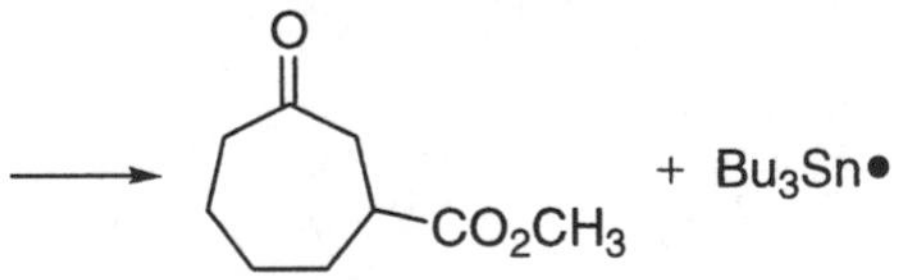

References

1. Dowd, P.; Choi, S.-C. *J. Am. Chem. Soc.* **1984**, *109*, 3493.
2. Beckwith, A. L. J.; O'Shea, D. M.; Gerba, S.; Westwood, S. W. *J. Chem. Soc., Chem. Commun.* **1987**, 666.

124

3. Beckwith, A. L. J.; O'Shea, D. M.; Gerba, S.; Westwood, S. W. *J. Am. Chem. Soc.*
    1988, *110*, 2565.
4. Dowd, P.; Choi, S.-C. *Tetrahedron* **1989**, *45*, 77.
5. Dowd, P.; Choi, S.-C. *Tetrahedron Lett.* **1989**, *30*, 6129.
6. Dowd, P.; Choi, S.-C. *Tetrahedron* **1991**, *47*, 4847.
7. Bowman, W. R.; Westlake, P. J. *Tetrahedron* **1992**, *48*, 4027.
8. Wang, C.; Gu, X.; Yu, M. S.; Curran, D. P. *Tetrahedron* **1998**, *54*, 8355.
9. Hasegawa, E.; Kitazume, T.; Suzuki, K.; Tosaka, E. *Tetrahedron Lett.* **1998**, *39*, 4059.
10. Hasegawa, E.; Yoneoka, A.; Suzuki, K.; Kato, T.; Kitazume, T.; Yanagi, K. *Tetrahedron* **1999**, *55*, 12957.
11. Kantorowski, E. J.; Kurth, M. J. *Tetrahedron* **2001**, *57*, 3859. (Review).
12. Sugi, M.; Togo, H. *Tetrahedron* **2002**, *57*, 3171.

Dutt–Wormall reaction

Azides from the reaction of sulfonamide and diazonium salts which may be derived from amines.

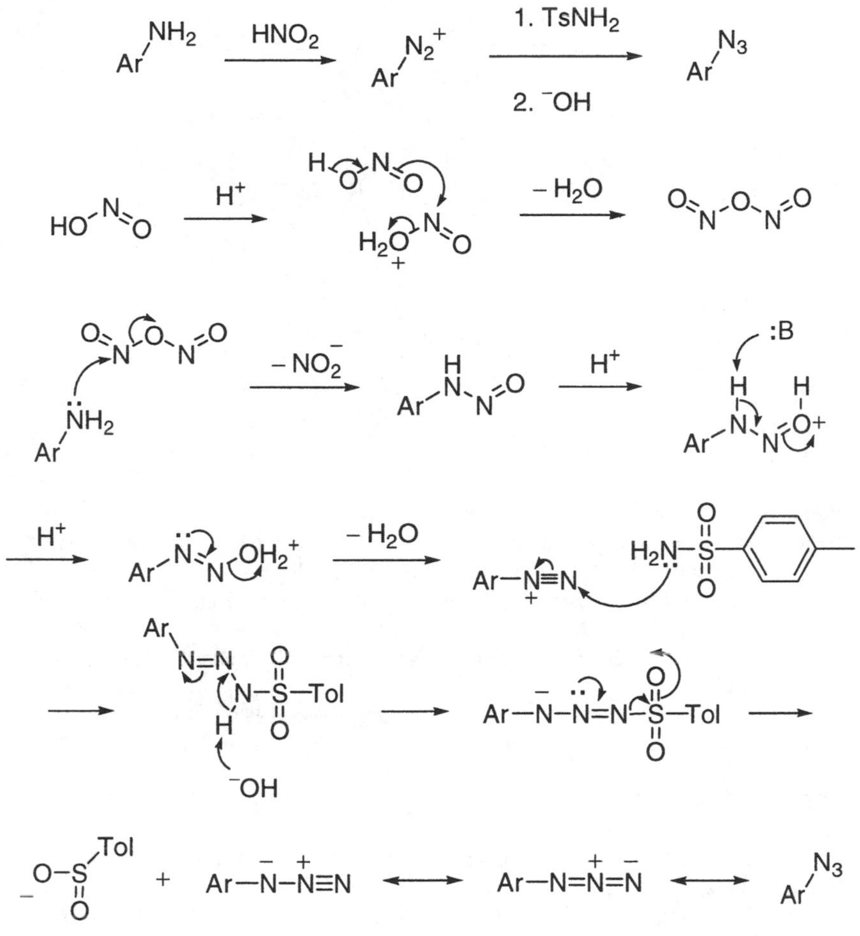

References

1. Dutt, J. C.; Whitehead, H. R.; Wormall, A. *J. Chem. Soc.* **1921**, *119*, 2088.
2. Bretschneider, H.; Rager, H. *Monatsh.* **1950**, *81*, 970.
3. Laing, I. G. In *Rodd's Chemistry of Carbon Compounds IIIC* **1973**, 107. (Review).

Eglinton reaction

Oxidative coupling of terminal alkynes mediated by stoichiometric (often excess) Cu(OAc)$_2$. A variant of the Glaser coupling reaction (page 160).

R$\equiv$H $\xrightarrow[\text{pyridine/MeOH}]{\text{Cu(OAc)}_2}$ R$\equiv$$\equiv$R

R$\equiv$H $\xrightarrow{\text{pyridine}}$ [pyridinium] + R$\equiv$$^-$ $\xrightarrow{\text{Cu(OAc)}_2}$

$\xrightarrow{\text{dimerization}}$ R$\equiv$$\equiv$R

References

1. Eglinton, G.; Galbraith, A. R. *Chem. Ind.* **1956**, 737.
2. Behr, O. M.; Eglinton, G.; Galbraith, A. R.; Raphael, R. A. *J. Chem. Soc.* **1960**, 3614.
3. Eglinton, G.; McRae, W. *Adv. Org. Chem.* **1963**, *4*, 225. (Review).
4. Altmann, M.; Friedrich, J.; Beer, F.; Reuter, R.; Enkelmann, V.; Bunz, U. H. F. *J. Am. Chem. Soc.* **1997**, *119*, 1427.
5. Srinivasan, R.; Devan, B.; Shanmugam, P.; Rajagopalan, K. *Indian J. Chem., Sect. B* **1997**, *36B*, 123.
6. Nakanishi, H.; Sumi, N.; Aso, Y.; Otsubo, T. *J. Org. Chem.* **1998**, *63*, 8632.
7. Müller, T.; Hulliger, J.; Seichter, W.; Weber, E.; Weber, T.; Wübbenhorst, M. *Chem. Eur. J.* **2000**, *6*, 54.
8. Märkl, G.; Zollitsch, T.; Kreimeier, P.; Prinzhorn, M.; Reithinger, S.; Eibler, E. *Chem. Eur. J.* **2000**, *6*, 3806.
9. Fabian, K. H. H.; Lindner, H.-J.; Nimmerfroh, N.; Hafner, K. *Angew. Chem., Int. Ed.* **2001**, *40*, 3402.
10. Siemsen, P.; Livingston, R. C.; Diederich, F. *Angew. Chem., Int. Ed.* **2000**, *39*, 2632. (Review).
11. Inouchi, K.; Kabashi, S.; Takimiya, K.; Aso, Y.; Otsubo, T. *Org. Lett.* **2002**, *4*, 2533.

Eschenmoser coupling reaction

Enamine from thiamide and alkyl halide.

$$\underset{\text{thiamide}}{\overset{\displaystyle S}{\underset{R^1}{\overset{\parallel}{\underset{N}{\overset{}{\bigg|}}}}}}\quad\xrightarrow[\text{2. base, thiophile}]{\text{1. } X\diagdown R^3}\quad R\overset{R^3}{\underset{R^1}{\underset{N}{\bigg|}}}R^2$$

e.g.

$$t\text{-BuO}_2C\underset{\underset{Bn}{N}}{\diagup}\diagdown S \xrightarrow[\text{2. PPh}_3,\ \text{Et}_3N,\ 90\%]{\text{1. BrCH}_2\text{CO}_2\text{CH}_3} t\text{-BuO}_2C\underset{\underset{Bn}{N}}{\diagup}\diagdown CO_2CH_3$$

References

1.   Roth, M.; Dubs, P.; Götschi, E.; Eschenmoser, A. *Helv. Chim. Acta* **1971**, *54*, 710.
2.   Peterson, J. S.; Fels, G.; Rapoport, H. *J. Am. Chem. Soc.* **1984**, *106*, 4539.
3.   Shiosaki, K. In *Comprehensive Organic Synthesis;* Trost, B. M.; Fleming, I., Eds.; Pergamon, **1991**, *Vol. 2*, 865–892. (Review).
4.   Levillain, J.; Vazeux, M. *Synthesis* **1995**, 56.
5.   Mulzer, J.; List, B.; Bats, J. W. *J. Am. Chem. Soc.* **1997**, *119*, 5512.
6.   Hodgkinson, T. J.; Kelland, L. R.; Shipman, M.; Vile, J. *Tetrahedron* **1998**, *54*, 6029.
7.   Ye, I.-H.; Choung, W.-K.; Kim, K. H.; Ha, D.-C. *Bull. Kor. Chem. Soc.* **2000**, *21*, 1169.

Eschenmoser–Tanabe fragmentation

Fragmentation of α,β-epoxyketones *via* the intermediacy of α,β-epoxy sulfonyl-hydrazones.

References

1. Eschenmoser, A.; Felix, D.; Ohloff, G. *Helv. Chim. Acta* **1967**, *50*, 708.
2. Tanabe, M.; Crowe, D. F.; Dehn, R. L. *Tetrahedron Lett.* **1967**, 3943.
3. Felix, D.; Müller, R. K.; Horn, U.; Joos, R.; Schreiber, J.; Eschenmoser, A. *Helv. Chim. Acta* **1972**, *55*, 1276.
4. Batzold, F. H.; Robinson, C. H. *J. Org. Chem.* **1976**, *41*, 313.
5. Chinn, L. J.; Lenz, G. R.; Choudary, J. B.; Nutting, E. F.; Papaioannou, S. E.; Metcalf, L. E.; Yang, P. C.; Federici, C.; Gauthier, M. *Eur. J. Org. Chem.* **1985**, *20*, 235.
6. Dai, W.; Katzenellenbogen, J. A. *J. Org. Chem.* **1993**, *58*, 1900.
7. Abad, A.; Arno, M.; Agullo, C.; Cunat, A. C.; Meseguer, B.; Zaragoza, R. J. *J. Nat. Prod.* **1993**, *56*, 2133.
8. Mueck-Lichtenfeld, C. *J. Org. Chem.* **2000**, *65*, 1366.

Étard reaction

Oxidation of an arylmethyl group to the corresponding aryl aldehyde using chromyl chloride.

Étard complex

References

1. Étard, A. L. *Compt. Rend.* **1880**, *90*, 524.
2. Hartford, W. H.; Darrin, M. *Chem. Rev.* **1958**, *58*, 1. (Review).
3. Roček, J. *Tetrahedron Lett.* **1962**, 135.
4. Wiberg, K. B.; Marshall, B.; Foster, G. *Tetrahedron Lett.* **1962**, 345.
5. Necsoiu, I.; Balaban, A. T.; Pascaru, I.; Sliam, E.; Elian, M.; Nenitzescu, C. D. *Tetrahedron* **1963**, *19*, 1133. (Mechanism Discussion).
6. Rentea, C. N.; Necsoiu, I.; Rentea, M.; Ghenciulescu, A.; Nenitzescu, C. D. *Tetrahedron* **1966**, *22*, 3501.
7. Schildknecht, H.; Hatzmann, G. *Angew. Chem., Int. Ed. Engl.* **1968**, *7*, 293.
8. Duffin, H. C.; Tucker, R. B. *Tetrahedron* **1968**, *24*, 6999.
9. Schiketanz, I. I.; Hanes, A.; Necsoiu, I. *Rev. Roum. Chim.* **1977**, *22*, 1097.
10. Schiketanz, I. I.; Badea, F.; Hanes, A; Necsoiu, I. *Rev. Roum. Chim.* **1984**, *29*, 353.
11. Luzzio, F. A.; Moore, W. J. *J. Org. Chem.* **1993**, *58*, 512.

130

Evans aldol reaction

Asymmetric aldol condensation of aldehyde and chiral acyl oxazolidinone, the Evans chiral auxiliary.

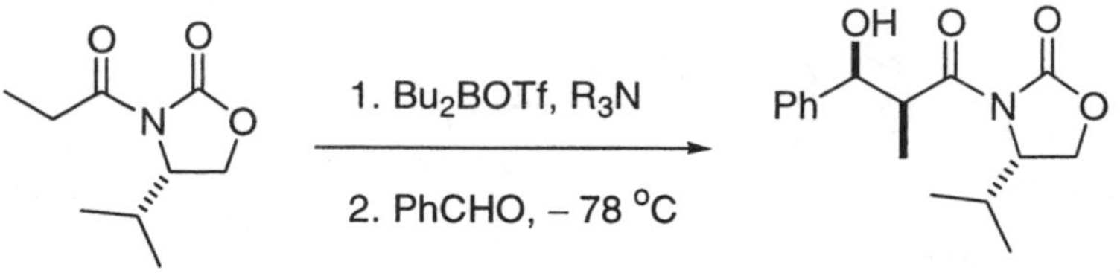

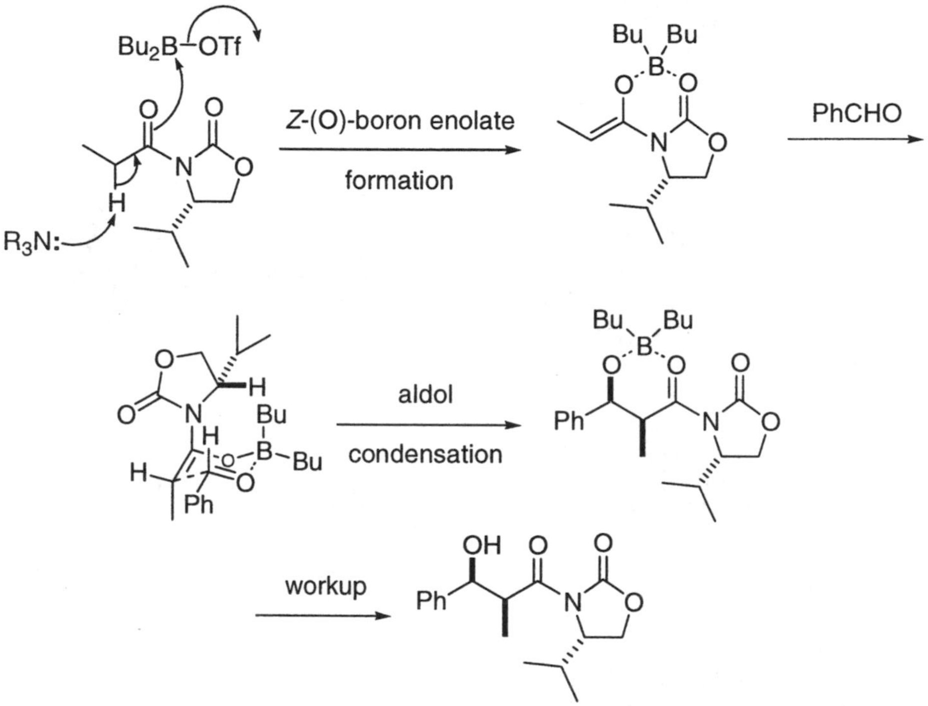

Refrences

1. Evans, D. A.; Bartroli, J.; Shih, T. L. *J. Am. Chem. Soc.* **1981**, *103*, 2127.
2. Evans, D. A.; McGee, L. R. *J. Am. Chem. Soc.* **1981**, *103*, 2876.
3. Allin, S. M.; Shuttleworth, S J. *Tetrahedron Lett.* **1996**, *37*, 8023.
4. Ager, D. J.; Prakash, I.; Schaad, D. R. *Aldrichimica Acta* **1997**, *30*, 3. (Review).
5. Braddock, D. C.; Brown, J. M. *Tetrahedron: Asymmetry* **2000**, *11*, 3591.
6. Lu, Y.; Schiller, P. W. *Synthesis* **2001**, 1639.
7. Li, G.; Xu, X.; Chen, D.; Timmons, C.; Carducci, M. D.; Headley, A. D. *Org. Lett.* **2003**, *5*, 329.

8. Williams, D. R.; Patnaik, S.; Clark, M. P. *J. Org. Chem* **2001**, *66*, 8463.
9. Matsushima, Y.; Itoh, H.; Nakayama, T.; Horiuchi, S.; Eguchi, T.; Kakinuma, K. *J. Chem. Soc., Perkin 1* **2002**, 949.
10. Guerlavais, V.; Carroll, P. J.; Joullie, M. M. *Tetrahedron: Asymmetry* **2002**, *13*, 675.

Favorskii rearrangement and Quasi-Favorskii rearrangement

Favorskii rearrangement

Transformation of α-haloketones to esters *via* base-catalyzed rearrangement.

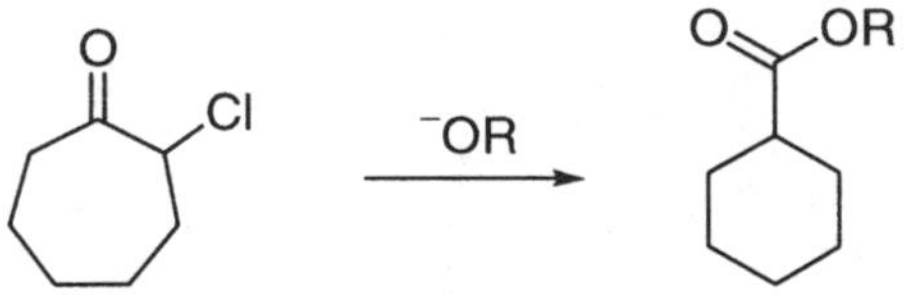

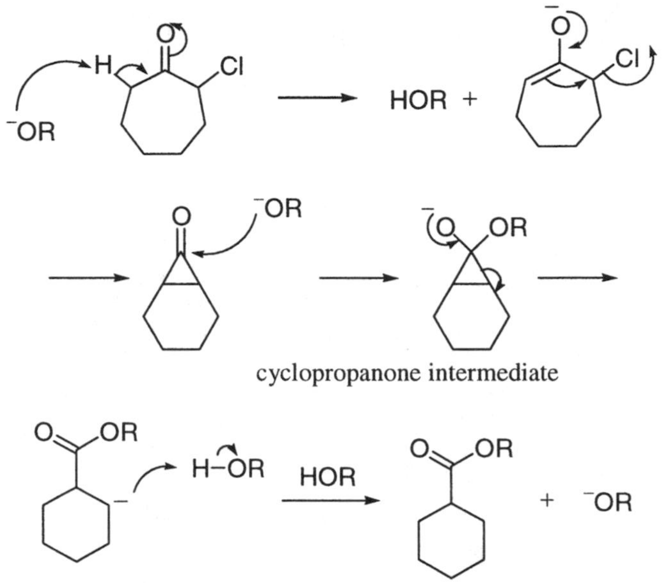

cyclopropanone intermediate

Quasi-Favorskii rearrangement

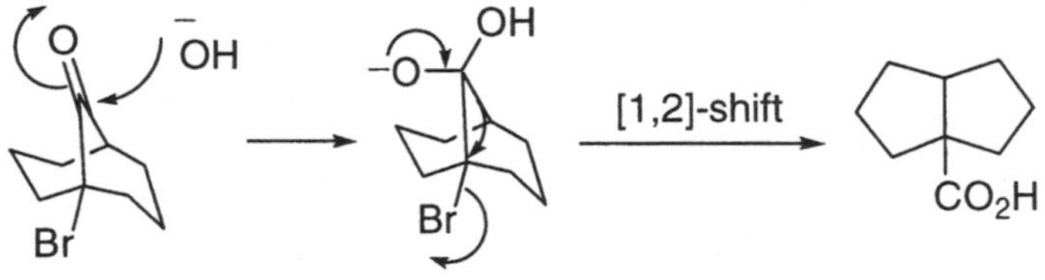

non-enolizable ketone

References

1. Favorskii, A. *J. Prakt. Chem.* **1895**, *51*, 533.
2. Chenier, P. J. *J. Chem. Educ.* **1978**, *55*, 286.

3. Barreta, A.; Waegill, B. In *Reactive Intermediates*; Abramovitch, R. A., ed.; Plenum Press: New York, **1982**, pp 527–585. (Review).

4. Gambacorta, A.; Turchetta, S.; Bovivelli, P.; Botta, M. *Tetrahedron* **1991**, *47*, 9097.

5. El-Wareth, A.; Sarhan, A. O.; Hoffmann, H. M. R. *J. Prakt. Chem./Chem.- Ztg.* **1997**, *339*, 390

6. Dhavale, D. D.; Mali, V. P.; Sudrik, S. G.; Sonawane, H. R. *Tetrahedron* **1997**, *53*, 16789.

7. Braverman, S.; Cherkinsky, M.; Kumar, E. V. K. S.; Gottlieb, H. E. *Tetrahedron* **2000**, *56*, 4521.

8. Mamedov, V. A.; Tsuboi, S.; Mustakimova, L. V.; Hamamoto, H.; Gubaidullin, A. T.; Litvinov, I. A.; Levin, Y. A. *Chem. Heterocycl. Compd.* **2001**, *36*, 911.

9. Muldgaard, L.; Thomsen, I. B.; Hazell, R. G.; Bols, M. *J. Chem. Soc., Perkin 1* **2002**, 1297.

10. Harmata, M.; Bohnert, G.; Kurti, L.; Barnes, C. L. *Tetrahedron Lett.* **2002**, *43*, 2347. (quasi-Favorskii rearrangement).

11. Harmata, M.; Bohnert, G. *Org. Lett.* **2003**, *5*, 59. (quasi-Favorskii rearrangement).

Feist–Bénary furan synthesis

α-Haloketones react with β-ketoesters in the presence of pyridine to fashion furans.

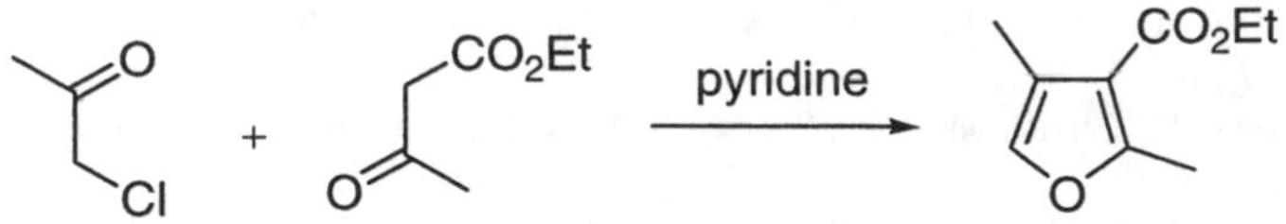

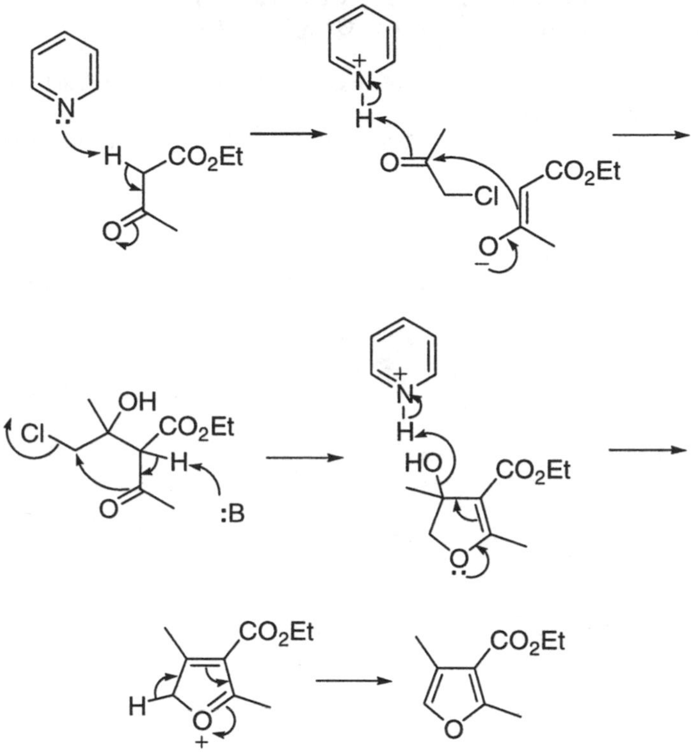

References

1. Feist, F. *Ber. Dtsch. Chem. Ges.* **1902**, *35*, 1537.
2. Bénary, E. *Ber. Dtsch. Chem. Ges.* **1911**, *44*, 489.
3. Bisagni, E.; Marquet, J. P.; Andre-Louisfert, J.; Cheutin, A.; Feinte, F. *Bull. Soc. Chim. Fr.* **1967**, 2796.
4. Cambie, R. C.; Moratti, S. C.; Rutledge, P. S.; Woodgate, P. D. *Synth. Commun.* **1990**, *20*, 1923.
5. Calter, M.; Zhu, C. *Abstr. Pap.-Am. Chem. Soc.* **2001**, 221st ORGN-574.
6. Calter, M.; Zhu, C. *Org. Lett.* **2002**, *4*, 205.
7. Calter, M.; Zhu, C.; Lachicotte, R. J. *Org. Lett.* **2002**, *4*, 20.

Ferrier rearrangement

Lewis-acid (such as $BF_3 \cdot OEt_2$, $SnCl_4$, etc.)-promoted rearrangement of unsaturated carbohydrates.

The axial addition is favored due to the anomeric effect.

References

1. Ferrier, R. J. *J. Chem. Soc. (C)* **1968**, 974.
2. Ferrier, R. J. *J. Chem. Soc., Perkin. Trans. 1* **1979**, 1455.
3. Fraser-Reid, B. *Acc. Chem. Res.* **1996**, *29*, 57.
4. Paquette, L. A. *Recent Res. Dev. Chem. Sci.* **1997**, *1*, 1.
5. Smith, A. B., III; Verhoest, P. R.; Minbiole, K. P.; Lim, J. J. *Org. Lett.* **1999**, *1*, 909.
6. Babu, B. S.; Balasubramanian, K. K. *Synth. Commun.* **1999**, *29*, 4299.
7. Taillefumier, C.; Chapleur, Y. *Can. J. Chem.* **2000**, *78*, 708.
8. Yadav, J. S.; Reddy, B. V. S.; Murthy, C. V. S. R.; Kumar, G. M. *Synlett* **2000**, 1450.
9. Abdel-Rahman, A. A.-H.; Winterfeld, G. A.; Takhi, M.; Schmidt, R. R. *Eur. J. Org. Chem.* **2002**, 713.
10. Swamy, N. R.; Venkateswarlu, Y. *Synthesis* **2002**, 598.
11. Lin, H.-C.; Yang, W.-B.; Gu, Y.-F.; Chen, C.-Y.; Wu, C.-Y.; Lin, C.-H. *Org. Lett.* **2003**, *5*, 1087.
12. Shimizu, M.; Iwasaki, Y.; Shibamoto, Y.; Sato, M.; DeLuca, H. F.; Yamada, S. *Bioorg. Med.Chem. Lett.* **2003**, *13*, 809.

Finkelstein reaction

S_N2 displacement of one alkyl halide with another halide.

$$R\diagup Cl \xrightarrow[\text{acetone, reflux}]{\text{excess KI}} R\diagup I \;+\; KCl$$

$$\overset{\_}{I}\diagdown\underset{R}{\overset{H\;H}{\diagup}}Cl \xrightarrow{\;S_N2\;} \underset{I\diagup R}{\overset{H\;H}{\diagup}} \;+\; Cl^{\_}$$

References

1. Finkelstein, H. *Ber. Dtsch. Chem. Ges.* **1910**, *43*, 1528.
2. Henne, A. L. *Org. React.* **1944**, *2*, 49–93. (Review).
3. "An abnormal Finkelstein reaction" Smith, W. B.; Branum, G. D. *Tetrahedron Lett.* **1983**, *22*, 2055.
4. Landin, D.; Albanese, D.; Mottadelli, S.; Penso, M. *J. Chem. Soc., Perkin Trans. 1* **1992**, 2309.
5. Zoller, T.; Uguen, D.; De Cian, A.; Fisher, J. *Tetrahedron Lett.* **1998**, *39*, 8089.
6. Mathews, D. P.; Green, J. E.; Shuker, A. J. *J. Comb. Chem.* **2000**, *2*, 119.
7. Creemers, A. F. L.; Lugtenburg, J. *J. Am. Chem. Soc.* **2002**, *124*, 6324.
8. "An aromatic Finkelstein reaction" Klapars, A.; Buchwald, S. L. *J. Am. Chem. Soc.* **2002**, *124*, 14844.

Fischer–Hepp rearrangement

Transformation of *N*-nitroso-anilines to the corresponding *para*-nitroso anilines. *Cf.* Orton rearrangement.

References

1. Fischer, O.; Hepp, E. *Ber. Dtsch. Chem. Ges.* **1886**, *19*, 2991.
2. Drake, N. L.; Winkler, H. J. S.; Kraebel, C. M.; Smith, T. D. *J. Org. Chem.* **1962**, *27*, 1026.
3. Baliga, B. T. *J. Org. Chem.* **1970**, *435*, 2031.
4. Williams, D. L. H. *Tetrahedron* **1975**, *31*, 1343.
5. Biggs, I. D.; Williams, D. L. H. *J. Chem. Soc., Perkin Trans. 2* **1976**, 691.
6. Biggs, I. D.; Williams, D. L. H. *J. Chem. Soc., Perkin Trans. 2* **1977**, 44.
7. Williams, D. L. H. *J. Chem. Soc., Perkin Trans. 2* **1982**, 801.
8. Lunn, G.; Sansone, E. B.; Keefer, L. K. *J. Org. Chem.* **1984**, *49*, 3470.
9. Kyziol, J. B. *J. Heterocycl. Chem.* **1985**, 1301.
10. Morris, P. I. *Chem. Ind.* **1999**, 968.

138

Fischer indole synthesis

Cyclization of arylhydrazones to indoles.

phenylhydrazine

phenylhydrazone

protonation

ene-hydrazine

[3,3]-sigmatropic

rearrangement

double imine

tautomerization

References

1. Fischer, E.; Jourdan, F. *Ber. Dtsch. Chem. Ges.* **1883**, *16*, 2241.
2. Robinson, B. *Chem. Rev.* **1969**, *69*, 227. (Review).
3. Ishii, H. *Acc. Chem. Res.* **1981**, *14*, 275. (Review).
4. Robinson, B. *The Fisher Indole Synthesis,* John Wiley & Sons, New York, NY, 1982. (Review).
5. Hughes, D. L.; Zhao, D. *J. Org. Chem.* **1993**, *58*, 228.
6. Hughes, D. L. *Org. Prep. Proc. Int.* **1993**, *25*, 607.
7. Bhattacharya, G.; Su, T.-L.; Chia, C.-M.; Chen, K.-T. *J. Org. Chem.* **2001**, *66*, 426.
8. Kozmin, S. A.; Iwama, T.; Huang, Y.; Rawal, V. H. *J. Am. Chem. Soc.* **2002**, *124*, 4628.
9. Pete, B.; Parlagh, G. *Tetrahedron Lett.* **2003**, *44*, 2537.

Fischer–Speier esterification

Often known as simply "Fischer esterification", protic acid-catalyzed esterification of acid and alcohol.

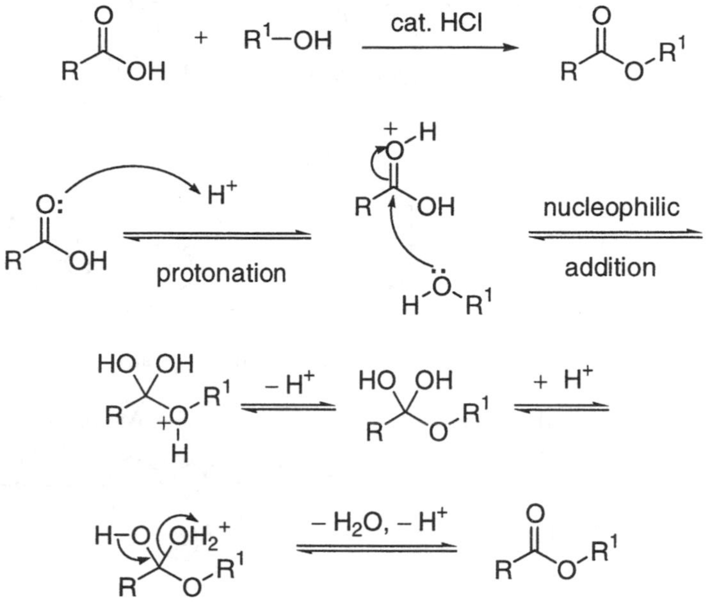

References

1. Fischer, E.; Speier, A. *Ber. Dtsch. Chem. Ges.* **1895**, *28*, 3252.
2. Hardy, J. P.; Kerrin, S. L.; Manatt, S. L. *J. Org. Chem.* **1973**, *38*, 4196.
3. Fujii, T.; Yoshifuji, S. *Chem. Pharm. Bull.* **1978**, *26*, 2253.
4. Pcolinski, M. J.; O'Mathuna, D. P.; Doskotch, R. W. *J. Nat. Prod.* **1978**, *58*, 209.
5. Kai, T.; Sun, X.-L.; Tanaka, M.; Takayanagi, H.; Furuhata, K. *Chem. Pharm. Bull.* **1996**, *44*, 208.
6. Birney, D. M.; Starnes, S. D. *J. Chem. Educ.* **1996**, *76*, 1560.
7. Cole, A. C., Jensen, J. L., Ntai, I.; Tran, K. L. T.; Weaver, K. J.; Forbes, D. C.; Davis, J. H., Jr. *J. Am. Chem. Soc.* **2002**, *124*, 5962.

Fleming oxidation

Cf. Tamao–Kumada oxidation.

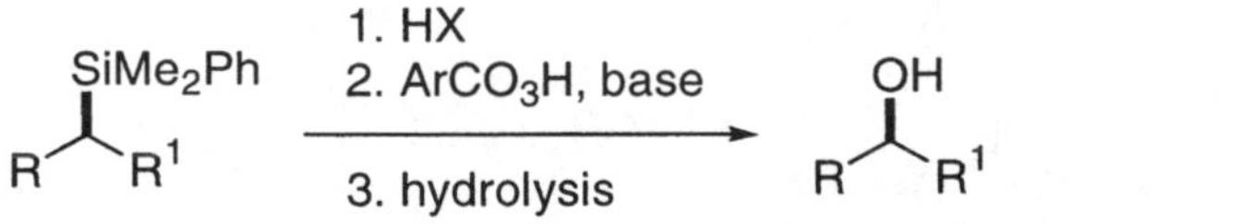

retention of configuration

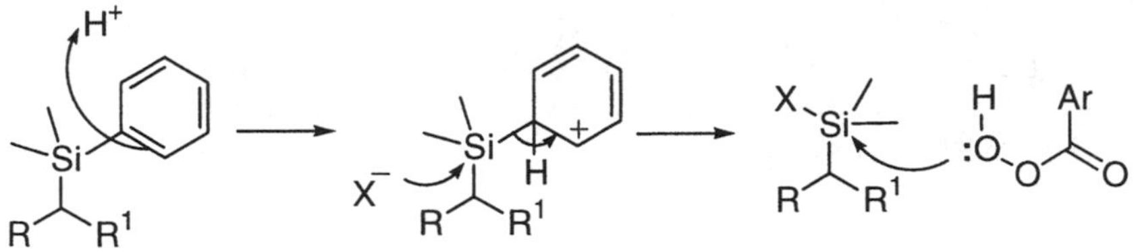

the β-carbocation is stabilized by the silicon group

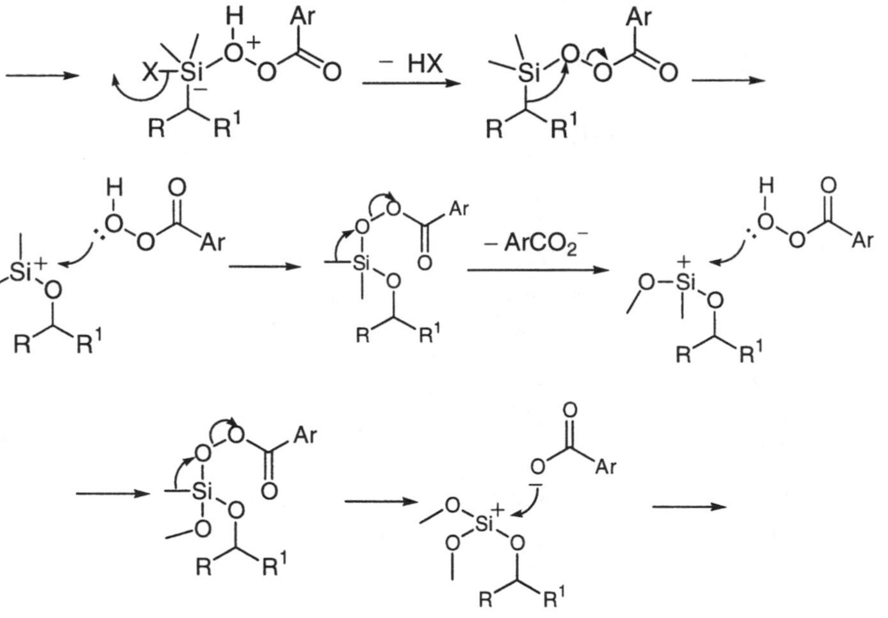

References

1. Fleming, I.; Henning, R.; Plaut, H. *J. Chem. Soc., Chem. Commun.* **1984**, 29.
2. Fleming, I.; Sanderson, P. E. J. *Tetrahedron Lett.* **1987**, *28*, 4229.
3. Fleming, I.; Dunogues, J.; Smithers, R. *Org. React.* **1989**, *37*, 57. (Review).
4. Jones, G. R.; Landais, Y. *Tetrahedron* **1996**, *52*, 7599.
5. Hunt, J. A.; Roush, W. R. *J. Org. Chem.* **1997**, *62*, 1112.
6. Knölker, H.-J.; Jones, P. G.; Wanzl, G. *Synlett* **1997**, 613.
7. Barrett, A. G. M.; Head, J.; Smith, M. L.; Stock, N. S.; White, A. J. P.; Williams, D. J. *J. Org. Chem.* **1999**, *64*, 6005.
8. Lee, T. W.; Corey, E. J. *Org. Lett.* **2001**, *3*, 3337.
9. Rubin, M.; Schwier, T.; Gevorgyan, V. *J. Org. Chem.* **2002**, *67*, 1936.
10. Boulineau, F. P.; Wei, A. *Org. Lett.* **2002**, *4*, 2281.
11. Jung, M. E.; Piizzi, G. *J. Org. Chem.* **2003**, *68*, 2572.

Forster reaction

α-Diazoketone formation from α-oximinoketones.

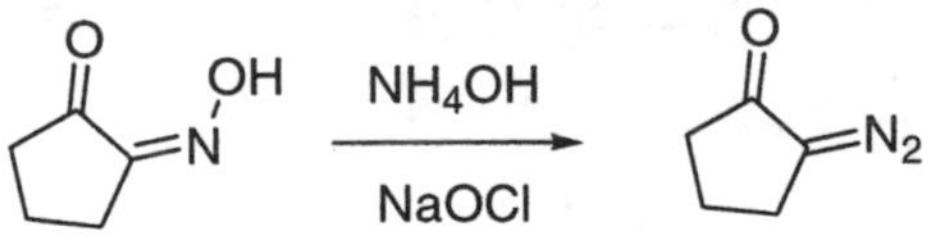

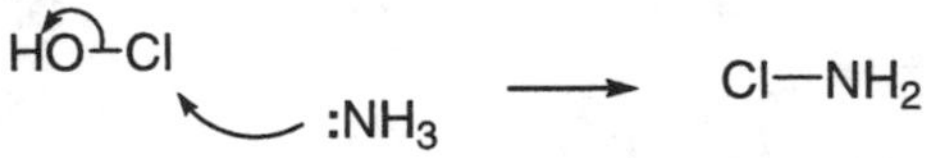

chloramine

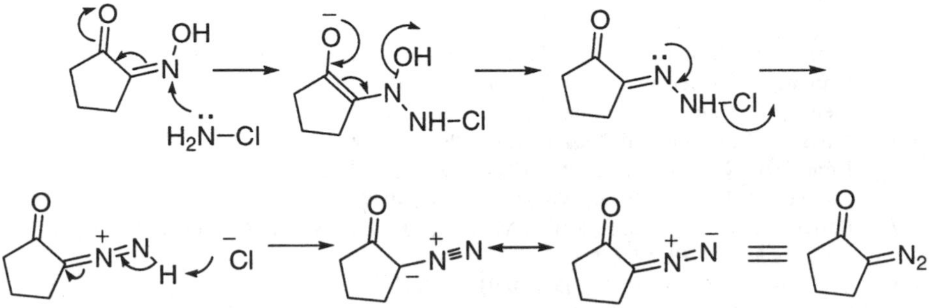

Alternatively:

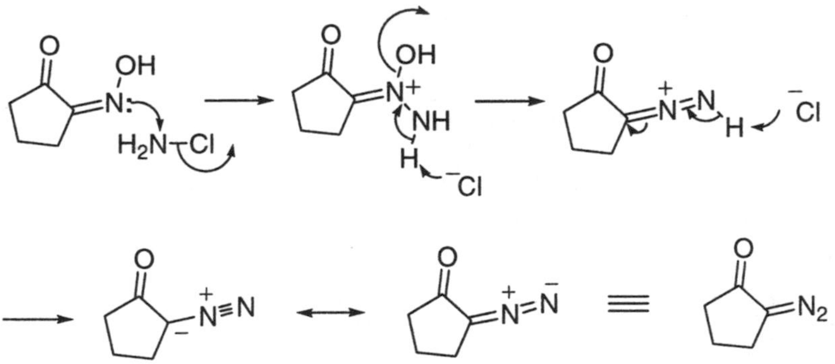

References

1. Forster, M. O. *J. Chem. Soc.* **1915**, *107*, 260.
2. Meinwald, J.; Gassman, P. G.; Miller, E. G. *J. Am. Chem. Soc.* **1959**, *81*, 4751.
3. Rundel, W. *Angew. Chem.* **1962**, *74*, 469.
4. Huneck, S. *Chem. Ber.* **1965**, *98*, 3204.

5. Overberger, C. G.; Anselme, J. P. *Tetrahedron Lett.* **1963**, 1405.
6. Van Leusen, A. M.; Strating, J.; Van Leusen, D. *Tetrahedron Lett.* **1973**, 5207.
7. L'abbe, G.; Dekerk, J. P.; Deketele, M. *J. Chem. Soc., Chem. Commun.* **1983**, 588.
8. L'abbe, G.; Luyten, I.; Toppet, S. *J. Heterocycl. Chem.* **1992**, *29*, 713.

Frater–Seebach alkylation

Asymmetric alkylation of β-hydroxyesters.

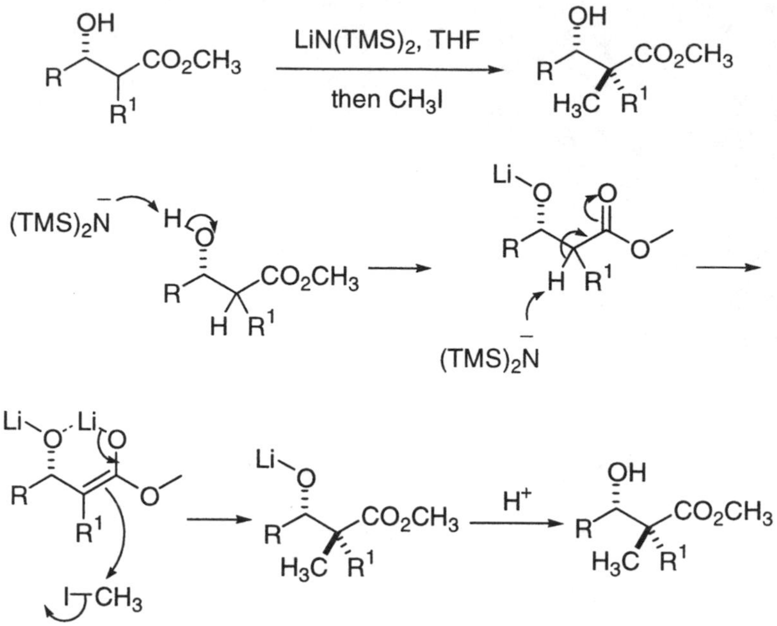

References

1. Frater, G.; Muller, U.; Gunter, W. *Tetrahedron* **1984**, *48*, 1269.
2. Seebach, D.; Imwinkelried, R.; Weber, T. *Modern Synth. Method* **1986**, *4*, 125. (Review).
3. Heathcock, C. H.; Kath, J. C.; Ruggeri, R. B. *J. Org. Chem.* **1995**, *60*, 1120.
4. Davenport, R. J.; Watson, R. J. *Tetrahedron Lett.* **2000**, *41*, 7983.
5. Sefkow, M. *Tetrahedron: Asymmetry* **2001**, *12*, 987.
6. Sefkow, M. *J. Org. Chem.* **2001**, *66*, 2343.
7. Sefkow, M.; Kelling, A.; Schilde, U. *Tetrahedron Lett.* **2001**, *42*, 5101.
8. Breit, B.; Zahn, S. K. *J. Org. Chem.* **2001**, *66*, 4870.

Friedel–Crafts reaction

Friedel–Crafts *acylation* reaction:

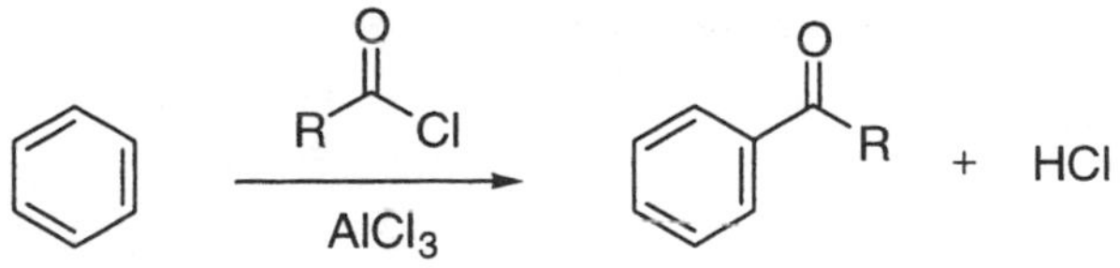

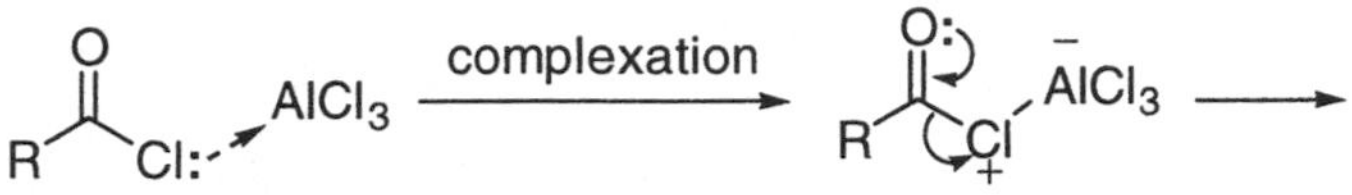

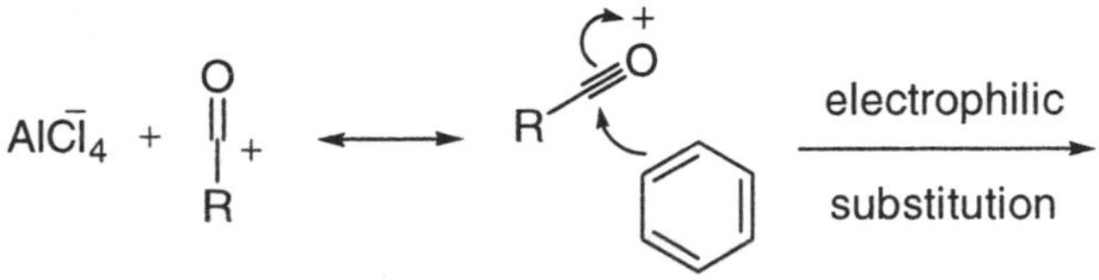

acylium ion

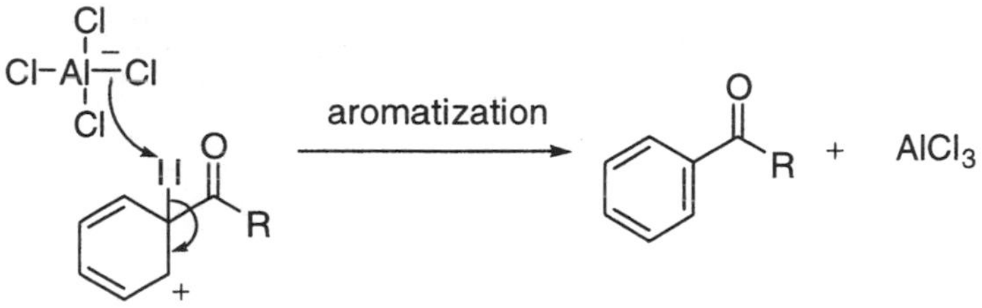

Friedel–Crafts *alkylation* reaction:

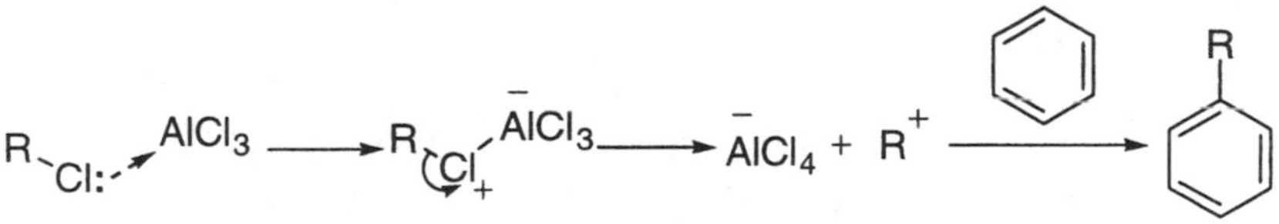

alkyl cation

References

1. Friedel, P.; Crafts, J. M. *Compt. Rend.* **1877**, *84*, 1392.
2. Pearson, D. E.; Buehler, C. A. *Synthesis* **1972**, 533.
3. Gore, P. H. *Chem. Ind.* **1974**, 727.
4. Chevrier, B.; Weis, R. *Angew. Chem.* **1974**, *86*, 12.
5. Schriesheim, A.; Kirshenbaum, I. *Chemtech* **1978**, *8*, 310.

6.   Ottoni, O.; Neder, A. V. F.; Dias, A. K. B.; Cruz, R. P. A.; Aquino, L. B. *Org. Lett.* **2000**, *3*, 1005.
7.   Fleming, I. *Chemtracts: Org. Chem.* **2001**, *14*, 405. (Review).
8.   Le Roux, C.; Dubac, J. *Synlett* **2002**, 181.
9.   Sefkow, M.; Buchs, J. *Tetrahedron Lett.* **2003**, *44*, 193.

Friedländer synthesis

Quinoline synthesis from the condensation of *o*-aminobenzaldehyde with aldehyde or ketone in the presence of NaOH.

References

1.   Friedländer, P. *Ber. Dtsch. Chem. Ges.* **1882**, *15*, 2572.
2.   Cheng, C.-C.; Yan, S.-J. *Org. Recat.* **1982**, *28*, 37. (Review).
3.   Thummel, R. P. *Synlett* **1992**, 1.
4.   Riesgo, E. C.; Jin, X.; Thummel, R. P. *J. Org. Chem.* **1996**, *61*, 3017.
5.   Mori, T.; Imafuku, K.; Piao, M.-Z.; Fujimori, K. *J. Heterocycl. Chem.* **1996**, *33*, 841.
6.   Ubeda, J. I.; Villacampa, M.; Avendano, C. *Synthesis* **1998**, 1176.
7.   Bu, X.; Deady, L. W. *Synth. Commun.* **1999**, *29*, 4223.
8.   Strekowski, L.; Czarny, A.; Lee, H. *J. Fluorine Chem.* **2000**, *104*, 281.
9.   Chen, J.; Deady, L. W.; Desneves, J.; Kaye, A. J.; Finlay, G. J.; Baguley, B. C.; Denny, W. A. *Bioorg. Med. Chem.* **2000**, *8*, 2461.

148

10. Gladiali, S.; Chelecci, G.; Mudadu, M. S.; Gastaut, M.-A.; Thummel, R. P. *J. Org. Chem.* **2001**, *66*, 400.
11. Hsiao, Y.; Rivera, N. R.; Yasuda, N.; Highes, D. L.; Reider, P. J. *Org. Lett.* **2002**, *4*, 1102, 1243.
12. Dormer, P. G.; Eng, K. K.; Farr, R. N.; Humphrey, G. R.; McWilliams, J. C.; Reider, P. J.; Sager, J. W.; Volante, R. P. *J. Org. Chem.* **2003**, *68*, 467.
13. Arcadi, A.; Chiarini, M.; Di Giuseppe, S.; Marinelli, F. *Synlett* **2003**, 203.

Fries rearrangement

Lewis acid-catalyzed rearrangement of phenol esters to 2- or 4-ketophenols.

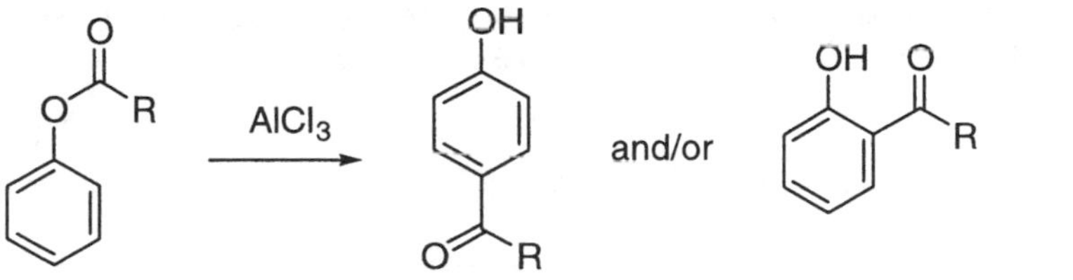

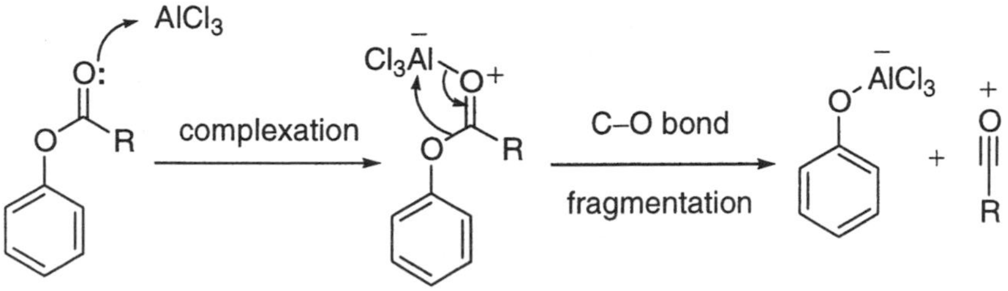

and/or

aluminum phenolate, acylium ion

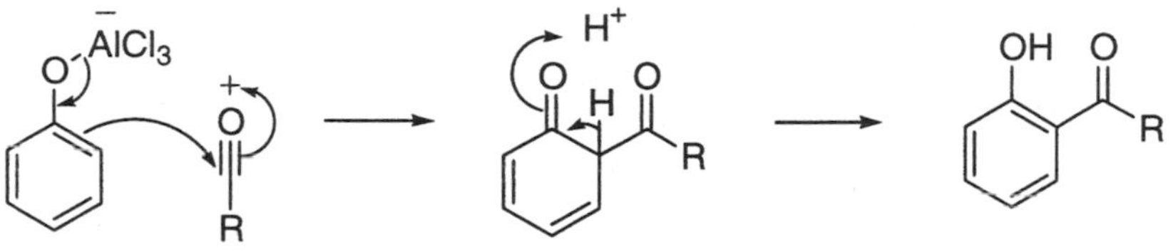

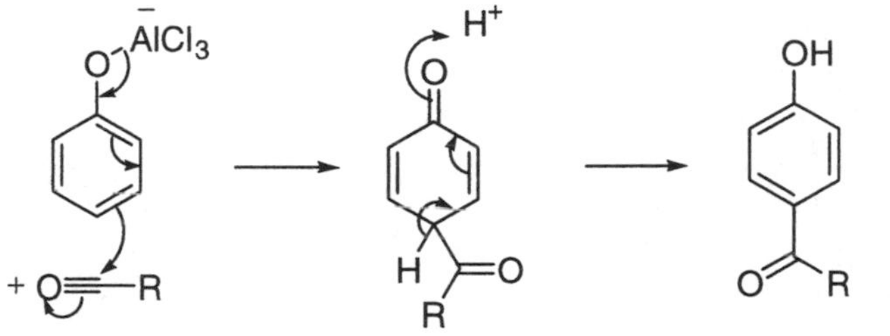

References

1. Fries, K.; Fink, G. *Ber. Dtsch. Chem. Ges.* **1908**, *41*, 4271.
2. Martin, R. *Bull. Soc. Chim. Fr.* **1974**, 983–8. (Review).
3. Martin, R. *Org. Prep. Proced. Int.* **1992**, *24*, 369.
4. Trehan, I. R.; Brar, J. S.; Arora, A. K.; Kad, G. L. *J. Chem. Educ.* **1997**, *74*, 324.
5. Harjani, J. R.; Nara, S. J.; Salunkhe, M. M. *Tetrahedron Lett.* **2001**, *42*, 1979.

6. Focken, T.; Hopf, H.; Snieckus, V.; Dix, I.; Jones, P. G. *Eur. J. Org. Chem.* **2001**, 2221.
7. Kozhevnikova, E. F.; Derouane, E. G.; Kozhevnikov, I. V. *Chem. Commun.* **2002**, 1178.
8. Clark, J. H.; Dekamin, M. G.; Moghaddam, F. M. *Green Chem.* **2002**, *4,* 366.
9. Sriraghavan, K.; Ramakrishnan, V. T. *Tetrahedron* **2003**, *59,* 1791.

Fritsch–Buttenberg–Wiechell rearrangement

Treatment of 1,1-diaryl-2-haloethylene with base affords diaryl acetylene *via* the intermediacy of alkylidene carbene.

e.g. a variant:

References

1. Fritsch, P. *Justus Liebigs Ann. Chem.* **1894**, *272*, 319.
2. Koebrich, G.; Merkel, D. *Angew. Chem., Int. Ed. Engl.* **1970**, *9*, 243.
3. Sket, B.; Zupan, M.; Pollak, A. *Tetrahedron Lett.* **1976**, 783.
4. Sket, B.; Zupan, M. *J. Chem. Soc., Perkin Trans. 1* **1979**, 752.
5. Hebda, C.; Szykula, J.; Orpiszewski, J.; Foehlisch, B. *Monat. Chem.* **1991**, *122,* 1029.
6. Creton, I.; Rezaei, H.; Marek, I.; Normant, J. F. *Tetrahedron Lett.* **1999**, *40*, 1899.
7. Rezaei, H.; Yamanoi, S.; Chemla, F.; Normant, J. F. *Org. Lett.* **2000**, *2*, 419.
8. Eisler, S.; Tykwinski, R. R. *J. Am. Chem. Soc.* **2000**, *122*, 10736.
9. Chernick, E. T.; Eisler, S.; Tykwinski, R. R. *Tetrahedron Lett.* **2001**, *42*, 8575.
10. Shi, S., Annabelle, L. K.; Chernick, E. T.; Eisler, S.; Tykwinski, R. R. *J. Org. Chem.* **2003**, *68*, 1339.

Fujimoto–Belleau reaction

Treatment of enol lactone with Grignard reagent leads to cyclic α-substituted α,β-unsaturated ketone. An alternative to the Robinson annulation.

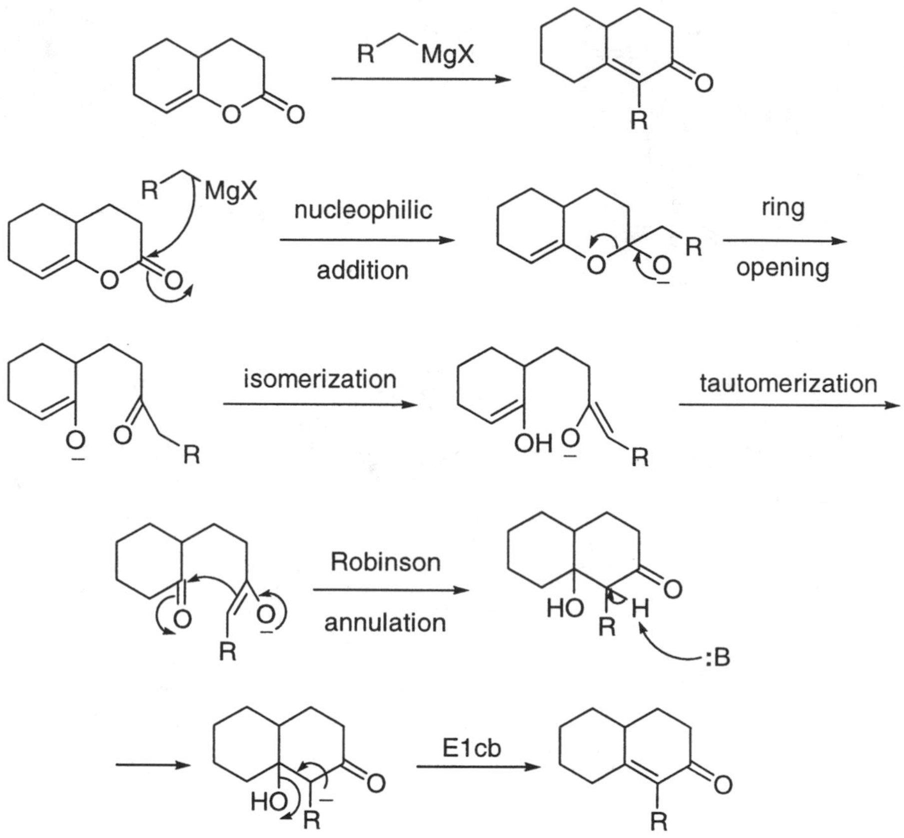

References

1. Fujimoto, C. I. *J. Am. Chem. Soc.* **1951**, *73*, 1856.
2. Weill-Raynal, J. *Synthesis* **1969**, 49.
3. Heys, J. R.; Senderoff, S. G. *J. Org. Chem.* **1989**, *54*, 4702.
4. Aloui, M.; Lygo, B.; Trabsa, H. *Synlett* **1994**, 115.
5. Revial, G.; Jabin, I.; Redolfi, M.; Pfau, M. *Tetrahedron: Asymmetry* **2001**, *12*, 1683.

Fukuyama amine synthesis

Transformation of a primary amine to a secondary amine using 2,4-dinitro-benzenesulfonyl chloride and alcohol.

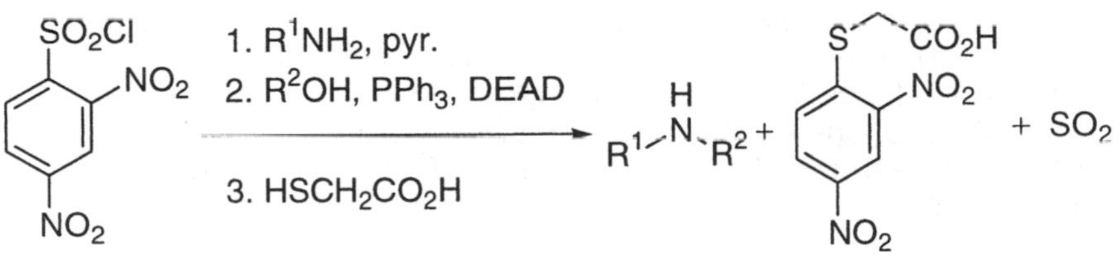

1. R^1NH_2, pyr.
2. R^2OH, PPh$_3$, DEAD
3. $HSCH_2CO_2H$

$+ SO_2$

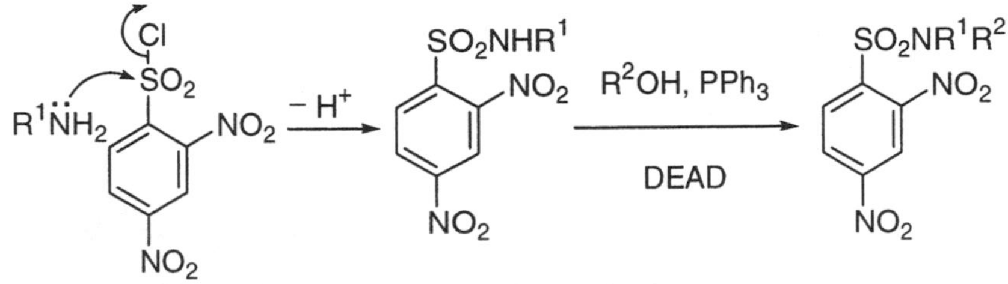

R^2OH, PPh$_3$

DEAD

See Mitsunobu reaction (page 265) for the mechanism.

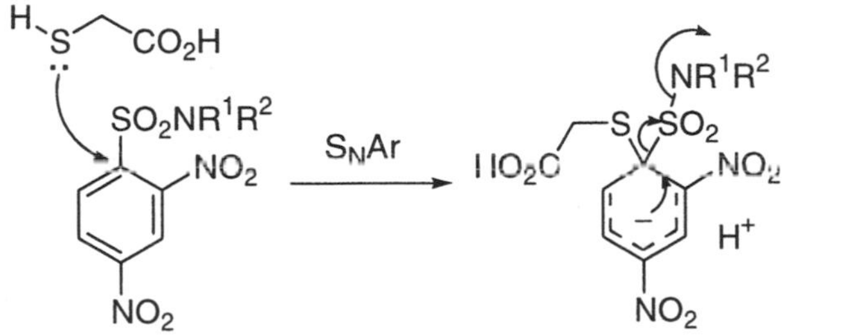

S_NAr

Meisenheimer complex

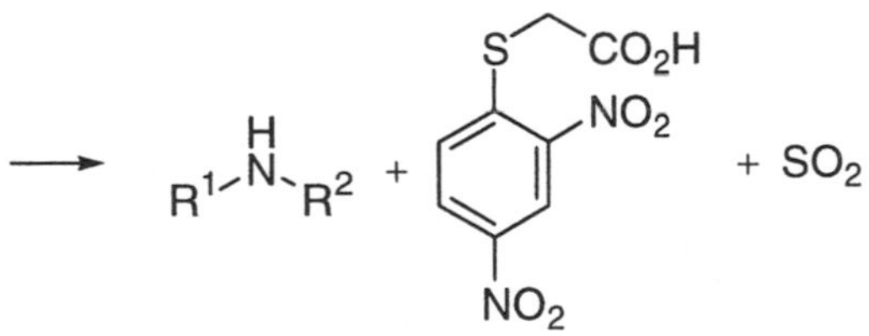

$+ SO_2$

154

References

1. Fukuyama, T.; Jow, C.-K.; Cheung, M. *Tetrahedron Lett.* **1995**, *36*, 6373.
2. Fukuyama, T.; Cheung, M.; Jow, C.-K.; Hidai, Y.; Kan, T. *Tetrahedron Lett.* **1997**, *38*, 5831.
3. Yang, L.; Chiu, K. *Tetrahedron Lett.* **1997**, *38*, 7307.
4. Piscopio, A. D.; Miller, J. F.; Koch, K. *Tetrahedron Lett.* **1998**, *39*, 2667.
5. Bolton, G. L.; Hodges, J. C. *J. Comb. Chem.* **1999**, *1*, 130.
6. Lin, X.; Dorr, H.; Nuss, J. M. *Tetrahedron Lett.* **2000**, *41*, 3309.
7. Amssoms, K.; Augustyns, K.; Yamani, A.; Zhang, M.; Haemers, A. *Synth. Commun.* **2002**, *32*, 319.

Gabriel synthesis

Synthesis of primary amines using potassium phthalimide and alkyl halides.

References

1. Gabriel, S. *Ber. Dtsch. Chem. Ges.* **1887**, *20*, 2224.
2. Press, J. B.; Haug, M. F.; Wright, W. B., Jr. *Synth. Commun.* **1985**, *15*, 837.
3. Slusarska, E.; Zwierzak, A. *Justus Liebigs Ann. Chem.* **1986**, 402.
4. Han, Y.; Hu, H. *Synthesis* **1990**, 122.
5. Ragnarsson, U.; Grehn, L. *Acc. Chem. Res.* **1991**, *24*, 285. (Review).
6. Toda, F.; Soda, S.; Goldberg, I. *J. Chem. Soc., Perkin Trans. 1* **1993**, 2357.
7. Khan, M. N. *J. Org. Chem.* **1996**, *61*, 8063.
8. Mamedov, V. A.; Tsuboi, S.; Mustakimova, L. V.; Hamamoto, H.; Gubaidullin, A. T.; Litvinov, I. A.; Levin, Y. A. *Chem. Heterocycl. Compd.* **2001**, *36*, 911.
9. Iida, K.; Tokiwa, S.; Ishii, T.; Kajiwara, M. *J. Labeled. Compd. Radiopharm.* **2002**, *45*, 569.

Gassman indole synthesis

Sulfur-substituted indole synthesis. The sulfur can be easily hydrogenolyzed off.

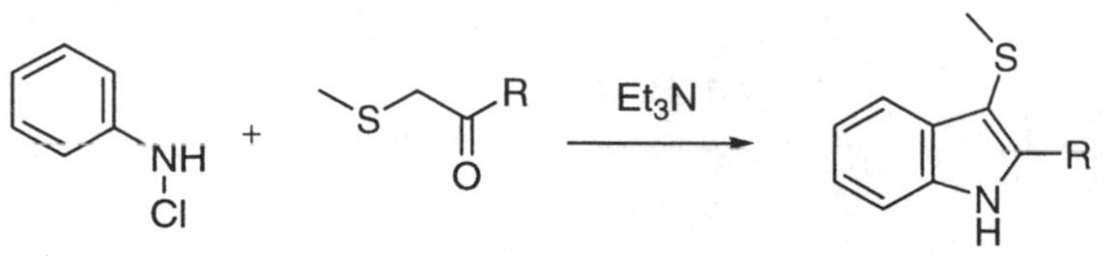

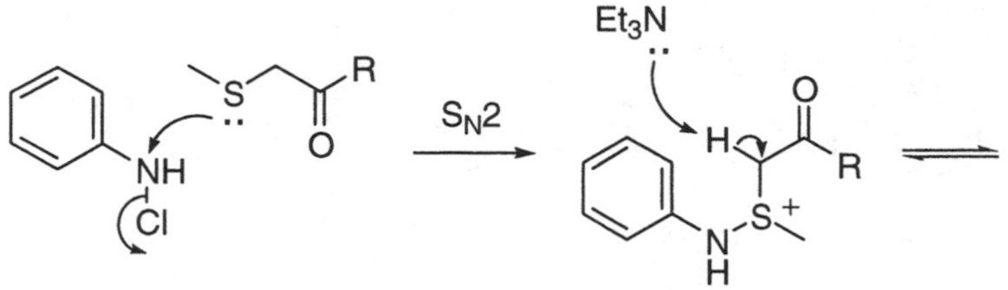

sulfonium ion

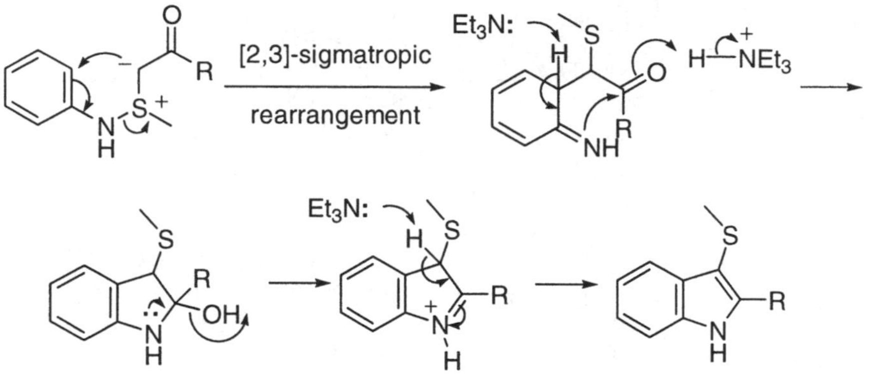

[2,3]-sigmatropic rearrangement

References

1. Gassman, P. G.; van Bergen, T. J.; Gilbert, D. P.; Cue, B. W. *J. Am. Chem. Soc.* **1974**, *96*, 5495, 5508, 5512.
2. Ishikawa, H.; Uno, T.; Miyamoto, H.; Ueda, H.; Tamaoka, H.; Tominaga, M.; Naka-gawa, K. *Chem. Pharm. Bull.* **1990**, *38*, 2459.
3. Wierenga, W. *J. Am. Chem. Soc.* **1981**, *103*, 5621.
4. Smith, A. B., III; Sunazuka, T.; Leenay, T. L.; Kingery-Wood, J. *J. Am. Chem. Soc.* **1990**, *112*, 8197.
5. Smith, A. B., III; Kingery-Wood, J.; Leenay, T. L.; Nolen, E. G.; Sunazuka, T. *J. Am. Chem. Soc.* **1992**, *114*, 1438.
6. Savall, B. M.; McWhorter, W. W.; Walker, E. A. *J. Org. Chem.* **1996**, *61*, 8696.

Gattermann–Koch reaction

Formylation of arenes using carbon monoxide and hydrogen chloride in the presence of aluminum chloride under high pressure.

acylium ion

References

1. Gattermann, L.; Koch, J. A. *Ber.***1897**, *30*, 1622.
2. Crounse, N. N. *Org. React.* **1949**, *5*, 290. (Review).
3. Tanaka, M.; Fujiwara, M.; Ando, H. *J. Org. Chem.* **1995**, *60*, 2106.
4. Tanaka, M.; Fujiwara, M.; Ando, H.; Souma, Y. *Chem. Commun.* **1996**, 159.
5. Tanaka, M.; Fujiwara, M.; Xu, Q.; Souma, Y.; Ando, H.; Laali, K. K. *J. Am. Chem. Soc.* **1997**, *119*, 5100.
6. Tanaka, M.; Fujiwara, M.; Xu, Q.; Ando, H.; Raeker, T J. *J. Org. Chem.* **1998**, *63*, 4408.
7. Kantlehner, W.; Vettel, M.; Gissel, A; Haug, E.; Ziegler, G.; Ciesielski, M.; Scherr, O.; Haas, R. *J. Prakt. Chem.* **2000**, *342*, 297.
8. Doana, M. I.; Ciuculescu, A.; Bruckner, A.; Pop, M.; Filip, P. *Rev. Roum. Chim.* **2002**, *46*, 345.

Gewald aminothiophene synthesis

Base-promoted aminothiophene formation from ketone, α-active methylene nitrile and elemental sulfur.

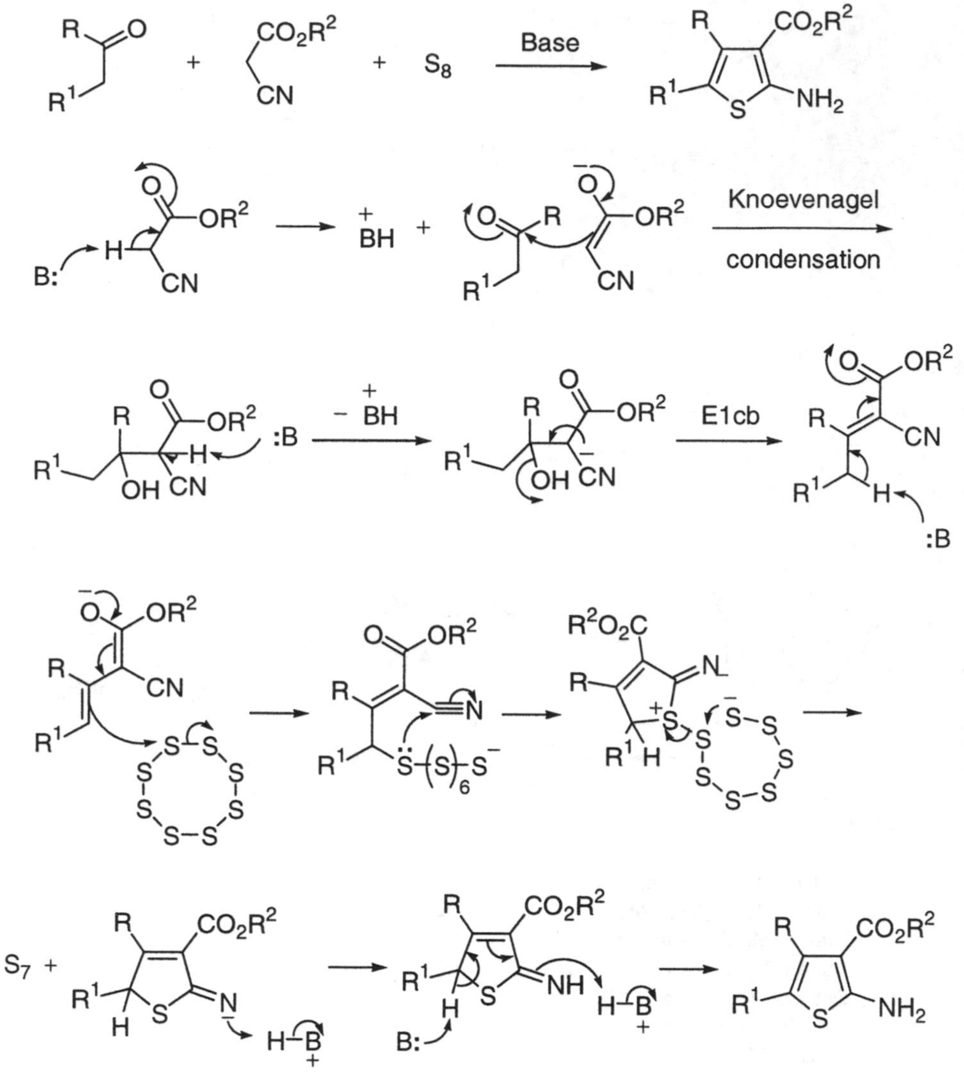

References

1. Gewald, K. *Chimia* **1980**, *34*, 101.
2. Peet, N. P.; Sunder, S.; Barbuch, R. J.; Vinogradoff, A. P. *J. Heterocycl. Chem.* **1986**, *23*, 129.
3. Guetschow, M.; Schroeter, H.; Kuhnle, G.; Eger, K. *Monatsh. Chem.* **1996**, *127*, 297.
4. Zhang, M.; Harper, R. W. *Bioorg. Med. Chem. Lett.* **1997**, *7*, 1629.
5. Sabnis, R. W.; Rangnekar, D. W.; Sonawane, N. D. *J. Heterocycl. Chem.* **1999**, *36*, 333.
6. Baraldi, P. G.; Zaid, A. Z.; Lampronti, I.; Fruttarolo, F. F.; Pavani, M. G.; Tabrizi, M. A.; Shryock, J. C. S.; Leung, E.; Romagnoli, R. *Bioorg. Med. Chem. Lett.* **2000**, *10*, 1953.
7. Pinto, I. L.; Jarvest, R. L.; Serafinowska, H. T. *Tetrahedron Lett.* **2000**, *41*, 1597.
8. Buchstaller, H.-P.; Siebert, C. D.; Lyssy, R. H.; Frank, I.; Duran, A.; Gottschlich, R.; Noe, C. R. *Monatsh. Chem.* **2001**, *132*, 279.
9. Hoener, A. P. F.; Henkel, B.; Gauvin, J.-C. *Synlett* **2003**, 63.

160

Glaser coupling

Oxidative homocoupling of terminal alkynes using copper catalyst.

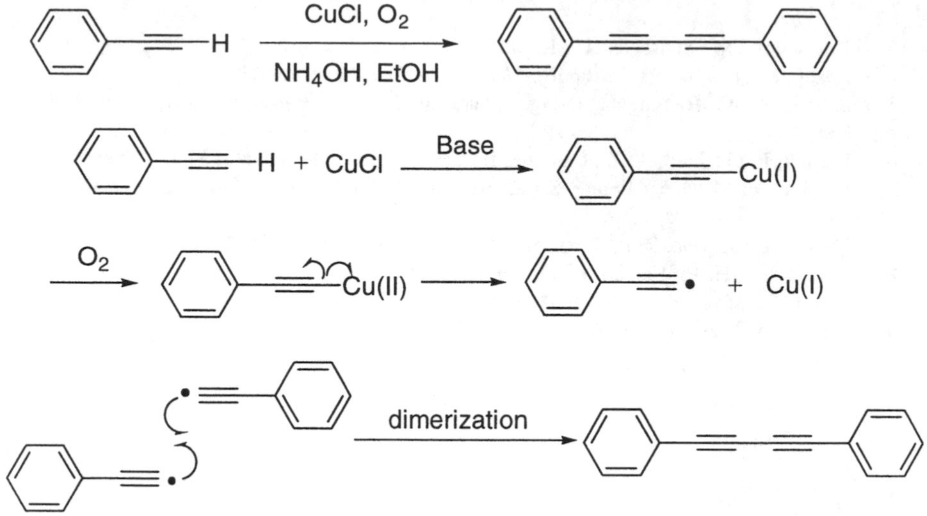

References

1. Glaser, C. *Ber. Dtsch. Chem. Ges.* **1869**, *2*, 422.
2. Hoeger, S.; Meckenstock, A.-D.; Pellen, H. *J. Org. Chem.* **1997**, *62*, 4556.
3. Li, J.; Jiang, H. *Chem. Commun.* **1999**, 2369.
4. Siemsen, P.; Livingston, R. C.; Diederich, F. *Angew. Chem., Int. Ed.* **2000**, *39*, 2632. (Review).
5. Setzer, W. N.; Gu, X.; Wells, E. B.; Setzer, M. C.; Moriarity, D. M. *Chem. Pharm. Bull.* **2001**, *48*, 1776.
6. Kabalka, G. W.; Wang, L.; Pagni, R. M. *Synlett* **2001**, 108.
7. Youngblood, W. J.; Gryko, D. T.; Lammi, R. K.; Bocian, D. F.; Holten, D.; Lindsey, J. S. *J. Org. Chem.* **2002**, *67*, 2111.

Gomberg–Bachmann reaction

Base-promoted radical coupling between an aryl diazonium salt and an arene to form a diaryl compound.

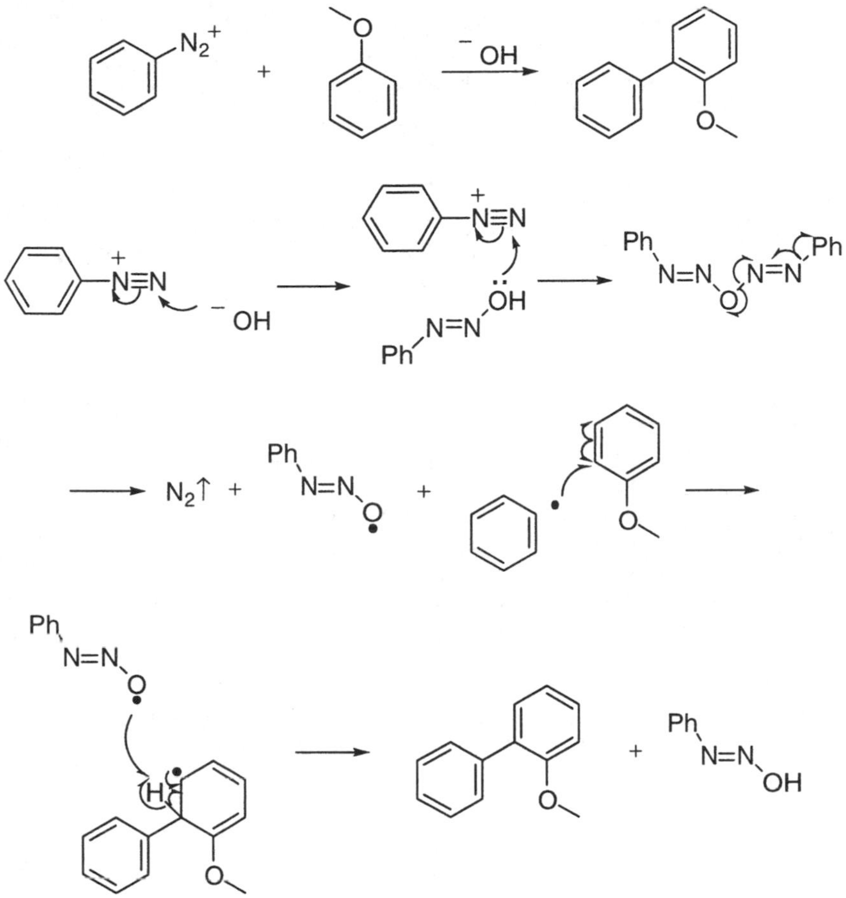

References

1. Gomberg, M.; Bachmann, W. E. *J. Am. Chem. Soc.* **1924**, *46*, 2339.
2. DeTar, D. F.; Kazimi, A. A. *J. Am. Chem. Soc.* **1955**, *77*, 3842.
3. Beadle, J. R.; Korzeniowski, S. H.; Rosenberg, D. E.; Garcia-Slanga, B. J.; Gokel, G. W. *J. Org. Chem.* **1984**, *49*, 1594.
4. McKenzie, T. C.; Rolfes, S. M. *J. Heterocycl. Chem.* **1987**, *24*, 859.
5. Gurczynski, M.; Tomasik, P. *Org. Prep. Proced. Int.* **1991**, *23*, 438.
6. Hales, N. J.; Heaney, H.; Hollinshead, J. H.; Sharma, R. P. *Tetrahedron* **1995**, *51*, 7403.
7. Lai, Y.-H.; Jiang, J. *J. Org. Chem.* **1997**, *62*, 4412.

Gribble indole reduction

Reduction of the indole double bond using sodium cyanoborohydride in glacial acetic acid. The use of sodium borohydride leads to reduction and *N*-alkylation.

References

1. Gribble, G. W.; Lord, P. D.; Skotnicki, J.; Dietz, S. E.; Eaton, J. T.; Johnson, J. L. *J. Am. Chem. Soc.* **1974**, *96*, 7812.
2. Gribble, G. W.; Hoffman, J. H. *Synthesis* **1977**, 859.
3. Gribble, G. W.; Nutaitis, C. F. *Org. Prep. Proc. Int.* **1985**, *17*, 317.
4. Rawal, V. H.; Jones, R. J.; Cava, M. P. *J. Org. Chem.* **1987**, *52*, 19.
5. Boger, D. L.; Coleman, R. S.; Invergo, B. L. *J. Org. Chem.* **1987**, *52*, 1521.
6. Siddiqui, M. A.; Snieckus, V. *Tetrahedron Lett.* **1990**, *31*, 1523.
7. Gribble, G. W. *ACS Symposium Series No. 641*, **1996**, pp 167–200.
8. Gribble, G. W. *Chem. Soc. Rev.* **1998**, *27*, 395. (Review).
9. He, F.; Foxman, B. M.; Snider, B. B. *J. Am. Chem. Soc.* **1998**, *120*, 6417.
10. Nicolaou, K. C.; Safina, B. S.; Winssinger, N. *Synlett* **2001**, 900

Gribble reduction of diaryl ketones

Reduction of diaryl ketones and diarylmethanols to diarylmethanes using sodium borohydride in trifluoroacetic acid. Also applicable to diheteroaryl ketones and alcohols.

$$Ar^1 \overset{O}{\underset{}{\diagup}} Ar^2 \xrightarrow[\text{CF}_3\text{CO}_2\text{H}]{\text{NaBH}_4} \left[Ar^1 \overset{OH}{\underset{}{\diagup}} Ar^2 \right] \longrightarrow Ar^1 \diagdown Ar^2$$

$$\underset{(CF_3CO_2)_3B-H}{Ar^1 \overset{\overset{\displaystyle H^+}{O:}}{\underset{}{\diagup}} Ar^2} \longrightarrow Ar^1 \overset{OH}{\underset{}{\diagup}} Ar^2 \xrightarrow{CF_3CO_2H}$$

$$Ar^1 \overset{\overset{+}{O}H_2}{\underset{}{\diagup}} Ar^2 \xrightarrow{-H_2O} Ar^1 \overset{+}{\diagdown} Ar^2 \xrightarrow{\overset{-}{H}B(OCOCF_3)_3} Ar_1 \diagdown Ar^2$$

References

1. Gribble, G. W.; Leese, R. M.; Evans, B. E. *Synthesis* **1977**, 172.
2. Gribble, G. W.; Kelly, W. J.; Emery, S. E. *Synthesis* **1978**, 763.
3. Gribble, G. W.; Nutaitis, C. F. *Org. Prep. Proc. Int.* **1985**, *17*, 317.
4. Kabalka, G. W.; Kennedy, T. P. *Org. Prep. Proc. Int.* **1989**, *21*, 348.
5. Daich, A.; Decroix, B. *J. Heterocycl. Chem.* **1992**, *29*, 1789.
6. Gribble, G. W. *ACS Symposium Series No. 641*, **1996**, pp 167–200. (Review).
7. Gribble, G. W. *Chem. Soc. Rev.* **1998**, *27*, 395. (Review).
8. Sattelkau, T.; Qandil, A. M.; Nichols, D. E. *Synthesis* **2001**, 267.

Grignard reaction

Addition of organomagnesium compounds (Grignard reagents), generated from organohalides and magnesium metal, to electrophiles.

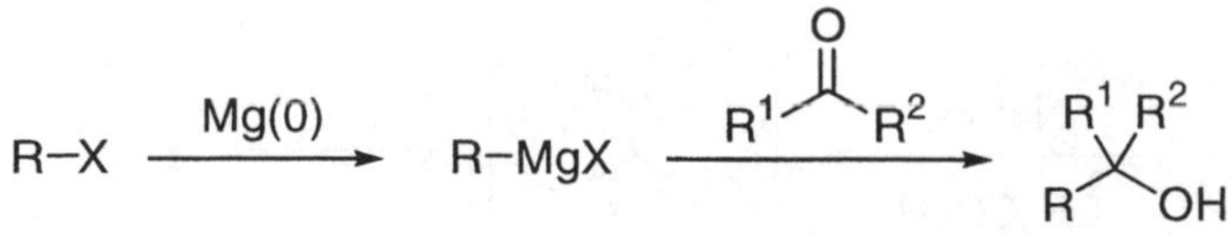

Formation of the Grignard reagent:

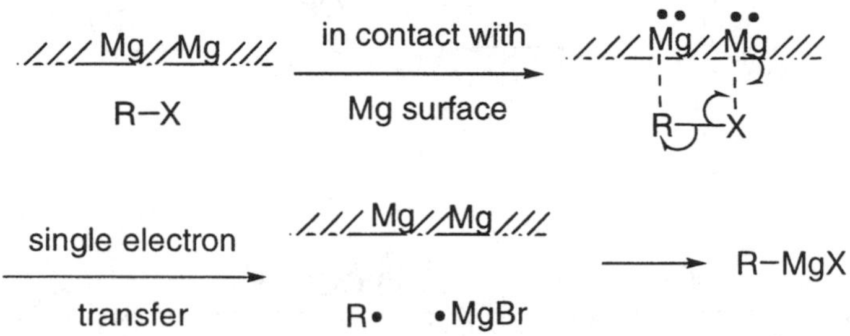

Grignard reaction:
Ionic mechanism,

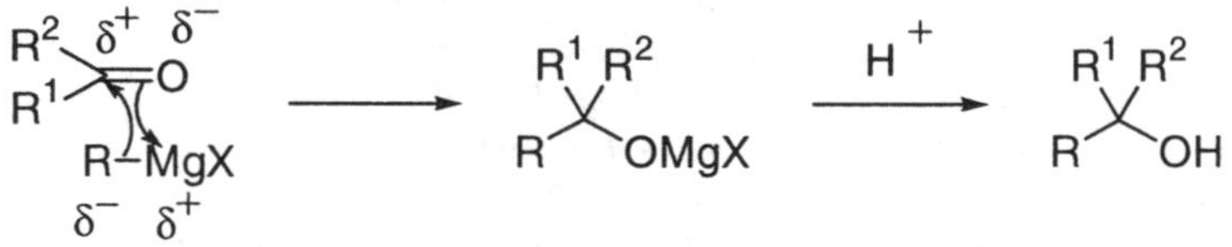

Radical mechanism,

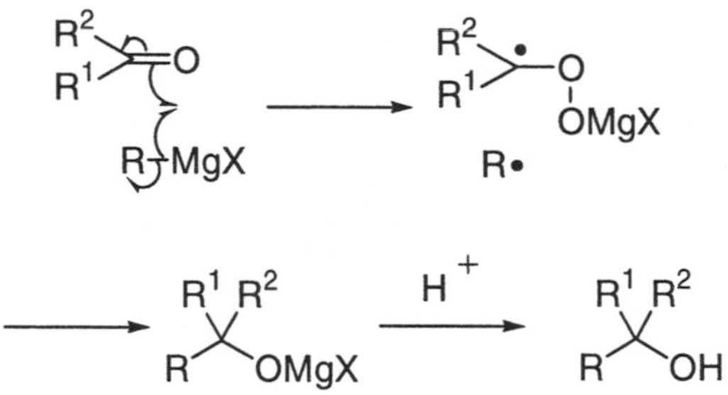

References

1. Grignard, V. *C. R. Acad. Sci.* **1900**, *130*, 1322.
2. Ashby, E. C.; Laemmle, J. T.; Neumann, H. M. *Acc. Chem. Res.* **1974**, *7*, 272. (Review).
3. Ashby, E. C.; Laemmle, J. T. *Chem. Rev.* **1975**, *75*, 521. (Review).
4. Lasperas, M.; Perez-Rubalcaba, A.; Quiroga-Feijoo, M. L. *Tetrahedron* **1980**, *36*, 3403.
5. Lund, T.; Pedersen, M. L.; Frandsen, L. A. *Tetrahedron Lett.* **1994**, *35*, 9225.
6. *Grignard Reagents* Richey, H. G., Jr., Ed.; Wiley: New York, 2000. (Review).
7. Holm, T.; Crossland, I. In *Grignard Reagents* Richey, H. G., Jr., Ed.; Wiley: New York, 2000, Chapter 1, pp 1–26. (Review).
8. Hoffmann, R. W.; Hölzer, B. *Chem. Commun.* **2001**, 491.
9. Toda, N.; Ori, M.; Takami, K.; Tago, K.; Kogen, H. *Org. Lett.* **2003**, *5*, 269.

Grob fragmentation

General scheme:

$$X = OH_2^+, OTs, I, Br, Cl; Y = O^-, NR_2$$

e.g.:

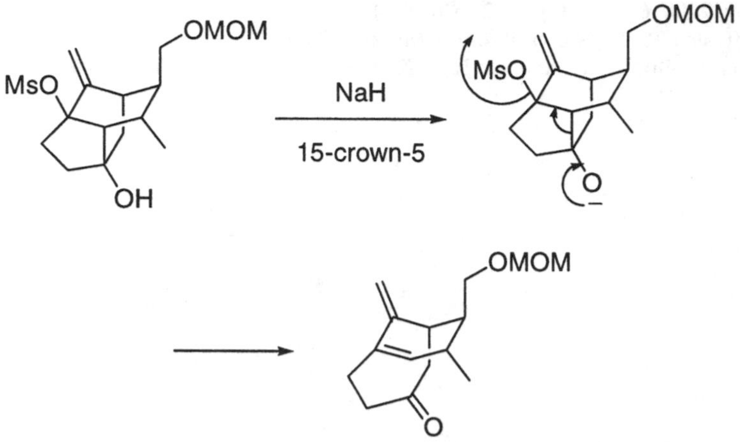

e.g.:

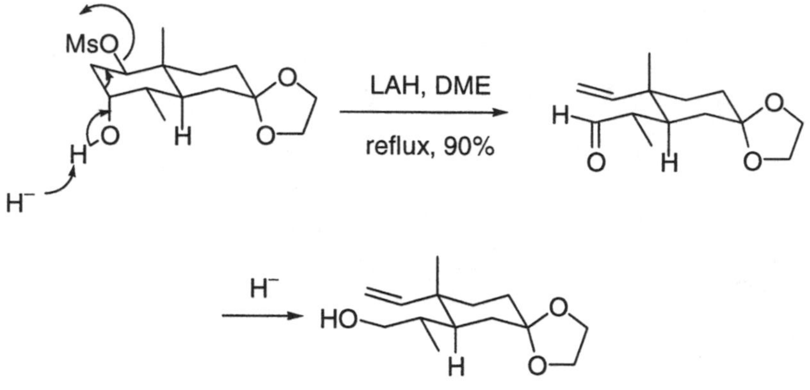

References

1. Grob, C. A.; Baumann, W. *Helv. Chim. Acta* **1955**, *38*, 594.
2. Grob, C. A.; Schiess, P. W. *Angew. Chem., Int. Ed. Engl.* **1967**, *6*, 1.
3. French, L. G.; Charlton, T. P. *Heterocycles* **1993**, *35*, 305.

4. Harmata, M.; Elahmad, S. *Tetrahedron Lett.* **1993**, *34*, 789.
5. Armesto, X. L.; Canle L., M.; Losada, M.; Santaballa, J. A. *J. Org. Chem.* **1994**, *59*, 4659.
6. Yoshimitsu, T.; Yanagiya, M.; Nagaoka, H. *Tetrahedron Lett.* **1999**, *40*, 5215.
7. Hu, W.-P.; Wang, J.-J.; Tsai, P.-C. *J. Org. Chem.* **2000**, *65*, 4208.
8. Molander, G. A.; Le Huerou, Y.; Brown, G. A. *J. Org. Chem.* **2001**, *66*, 4511.
9. Alder, R. W.; Harvey, J. N.; Oakley, M. T. *J. Am. Chem. Soc.* **2002**, *124*, 4960.
10. Paquette, L. A.; Yang, J.; Long, Y. O. *J. Am. Chem. Soc.* **2002**, *124*, 6542.
11. Barluenga, J.; Alvarez-Perez, M.; Wuerth, K.; Rodriguez, F.; Fananas, F. *J. Org. Lett.* **2003**, *5*, 905.

Guareschi–Thorpe condensation

2-Pyridone formation from the condensation of cyanoacetic ester with acetoacetic ester in the presence of ammonia.

References

1. Baron, H.; Renfry, F. G. P.; Thorpe, J. F. *J. Chem. Soc.* **1904**, *85*, 1726.
2. Brunskill, J. S. A. *J. Chem. Soc. (C)* **1968**, 960.
3. Brunskill, J. S. A. *J. Chem. Soc., Perkin Trans. 1* **1972**, 2946.
4. Krstic, V.; Misic-Vukovic, M.; Radojkovic-Velickovic, M. *J. Chem. Res. (S)* **1991**, 82.
5. Narsaiah, B.; Sivaprasad, A.; Venkataratnam, R. V. *Org. Prep. Proced. Int.* **1993**, *25*, 116.
6. Mijin, D. Z.; Misic-Vukovic, M M. *Indian J. Chem., Sect. B* **1995**, *34B*, 348.
7. Mijin, D. Z.; Misic-Vukovic, M M. *Indian J. Chem., Sect. B* **1998**, *37B*, 988.
8. Al-Omran, F.; El-Khair, A. A. Mijin, D. Z.; Misic-Vukovic, M M. *Indian J. Chem., Sect. B* **2001**, *40B*, 608.

Hajos–Wiechert reaction

Asymmetric Robinson annulation catalyzed by (S)-(−)-proline.

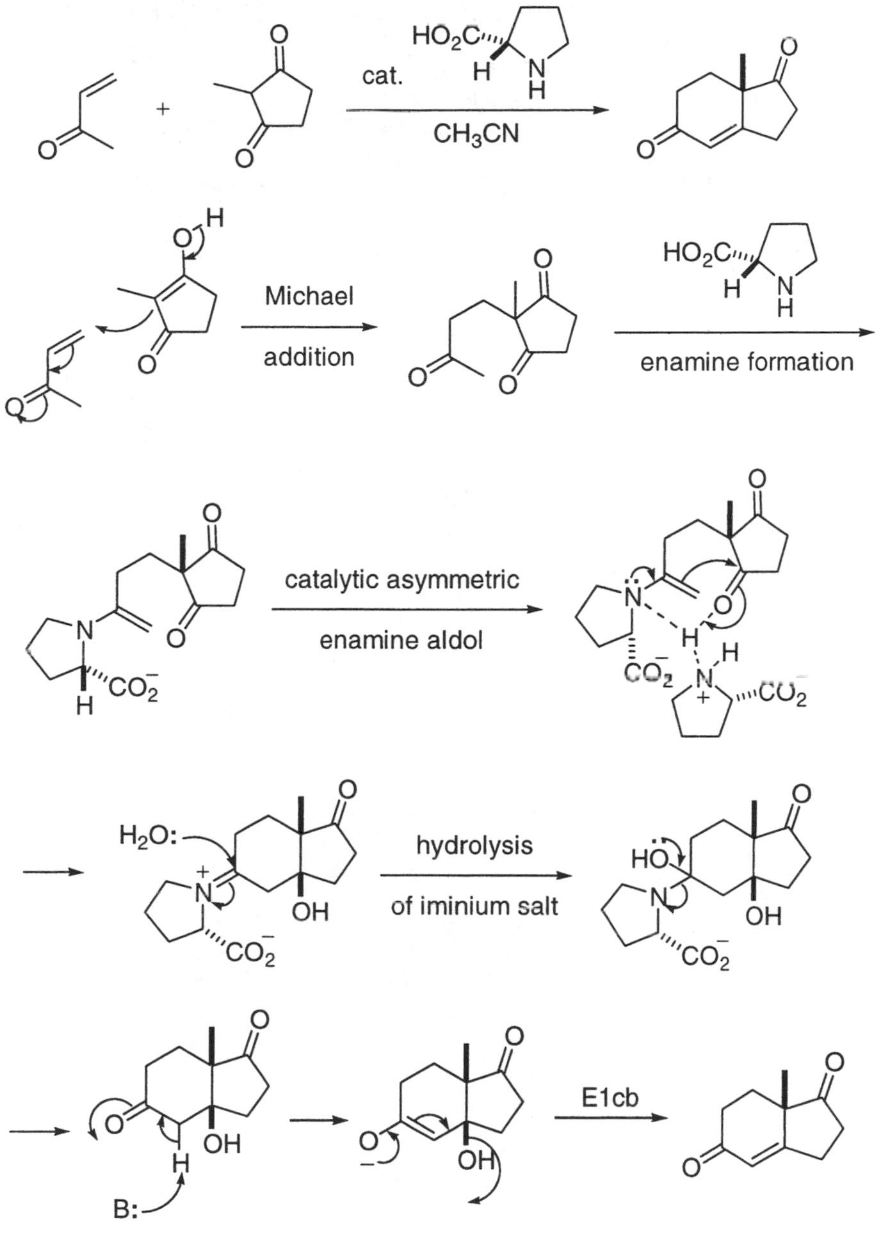

170

References

1. Hajos, Z. G.; Parrish, D. R. *J. Org. Chem.* **1974**, *39*, 1615.
2. Eder, U.; Sauer, G.; Wiechert, R. *Angew. Chem., Int. Ed. Engl.* **1971**, *10*, 496.
3. Brown, K. L.; Dann, L.; Duntz, J. D.; Eschenmoser, A.; Hobi, R.; Kratky, C. *Helv. Chim. Acta* **1978**, *61*, 3108.
4. Agami, C. *Bull. Soc. Chim. Fr.* **1988**, 499.
5. Nelson, S. G. *Tetrahedron: Asymmetry* **1998**, *9*, 357.
6. List, B.; Lerner, R. A.; Barbas, C. F., III. *J. Am. Chem. Soc.* **2000**, *122*, 2395.
7. List, B.; Pojarliev, P.; Castello, C. *Org. Lett.* **2001**, *3*, 573.
8. Hoang, L.; Bahmanyar, S.; Houk, K. N.; List, B. *J. Am. Chem. Soc.* **2003**, *125*, 16.

Haller–Bauer reaction

Base-induced cleavage of non-enolizable ketones leading to carboxylic amide derivative and a neutral fragment in which the carbonyl group is replaced by a hydrogen.

non-enolizable ketone

References

1. Haller, A.; Bauer, E. *Compt. Rend.* **1908**, *147*, 824.
2. Paquette, L. A.; Gilday, J. P.; Maynard, G. D. *J. Org. Chem.* **1989**, *54*, 5044.
3. Paquette, L. A.; Gilday, J. P. *Org. Prep. Proc. Int.* **1990**, *22*, 167.
4. Mehta, G.; Praveen, M. *J. Org. Chem.* **1995**, *60*, 279.
5. Mehta, G.; Reddy, K. S.; Kunwar, A. C. *Tetrahedron Lett.* **1996**, *37*, 2289.
6. Mehta, G.; Reddy, K. S. *Synlett* **1996**, 229.
7. Mittra, A.; Bhowmik, D. R.; Venkateswaran, R. V. *J. Org. Chem.* **1998**, *63*, 9555.
8. Mehta, G.; Venkateswaran, R. V. *Tetrahedron* **2000**, *56*, 1399.
9. Arjona, O.; Medel, R.; Plumet, J. *Tetrahedron Lett.* **2001**, *42*, 1287.

Hantzsch pyridine synthesis

Dihydropyridine from the condensation of aldehyde, β-ketoester and ammonia.

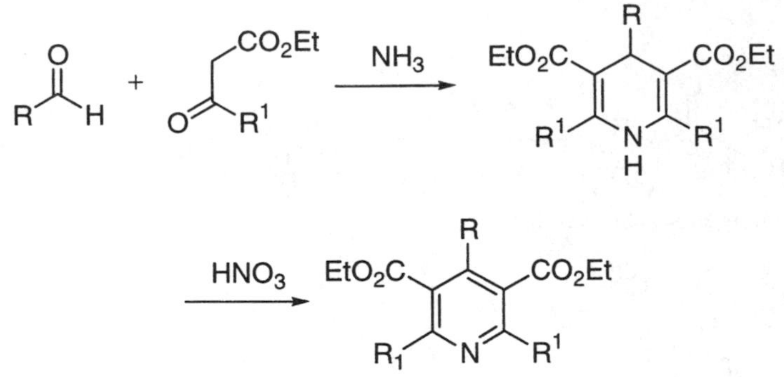

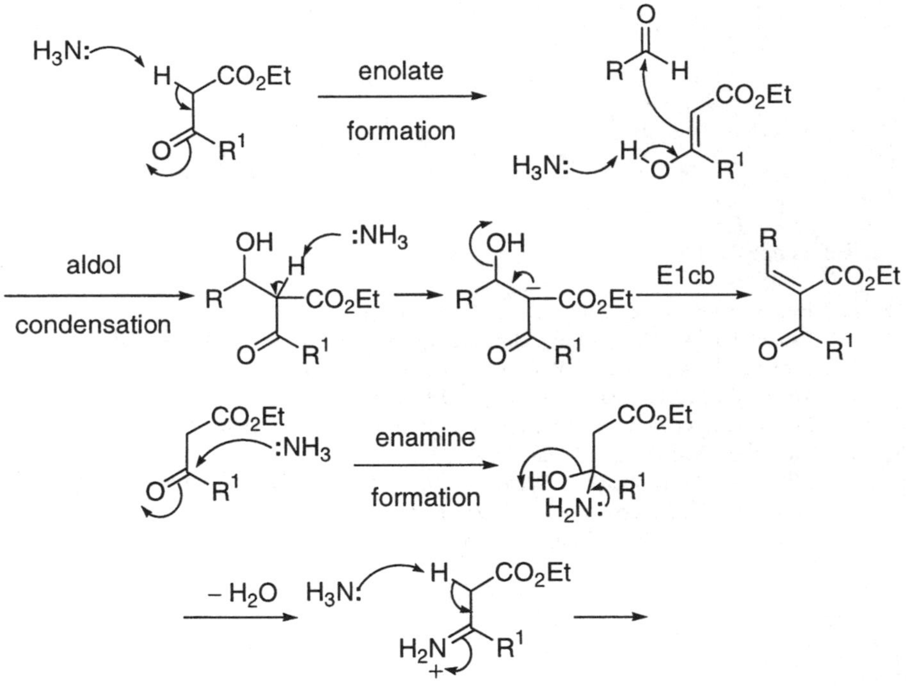

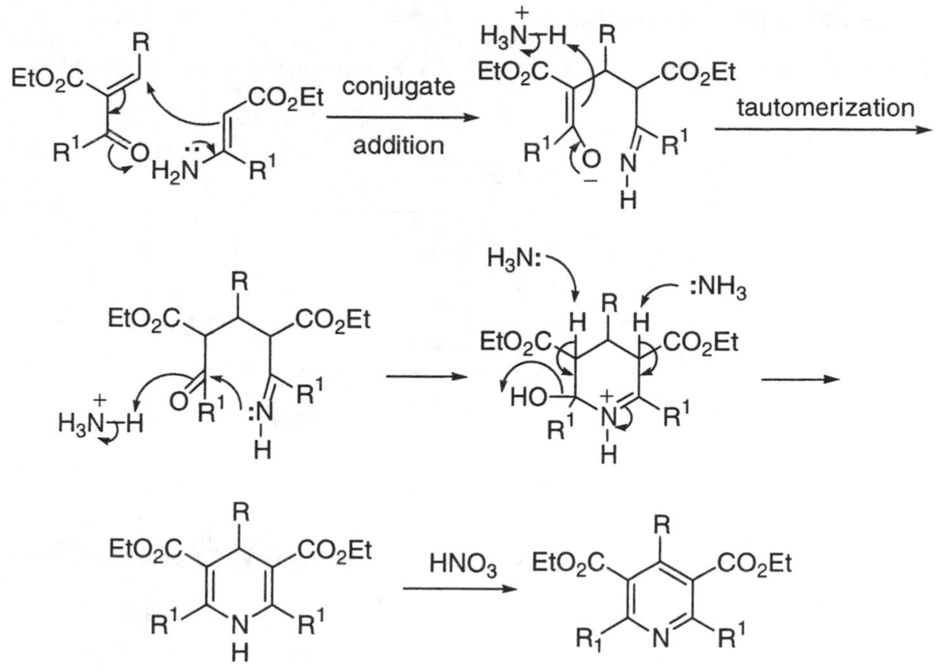

References

1. Hantzsch, A. *Ann.* **1882**, *215*, 1.
2. Balogh, M.; Hermecz, I.; Naray-Szabo, G.; Simon, K.; Meszaros, Z. *J. Chem. Soc., Perkin Trans. 1* **1986**, 753.
3. Katritzky, A. R.; Ostercamp, D. L.; Yousaf, T. I. *Tetrahedron* **1986**, *42*, 5729.
4. Shah, A. C.; Rehani, R.; Arya, V. P. *J. Chem. Res., (S)* **1994**, 106.
5. Menconi, I.; Angeles, E.; Martinez, L.; Posada, M. E.; Toscano, R. A.; Martinez, R. *J. Heterocycl. Chem.* **1995**, *32*, 831.
6. Goerlitzer, K.; Heinrici, C.; Ernst, L. *Pharmazie* **1999**, *54*, 35.
7. Raboin, J.-C.; Kirsch, G.; Beley, M. *J. Heterocycl. Chem.* **2000**, *37*, 1077.
8. Sambongi, Y.; Nitta, H.; Ichihashi, K.; Futai, M.; Ueda, I. *J. Org. Chem.* **2002**, *67*, 3499.

174

Hantzsch pyrrole synthesis

Reaction of α-chloromethyl ketones with β-ketoesters and ammonia to assemble pyrroles.

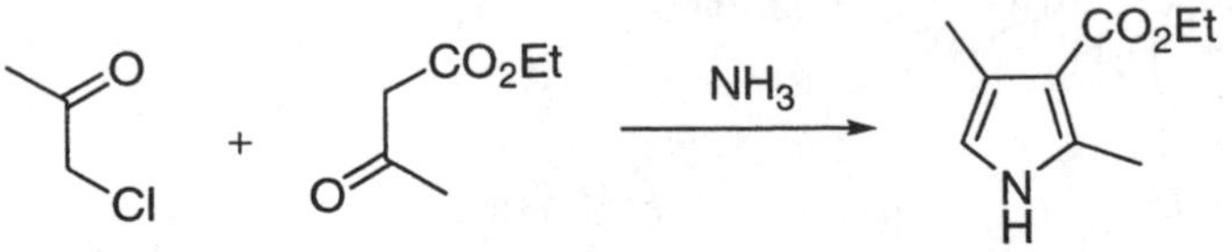

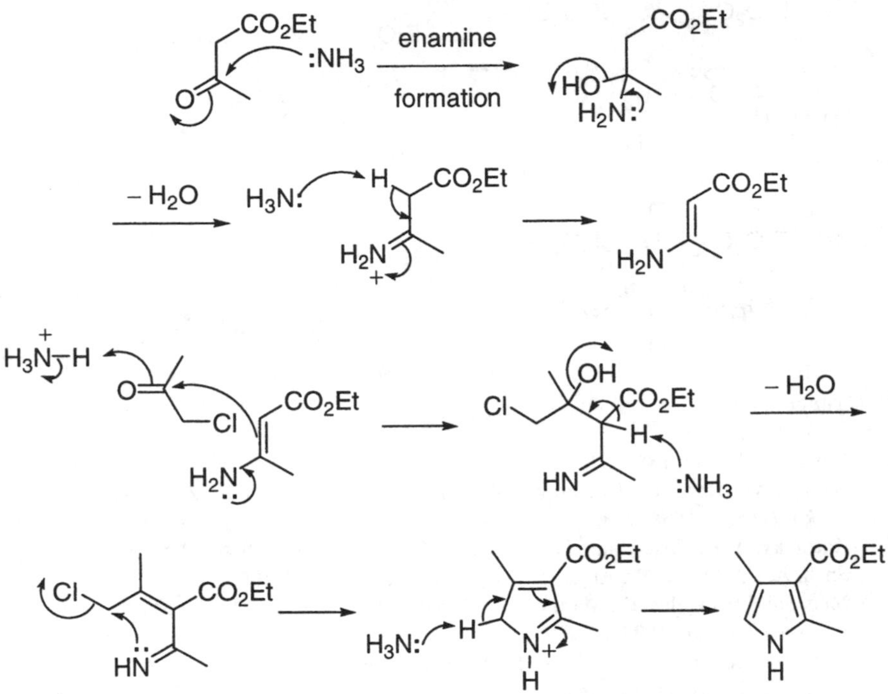

References

1. Hantzsch, A. *Ber. Dtsch. Chem. Ges.* **1890**, *23*, 1474.
2. Hort, E. V.; Anderson, L. R. *Kirk-Othmer Encycl. Chem. Technol.*; 3rd Ed.; **1982**, *19*, 499. (Review).
3. Katritzky, A. R.; Ostercamp, D. L.; Yousaf, T. I. *Tetrahedron* **1987**, *43*, 5171.
4. Kirschke, K.; Costisella, B.; Ramm, M.; Schulz, B. *J. Prakt. Chem.* **1990**, *332*, 143.
5. Trautwein, A. W.; Süßmuth, R. D.; Jung, G. *Bioorg. Med. Chem. Lett.* **1998**, *8*, 2381.
6. Ferreira, V. F.; De Souza, M. C. B. V.; Cunha, A. C.; Pereira, L. O. R.; Ferreira, M. L. G. *Org. Prep. Proceed. Int.* **2002**, *33*, 411.

Haworth reaction

Friedel–Crafts reaction of an arene with succinic anhydride is followed by reduction and an additional intramolecular Friedel–Crafts reaction to give tetralone. The entire process is called the Haworth reaction.

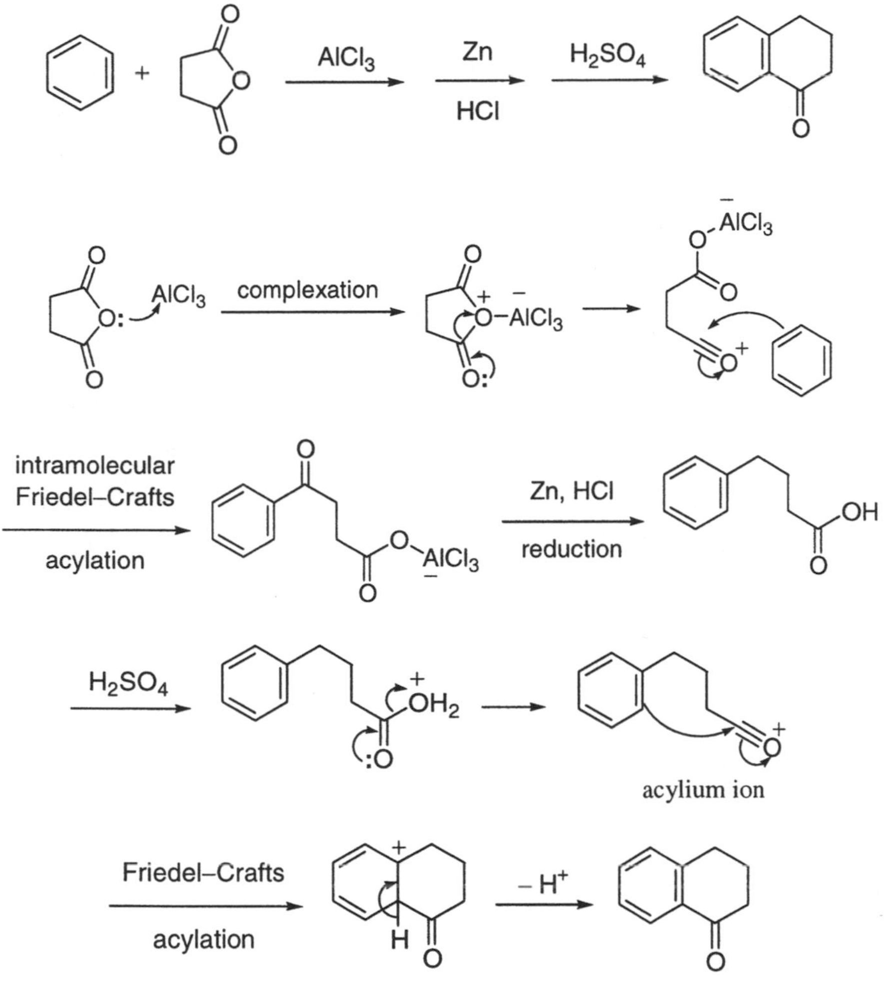

References

1. Haworth, R. D. *J. Chem. Soc.* **1932**, 1125.
2. Agranat, I.; Shih, Y. *J. Chem. Educ.* **1976**, *53*. 488.
3. Silveira, A., Jr.; McWhorter, E. J. *J. Org. Chem.* **1972**, *37*. 3687.
4. Aichaoui, H.; Poupaert, J. H.; Lesieur, D.; Henichart, J. P. *Bull. Soc. Chim. Belg.* **1992**, *101*. 1053.

5. Auger, P.; Malaiyandi, M.; Wightman, R. H.; Williams, D. T. *J. Labeled. Compd. Radiopharm.* **1993**, *33*, 263.
6. Wipf, P.; Jung, J.-K. *J. Org. Chem.* **2000**, *65*, 6319.
7. Kadam, A. J.; Baraskar, U. K.; Mane, R. B. *Indian J. Chem., Sect. B* **2000**, *39B*, 822.

Hayashi rearrangement

Rearrangement of *o*-benzoylbenzoic acids in the presence of sulfuric acid or phosphorus pentoxide.

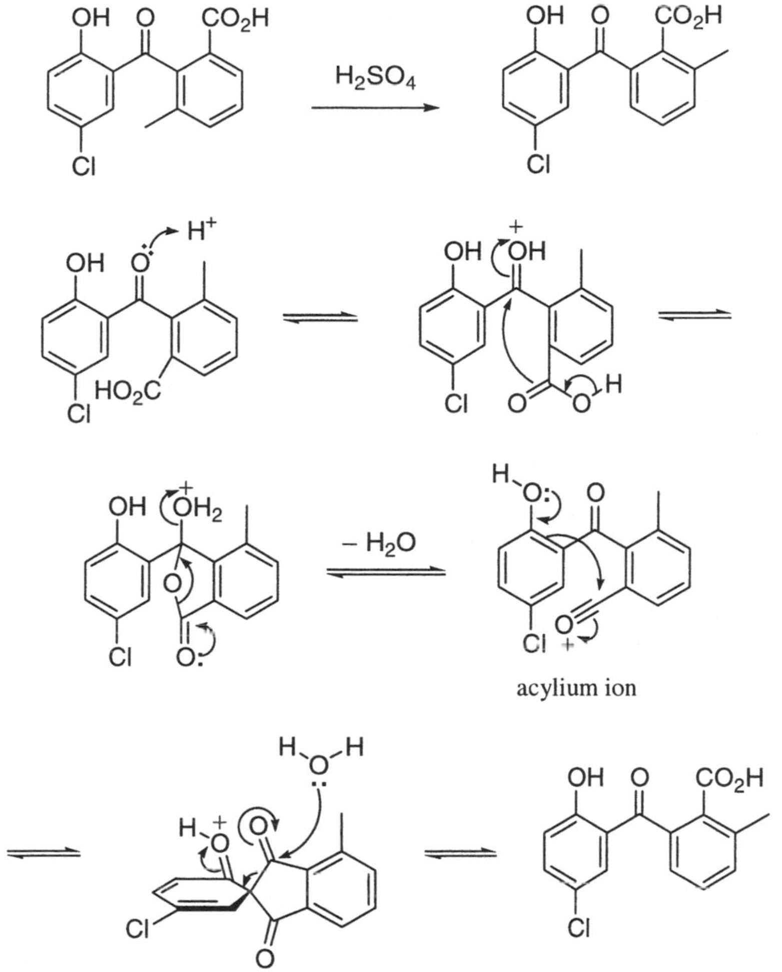

acylium ion

spirocyclic intermediate

References

1. Hayashi, M. *J. Chem. Soc.* **1927**, 2516.
2. Sandin, R. B.; Melby, R.; Crawford, R.; McGreer, D. G. *J. Am. Chem. Soc.* **1956**, *78*, 3817.
3. Newman, M. S.; Ihrman, K. G. *J. Am. Chem. Soc.* **1958**, *80*, 3652.
4. Cristol, S. J.; Caspar, M. L. *J. Org. Chem.* **1968**, *33*, 2020.

178

5.   Cadogan, J. I. G.; Kulik, S.; Tood, M. J. *J. Chem. Soc., Chem. Commun.* **1968**, 736.
6.   Newmann, M. S. *Acc. Chem. Res.* **1972**, *5*, 354.
7.   Cushman, M.; Choong, T.-C.; Valko, J. T.; Koleck, M. P. *J. Org. Chem.* **1980**, *45*, 5067.
8.   Opitz, A.; Roemer, E.; Haas, W.; Gorls, H.; Werner, W.; Grafe, U. *Tetrahedron* **2000**, *56*, 5147.

Heck reaction

Palladium-catalyzed coupling between organohalides or triflates with olefins.

$$R-X \xrightarrow{\text{Pd(0)}} R\diagup\diagdown Z$$

X = I, Br, OTf, *etc.*
Z = H, R, Ar, CN, CO$_2$R, OR, OAc, NHAc, *etc.*

References

1. Heck, R. F.; Nolley, J. P., Jr. *J. Am. Chem. Soc.* **1968**, *90*, 5518.
2. Heck, R. F. *Acc. Chem. Res.* **1979**, *12*, 146. (Review).
3. Heck, R. F. *Org. React.* **1982**, *27*, 345. (Review).
4. Heck, R. F. *Palladium Reagents in Organic Synthesis,* Academic Press, London, **1985**. (Review).
5. Akita, Y.; Inoue, A.; Mori, Y.; Ohta, A. *Heterocycles* **1986**, *24*, 2093.
6. Hegedus, L. S. *Transition Metals in the Synthesis of Complex Organic Molecule* **1994**, University Science Books: Mill Valley, CA, pp 103–113. (Review).
7. Beletskaya, I. P.; Cheprakov, A. V. *Chem. Rev.* **2000**, *100*, 3009. (Review).
8. Amatore, C.; Jutand, A. *Acc. Chem. Res.* **2000**, *33*, 314. (Review).
9. Franzen, R. *Can. J. Chem.* **2000**, *78*, 957.
10. Mayasundari, A.; Young, D. G. J. *Tetrahedron Lett.* **2001**, *42*, 203.
11. Haeberli, A.; Leumann, C. J. *Org. Lett.* **2001**, *3*, 489.

12. Gilbertson, S. R.; Fu, Z.; Xie, D. *Tetrahedron Lett.* **2001**, *42*, 365.
13. Andrus, M. B.; Song, C.; Zhang, J. *Org. Lett.* **2002**, *4*, 2079.
14. Reddy, P. R.; Balraju, V.; Madhavan, G. R.; Banerji, B.; Iqbal, J. *Tetrahedron Lett.* **2003**, *44*, 353.

Hegedus indole synthesis

Stoichiometric Pd(II)-mediated oxidative cyclization of alkenyl anilines to indoles. *Cf.* Wacker oxidation.

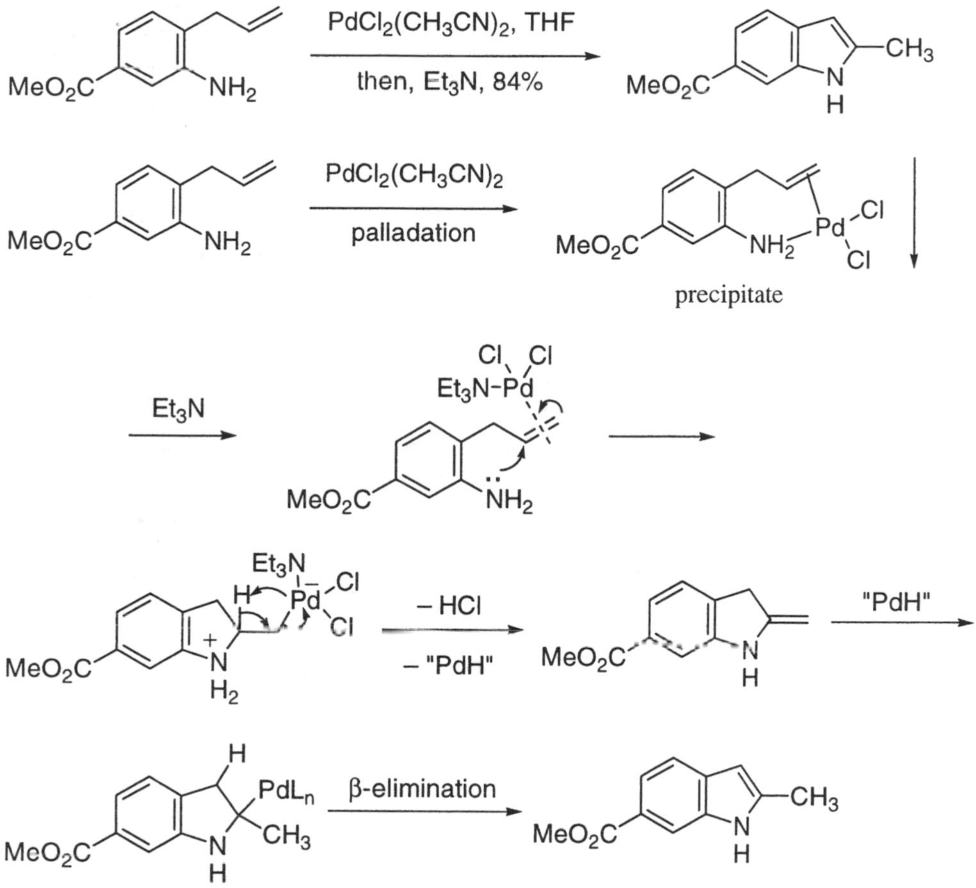

References

1. Hegedus, L. S.; Allen, G. F.; Waterman, E. L. *J. Am. Chem. Soc.* **1976**, *98*, 2674.
2. Hegedus, L. S.; Allen, G. F.; Bozell, J. J.; Waterman, E. L. *J. Am. Chem. Soc.* **1978**, *100*, 5800.
3. Hegedus, L. S.; Winton, P. M.; Varaprath, S. *J. Org. Chem.* **1981**, *46*, 2215.
4. Hegedus, L. S. *Angew. Chem., Int. Ed. Engl.* **1988**, *27*, 1113.
5. A ruthenium variant: Kondo, T.; Okada, T.; Mitsudo, T. *J. Am. Chem. Soc.* **2002**, *124*, 186.

Hell–Volhardt–Zelinsky reaction

α-Bromination of carboxylic acids using Br_2/PBr_3.

α-bromoacid

References

1. Hell, C. *Ber. Dtsch. Chem. Ges.* **1881**, *14*, 891.
2. Little, J. C.; Sexton, A. R.; Tong, Y.-L. C.; Zurawic, T. E. *J. Am. Chem. Soc.* **1969**, *91*, 7098.
3. Chatterjee, N. R. *Indian J. Chem., Sect. B* **1978**, *16B*, 730.
4. Kortylewicz, Z. P.; Galardy, R. E. *J. Med. Chem.* **1990**, *33*, 263.
5. Kolasa, T.; Miller, M. J. *J. Org. Chem.* **1990**, *55*, 4246.
6. Krasnov, V. P.; Bukrina, I. M.; Zhdanova, E. A.; Kodess, M. I.; Korolyova, M. A. *Synthesis* **1994**, 961.
7. Zhang, L. H.; Duan, J.; Xu, Y.; Dolbier, W. R., Jr. *Tetrahedron Lett.* **1996**, *39*, 9621.
8. Sharma, A.; Chattopadhyay, S. *J. Org. Chem.* **1999**, *64*, 8059.
9. Stack, D. E.; Hill, A. L.; Differdaffer, C. B.; Burns, N. M. *Org. Lett.* **2002**, *4*, 4487.

Henry reaction (nitroaldol reaction)

The nitroaldol condensation reaction involving aldehydes and nitronates, derived from deprotonation of nitroalkanes by bases.

nitronate

References

1. Henry, L. *Compt. Rend.* **1895**, *120*, 1265.
2. Matsumoto, K. *Angew. Chem.* **1984**, *96*, 599.
3. Sakanaka, O.; Ohmori, T.; Kozaki, S.; Suami, T.; Ishii, T.; Ohba, S.; Saito, Y. *Bull. Chem. Soc. Jpn.* **1986**, *59*, 1753.
4. Rosini, G. In *Comprehensive Organic Synthesis;* Trost, B. M.; Fleming, I., Eds.; Pergamon, **1991**, *2*, 321–340. (Review).
5. Barrett, A. G. M.; Robyr, C.; Spilling, C. D. *J. Org. Chem.* **1989**, *54*, 1233.
6. Bandgar, B. P.; Uppalla, L. S. *Synth. Commun.* **2000**, *30*, 2071.
7. Luzzio, F. A. *Tetrahedron* **2001**, *57*, 915. (Review).
8. Trost, B. M.; Yeh, V. S. C. *Angew. Chem., Int. Ed.* **2002**, *41,* 861.
9. Ma, D.; Pan, Q.; Han, F. *Tetrahedron Lett.* **2002**, *43*, 9401.
10. Westermann, B. *Angew. Chem., Int. Ed.* **2003**, *42*, 151. (Review on aza-Henry reaction).
11. Risgaard, T.; Gothelf, K. V.; Jøgensen, K. A. *Org. Biomol. Chem.* **2003**, *1*, 153.
12. Ballini, R.; Bosica, G.; Livi, D.; Palmieri, A.; Maggi, R.; Sartori, G. *Tetrahedron Lett.* **2003**, *44*, 2271.

Herz reaction

Treatment of anilines with sulfur monochloride to produce thiazothionium chlorides, which upon treatment with base give α-aminothiophenols.

References

1. Herz, R. Ger. Pat. 360,690, **1914**.
2. Ried, W.; Valentin, J. *Justus Justus Liebigs Ann. Chem.* **1966**, *699*, 183.
3. Hope, P.; Wiles, L. Λ. *J. Chem. Soc. (C)* **1966**, 1642.
4. Schneller, S. W. *Int. J. Sulfur Chem. B* **1972**, *7*, 155.
5. Chenard, B. L. *J. Org. Chem.* **1984**, *49*, 1224.
6. Belica, P. S.; Manchand, P. S. *Synthesis* **1990**, 539.
7. Grandolini, G.; Perioli, L.; Ambrogi, V. *Gazz. Chim. Ital.* **1997**, *127*, 411.
8. Koutentis, P. A.; Rees, C. W. *J. Chem. Soc., Perkin Trans. 1* **2002**, 315.

Heteroaryl Heck reaction

Intermolecular or intramolecular Heck reaction that occurs onto a heteroaryl recipient.

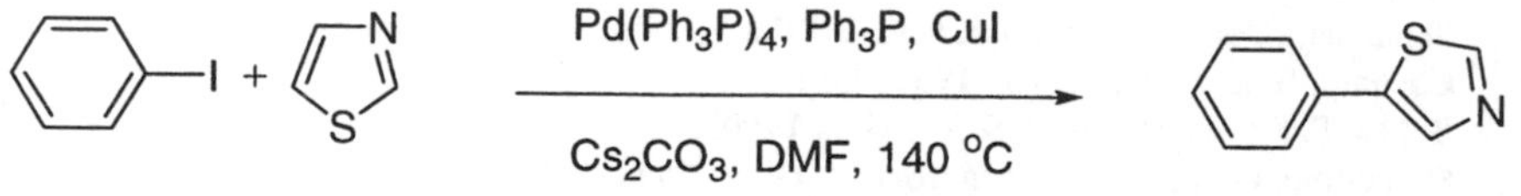

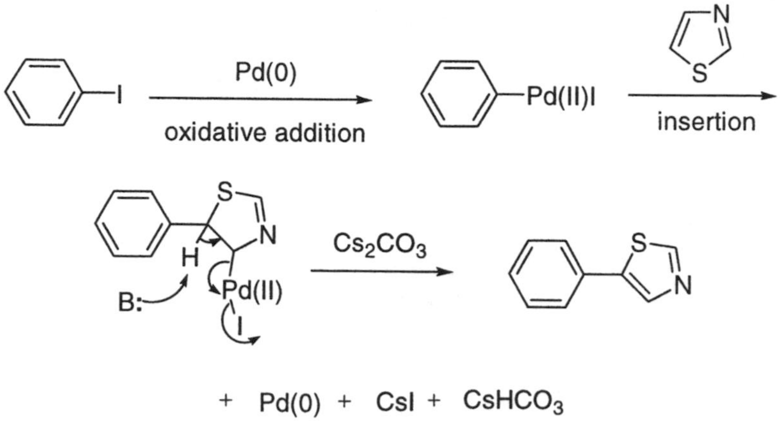

$$+ \quad Pd(0) \quad + \quad CsI \quad + \quad CsHCO_3$$

References

1. Ohta, A.; Akita, Y.; Ohkuwa, T.; Chiba, M.; Fukunaka, R.; Miyafuji, A.; Nakata, T.; Tani, N. Aoyagi, Y. *Heterocycles* **1990**, *31*, 1951.
2. Aoyagi, Y.; Inoue, A.; Koizumi, I.; Hashimoto, R.; Tokunaga, K.; Gohma, K.; Komatsu, J.; Sekine, K.; Miyafuji, A.; Konoh, J. Honma, R. Akita, Y.; Ohta, A. *Heterocycles* **1992**, *33*, 257.
3. Proudfoot, J. R. *et al. J. Med. Chem.* **1995**, *38*, 4930.
4. Pivsa-Art, S.; Satoh, T.; Kawamura, Y.; Miura, M.; Nomura, M. *Bull. Chem. Soc. Jpn.* **1998**, *71*, 467.
5. Li, J. J.; Gribble, G. W. In *Palladium in Heterocycl. Chemistry*; **2000**, Pergamon: Oxford, p16. (Review).

Hiyama cross-coupling reaction

Palladium-catalyzed cross-coupling reaction of organosilicons with organic halides, triflates, *etc.* in the presence of an activating agent such as fluoride or hydroxide (transmetallation is reluctant to occur without the effect of an activating agent). For the catalytic cycle, see the Kumada coupling on page 234.

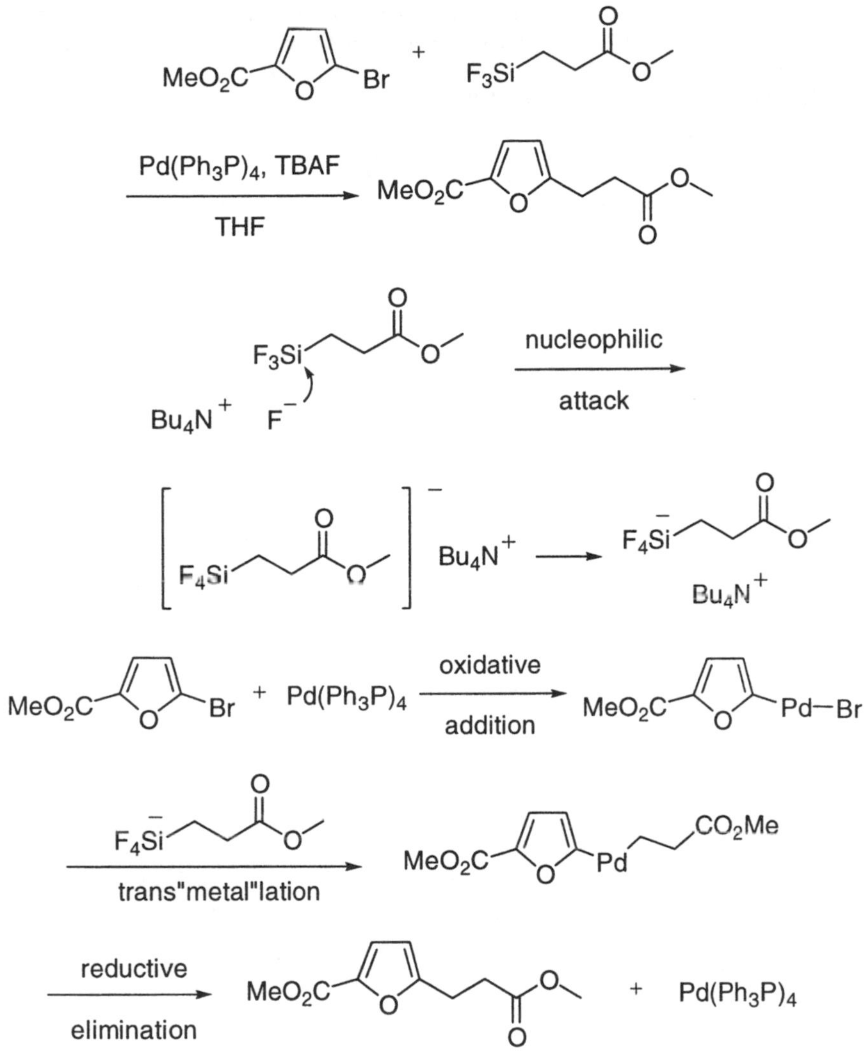

188

References

1. Hiyama, T.; Hatanaka, Y. *Pure Appl. Chem.* **1994**, *66*, 1471.
2. Matsuhashi, H.; Kuroboshi, M.; Hatanaka, Y.; Hiyama, T. *Tetrahedron Lett.* **1994**, *35*, 6507.
3. Mateo, C.; Fernandez-Rivas, C.; Echavarren, A. M.; Cardenas, D. J. *Organometallics* **1997**, *16*, 1997.
4. Hiyama, T. In *Metal-Catalyzed Cross-Coupling Reactions;* **1998**, Diederich, F.; Stang, P. J., Eds.; Wiley–VCH Verlag GmbH: Weinheim, Germany, 421–53. (Review).
5. Denmark, S. E.; Wang, Z. *J. Organomet. Chem.* **2001**, *624*, 372.
6. Hiyama, T. *J. Organomet. Chem.* **2002**, *653*, 58.

Hoch–Campbell aziridine synthesis

Treatment of ketoximes with Grignard reagents and subsequent hydrolysis of the organometallic complex to form aziridines.

References

1. Hoch, J. *Compt. Rend. Acad. Sci.* **1934**, *198*, 1865.
2. Campbell, K. N.; McKenna, J. F. *J. Org. Chem.* **1939**, *4*, 198.
3. Kotera, K.; Kitahonoki, K. *Org. Prep. Proced. Int.* **1952**, *1*, 305. (Review).
4. Eguchi, S.; Ishii, Y. *Bull. Chem. Soc. Jpn.* **1963**, *36*, 1434.
5. Dermer, O. C.; Ham, G. E. *Ethyleneimine and Other Aziridines,* Academic Press: New York, **1969**, pp65–68. (Review).
6. Alvernhe, G.; Laurent, A. *Bull. Soc. Chim. Fr.* **1970**, 3003.
7. Freeman, J. P. *Chem. Rev.* **1973**, *73*, 283. (Review).
8. Chaabouni, R.; Laurent, A. *Bull. Soc. Chim. Fr.* **1973**, 2680.
9. Bartnik, R.; Laurent, A. *Bull. Soc. Chim. Fr.* **1975**, 173.

10. Tzikas, A.; Tamm, C.; Boller, A.; Fürst, A. *Helv. Chim. Acta* **1976**, *59*, 1850.
11. Sasaki, T.; Eguchi, S.; Hattori, S. *Heterocycles* **1978**, *11*, 235.
12. Alvernhe, G.; Laurent, A. *J. Chem. Soc. (S)* **1978**, 28.
13. Laurent, A.; Marsura, A.; Pierre, J.-L. *J. Heterocycl. Chem.* **1980**, *17*, 1009.
14. Quinze, K.; Laurent, A.; Mison, P. *J. Fluorine Chem.* **1989**, *44*, 211.

Hodges–Vedejs metallation of oxazoles

Metallation of an oxazole followed by treatment with benzaldehyde results in a 4-substituted oxazole as the major product [1]:

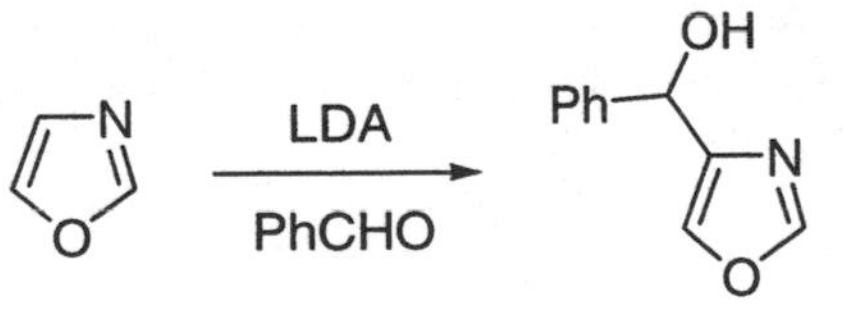

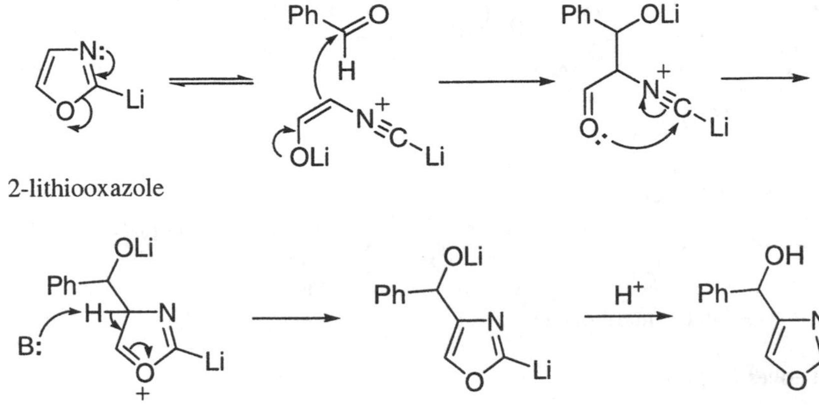

2-lithiooxazole

However, the ring-opening process can be prevented by addition of boranes [3]:

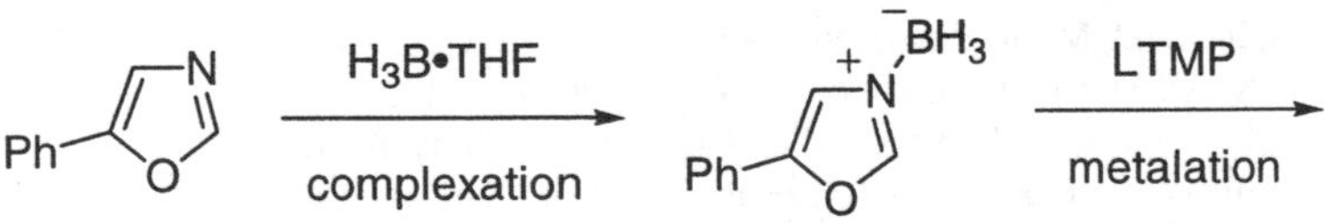

LTMP = lithium tetramethylpiperidine

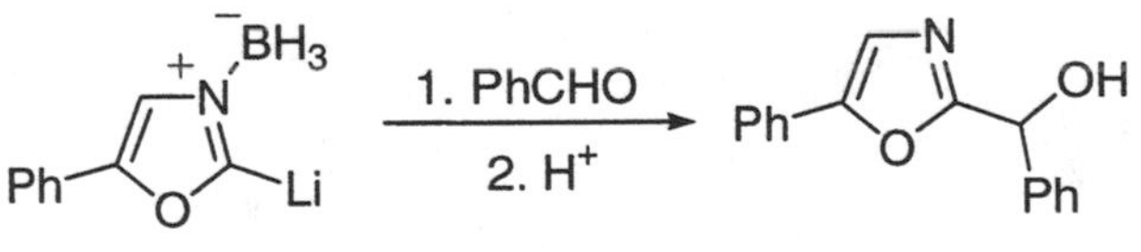

References

1. Hodges, J. C.; Patt, W. C.; Connolly, C. J. *J. Org. Chem.* **1991**, *56*, 449.
2. Iddon, B. *Heterocycles* **1994**, *37*, 1321.
3. Vedejs, E.; Monahan, S. D. *J. Org. Chem.* **1996**, *61*, 5192.
4. Vedejs, E.; Luchetta, L. M. *J. Org. Chem.* **1999**, *64*, 1011.

Hofmann rearrangement (Hofmann degradation reaction)

Upon treatment of primary amides with hypohalites, primary amines with one less carbon are obtained *via* the intermediacy of isocyanate.

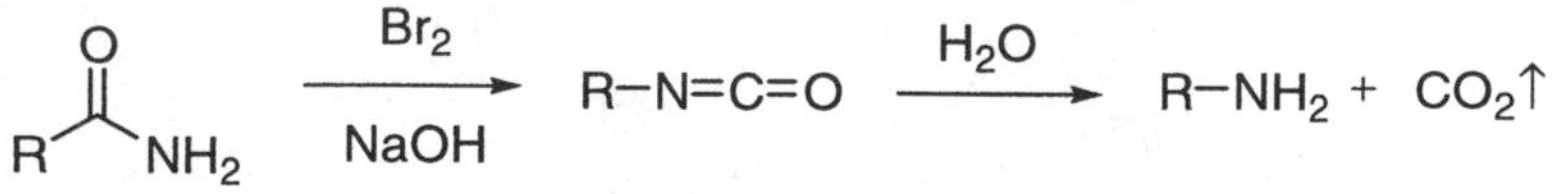

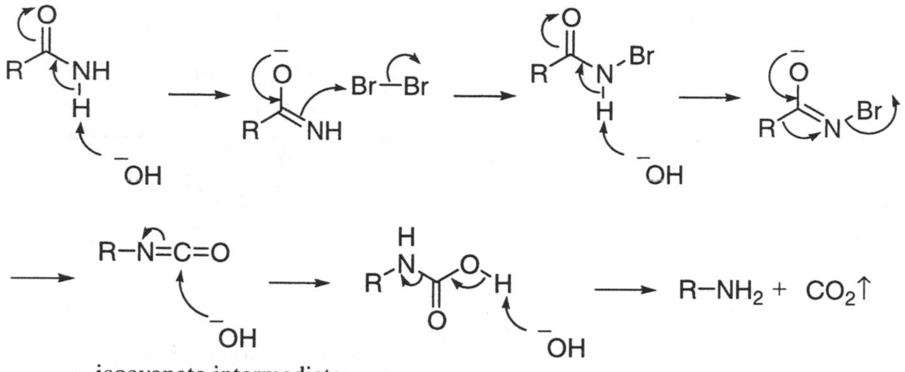

isocyanate intermediate

References

1. Hofmann, A. W. *Ber. Dtsch. Chem. Ges.* **1881**, *14*, 2725.
2. Grillot, G. F. *Mech. Mol. Migr.* **1971**, 237.
3. Jew, S.-s.; Kang, M.-h. *Arch. Pharmacal Res.* **1994**, *17*, 490.
4. Huang, X.; Seid, M.; Keillor, J. W. *J. Org. Chem.* **1997**, *62*, 7495.
5. Monk, K. A.; Mohan, R. S. *J. Chem. Educ.* **1999**, *76*, 1717.
6. Togo, H.; Nabana, T.; Yamaguchi, K. *J. Org. Chem.* **2000**, *65*, 8391.
7. Yu, C.; Jiang, Y.; Liu, B.; Hu, L. *Tetrahedron Lett.* **2001**, *42*, 1449.
8. Lopez-Garcia, M.; Alfonso, I.; Gotor, V. *J. Org. Chem.* **2003**, *68*, 648.

Hofmann–Löffler–Freytag reaction

Formation of pyrrolidines or piperidines by thermal or photochemical decomposition of protonated N-haloamines.

chloroammonium salt

nitrogen radical cation

References

1. Hofmann, A. W. *Ber. Dtsch. Chem. Ges.* **1879**, *12*, 984.
2. Löffler, K.; Freytag, C. *Ber. Dtsch. Chem. Ges.* **1909**, *42*, 3727.
3. Wolff, M. E. *Chem. Rev.* **1963**, *63*, 55. (Review).
4. Furstoss, R.; Teissier, P.; Waegell, B. *Tetrahedron Lett.* **1970**, 1263.
5. Deshpande, R. P.; Nayak, U. R. *Indian J. Chem., Sect. B* **1979**, *17B*, 310.
6. Hammerum, S. *Tetrahedron Lett.* **1981**, *22*, 157.
7. Uskokovic, M. R.; Henderson, T.; Reese, C.; Lee, H. L.; Grethe, G.; Gutzwiller, J. *J. Am. Chem. Soc.* **1978**, *100*, 571.
8. Majetich, G.; Wheless, K. *Tetrahedron* **1995**, *51*, 7095.
9. Madsen, J.; Viuf, C.; Bols, M. *Chem. Eur. J.* **2000**, *6*, 1140.
10. Togo, H.; Katohgi, M. *Synlett* **2001**, 565.

Hofmann–Martius reaction

Heating the HCl salts of arylalkylamines renders intermolecular migration of the alkyl groups.

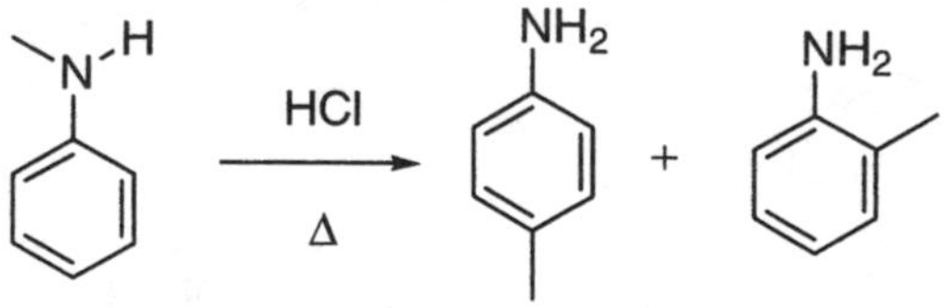

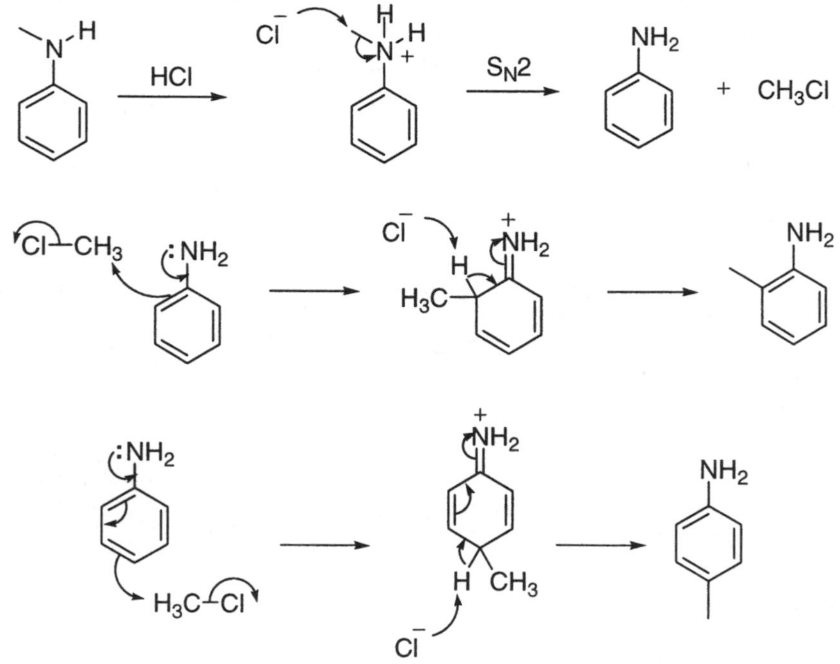

Reilly–Hickinbottom rearrangement is a variation of the Hofmann–Martius reaction in which a Lewis acid is used instead of a protic acid. The reaction follows an analogous pathway:

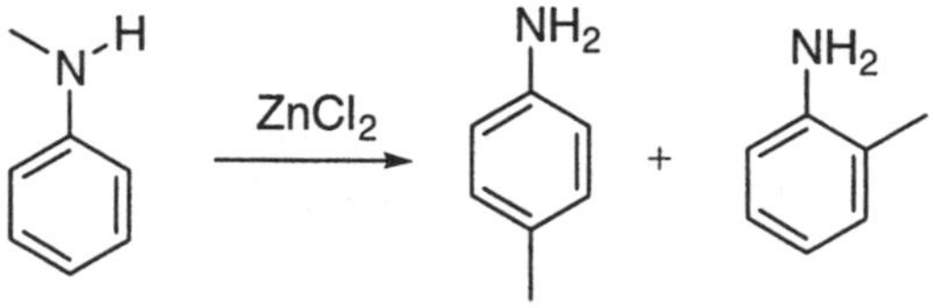

References

1. Hofmann, A. W.; Martius, C. A. *Ber.* **1964**, *20*, 2717.
2. Ogata, Y.; *et al. Tetrahedron* **1964**, 1263.
3. Ogata, Y.; *et al. J. Org. Chem.* **1970**, *35*, 1642.
4. Grillot, G. F. *Mech. Mol. Migr.* **1971**, *3*, 237.
5. Giumanini, A. G.; Roveri, S.; Del Mazza, D. *J. Org. Chem.* **1975**, *40*, 1677.
6. Hori, M.; Kataoka, T.; Shimizu, H.; Hsu, C. F.; Hasegawa, Y.; Eyama, N. *J. Chem. Soc., Perkin Trans. 1* **1988**, 2271.
7. Siskos, M. G.; Tzerpos, N.; Zarkadis, A. *Bull. Soc. Chim. Belg.* **1996**, *105*, 759.

Hooker oxidation

Oxidation of 2-hydroxy-3-alkyl-1,4-quinones with potassium permanganate with shortening of the alkyl side chain by one methylene and the position exchange between hydroxyl and alkyl groups.

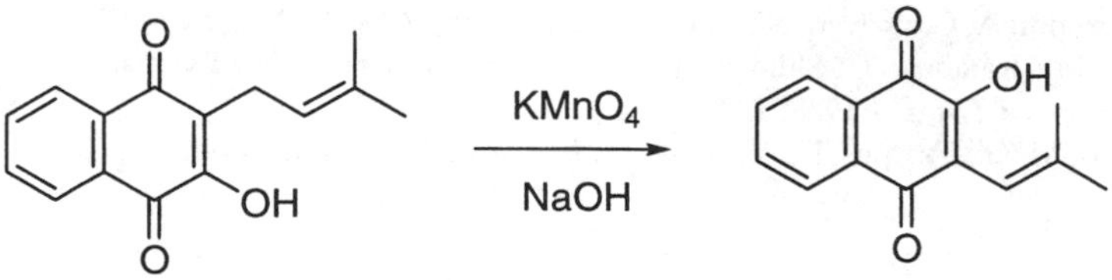

lapachol

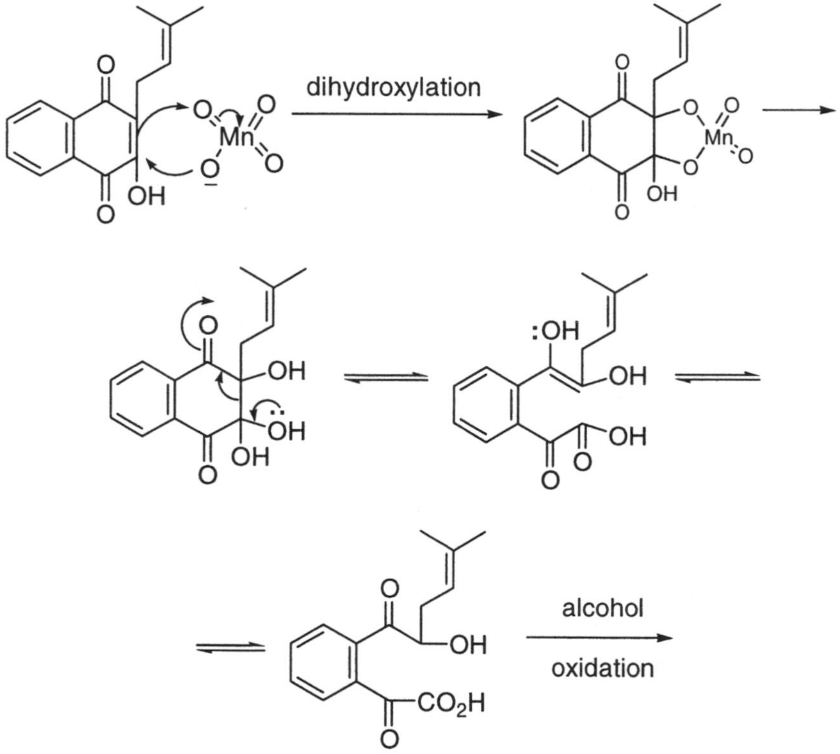

dihydroxylation

alcohol

oxidation

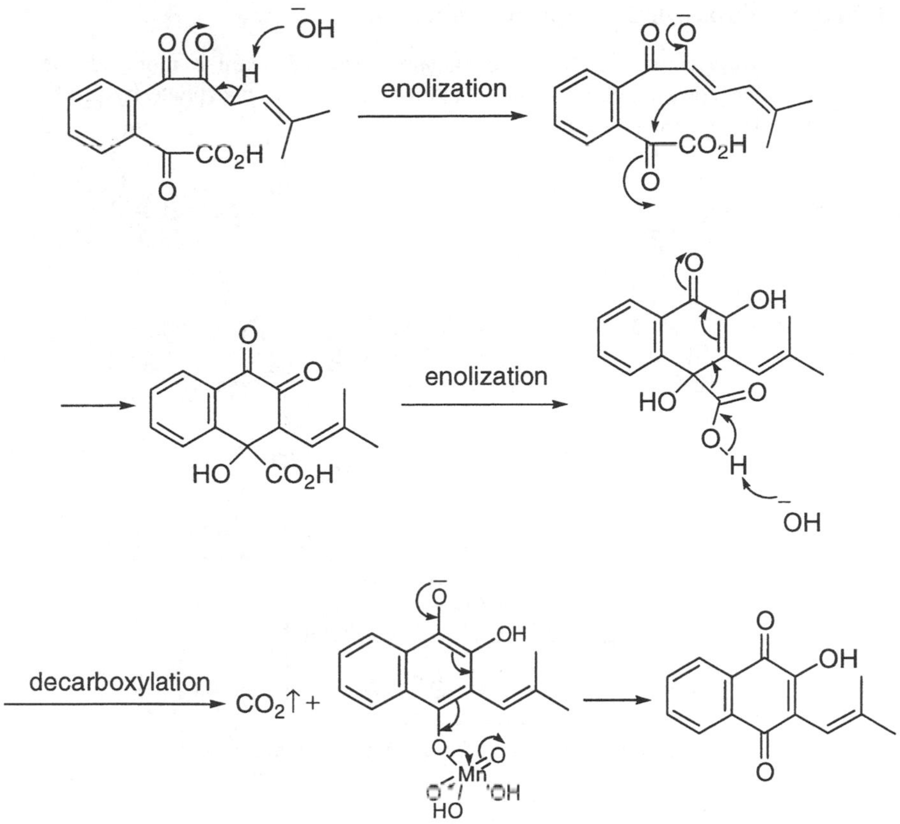

References

1. Hooker, S. C. *J. Am. Chem. Soc.* **1936**, *58,* 1174.
2. Fieser, L. F.; Hartwell, J. L.; Seligman, A. M. *J. Am. Chem. Soc.* **1936**, *58,* 1223.
3. Paulshock, M.; Moser, C. M. *J. Am. Chem. Soc.* **1950**, *72,* 5073.
4. Fawaz, G.; Fieser, L. F. *J. Am. Chem. Soc.* **1950**, *72,* 996.
5. Shchukina, L. A. *J. Gen. Chem. U.S.S.R.* **1956**, *26,* 1907.
6. Pratt, Y. T.; Drake, N. L. *J. Am. Chem. Soc.* **1957**, *79,* 5024.
7. Fieser, L. F.; Sachs, D. H. *J. Am. Chem. Soc.* **1968**, *90,* 4129.
8. Lee, K. Hee; Moore, H. W. *Tetrahedron Lett.* **1993**, *34,* 235.
9. Lee, K.; Turnbull, P.; Moore, H. W. *J. Org. Chem.* **1995**, *60,* 461.

Horner–Wadsworth–Emmons reaction

Olefin formation from aldehydes and phosphonates. Workup is more advantageous than the corresponding Wittig reaction because the phosphate by-product can be washed away with water.

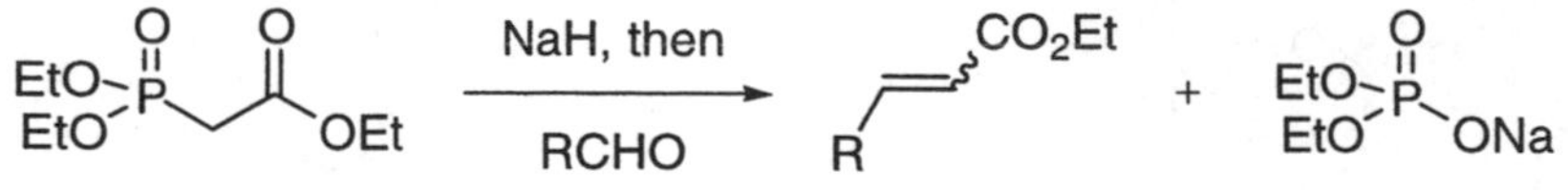

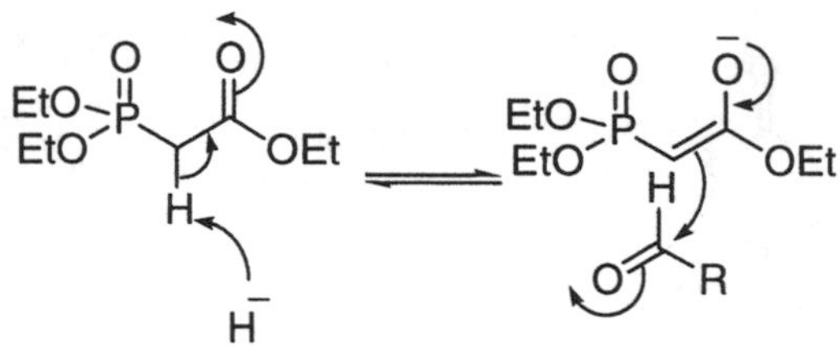

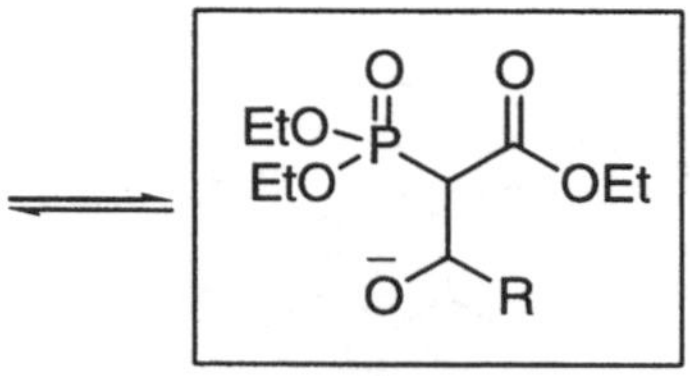

erythro (kinetic) or *threo* (thermodynamic)

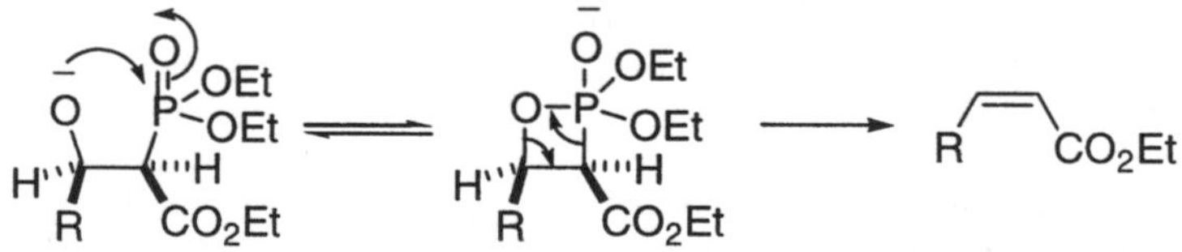

erythro, kinetic adduct

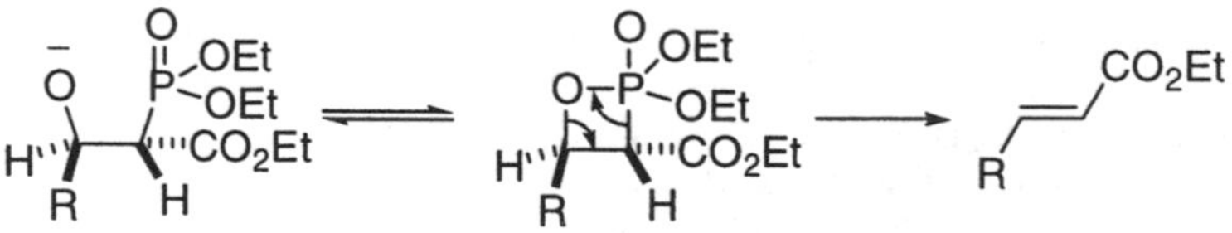

threo, thermodynamic adduct

References

1. Horner, L.; Hoffmann, H.; Wippel, H. G.; Klahre, G. *Chem. Ber.* **1959**, *92*, 2499.
2. Wadsworth, W. S., Jr.; Emmons, W. D. *J. Am. Chem. Soc.* **1961**, *62*, 1733.

3. Wadsworth, D. H.; Schupp, O. E.; Seus, E. J.; Ford, J. A., Jr. *J. Org. Chem.* **1965**, *30*, 680.

4. Maryanoff, B. E.; Reitz, A. B. *Chem. Rev.* **1989**, *89*, 863. (Review).

5. Ando, K. *J. Org. Chem.* **1997**, *62*, 1934.

6. Ando, K. *J. Org. Chem.* **1999**, *64*, 6815.

7. Simoni, D.; Rossi, M.; Rondanin, R.; Mazzali, A.; Baruchello, R.; Malagutti, C.; Roberti, M.; Invidiata, F. P. *Org. Lett.* **2000**, *2*, 3765.

8. Reiser, U.; Jauch, J. *Synlett* **2001**, 90.

9. Comins, D. L.; Ollinger, C. G. *Tetrahedron Lett.* **2001**, *42*, 4115.

10. Harusawa, S.; Koyabu, S.; Inoue, Y.; Sakamoto, Y.; Araki, L.; Kurihara, T. *Synthesis* **2002**, 1072.

11. Lattanzi, A.; Orelli, L. R.; Barone, P.; Massa, A.; Iannece, P.; Scettri, A. *Tetrahedron Lett.* **2003**, *44*, 1333.

Houben–Hoesch reaction

Acid-catalyzed acylation of phenols using nitriles.

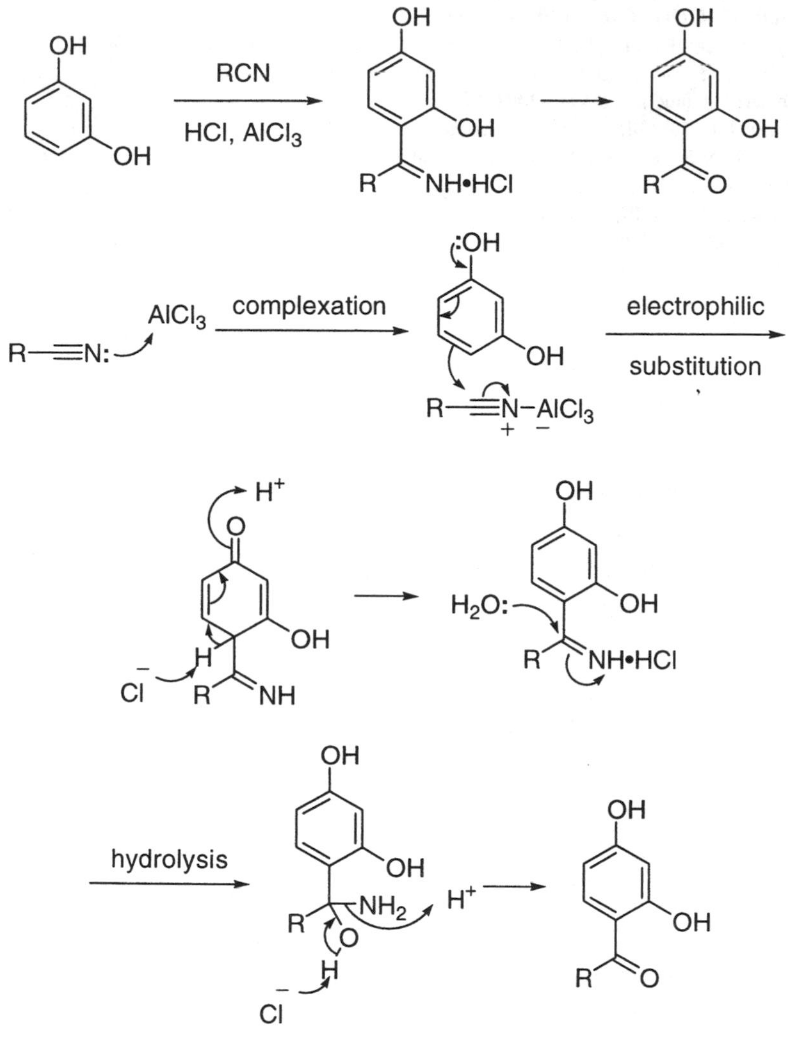

References

1. Hoesch, K. *Ber. Dtsch. Chem. Ges.* **1915**, *48*, 1122.
2. Amer, M. I.; Booth, B. L.; Noori, G. F. M.; Proenca, M. F. J. R. P. *J. Chem. Soc., Perkin Trans. 1* **1983**, 1075.

3.   Yato, M.; Ohwada, T.; Shudo, K. *J. Am. Chem. Soc.* **1991**, *113*, 691.
4.   Sato, Y.; Yato, M.; Ohwada, T.; Saito, S.; Shudo, K. *J. Am. Chem. Soc.* **1995**, *117*, 3037.
5.   Kawecki, R.; Mazurek, A. P.; Kozerski, L.; Maurin, J. K. *Synthesis* **1999**, 751.
6.   Zhu, H.-Y.; Zhang, C.-M.; Liu, F.-C. *Hecheng Huaxue* **2000**, *8*, 284.
7.   Udwary, D. W.; Casillas, L. K.; Townsend, C. A. *J. Am. Chem. Soc.* **2002**, *124*, 5294.

Hunsdiecker reaction

Conversion of silver carboxylate to halide.

$$R \overset{O}{\underset{O}{\parallel}}\text{-} \overset{+}{Ag} \xrightarrow{X_2} R\text{-}X + CO_2\uparrow + AgX$$

$$\xrightarrow{} AgX + R \overset{O}{\underset{O}{\parallel}} X \xrightarrow[\text{cleavage}]{\text{homolytic}}$$

$$X\bullet + R \overset{O}{\underset{O}{\parallel}} \bullet \longrightarrow CO_2\uparrow + R\bullet \xrightarrow{R \overset{O}{\underset{O}{\parallel}} X} R\text{-}X + R \overset{O}{\underset{O}{\parallel}} \bullet$$

References

1. Borodin, B. *Justus Liebigs Ann. Chem.* **1861**, *119*, 121.
2. Hunsdiecker, H.; Hunsdiecker, C. *Ber. Dtsch. Chem. Ges.* **1942**, *75*, 291.
3. Sheldon, R. A.; Kochi, J. K. *Org. React.* **1972**, *19*, 326. (Review).
4. Barton, D. H. R.; Crich, D.; Motherwell, W. B. *Tetrahedron Lett.* **1983**, *24*, 4979.
5. Crich, D. In *Comprehensive Organic Synthesis;* Trost, B. M.; Steven, V. L., Eds.; Pergamon, **1991**, *Vol. 7*, 723–734.
6. Naskar, D.; Chowdhury, S.; Roy, S. *Tetrahedron Lett.* **1998**, *39*, 699.
7. Camps, P.; Lukach, A. E.; Pujol, X.; Vazquez, S. *Tetrahedron* **2000**, *56*, 2703.
8. De Luca, L.; Giacomelli, G.; Porcu, G.; Taddei, M. *Org. Lett.* **2001**, *3*, 855.
9. Das, J. P.; Roy, S. *J. Org. Chem.* **2002**, *67*, 7861.

Ing–Manske procedure

A variant of Gabriel amine synthesis where hydrazine is used to release the amine from the corresponding phthalimide:

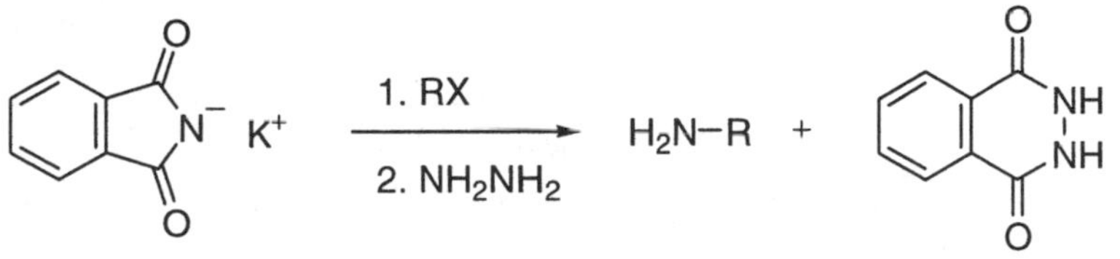

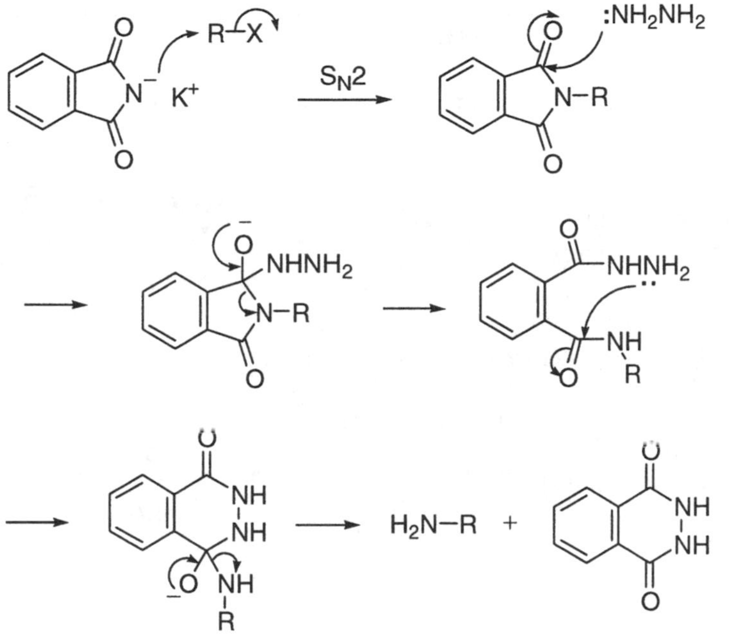

References

1. Ing, H. R.; Manske, R. H. F. *J. Chem. Soc.* **1926**, 2348.
2. Ueda, T.; Ishizaki, K. *Chem. Pharm. Bull.* **1967**, *15*, 228.
3. Khan, M. N. *J. Org. Chem.* **1995**, *60*, 4536.
4. Hearn, M. J.; Lucas, L. E. *J. Heterocycl. Chem.* **1984**, *21*, 615.
5. Khan, M. N. *J. Org. Chem.* **1996**, *61*, 8063.

Jacobsen–Katsuki epoxidation

Manganese(III)-catalyzed asymmetric epoxidation of (Z)-olefins.

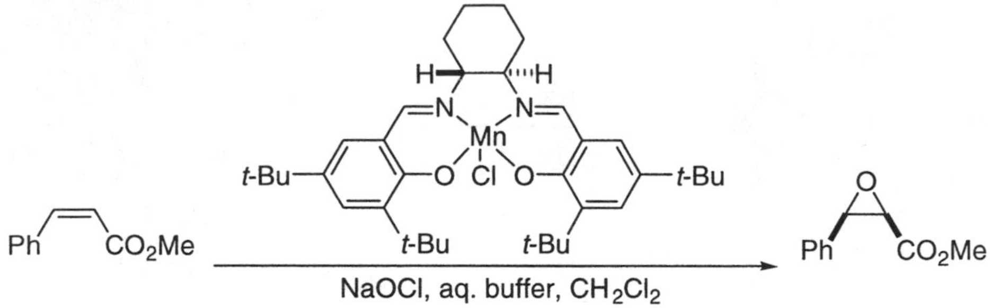

1. Concerted oxygen transfer (*cis*-epoxide):

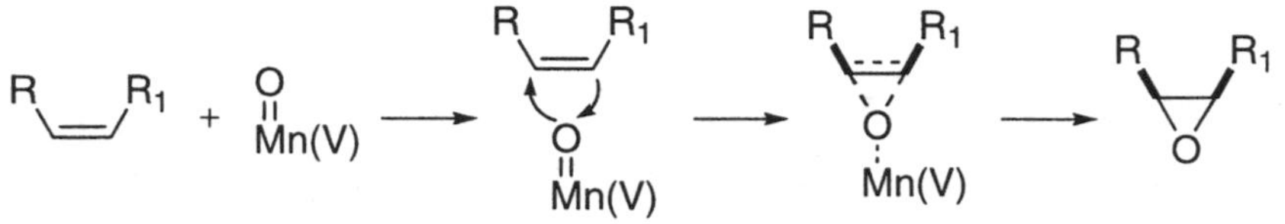

2. Oxygen transfer *via* radical intermediate (*trans*-epoxide):

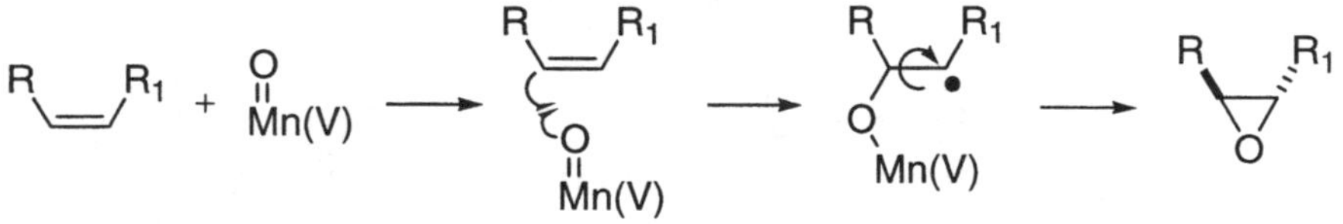

3. Oxygen transfer *via* manganaoxetane intermediate (*cis*-epoxide):

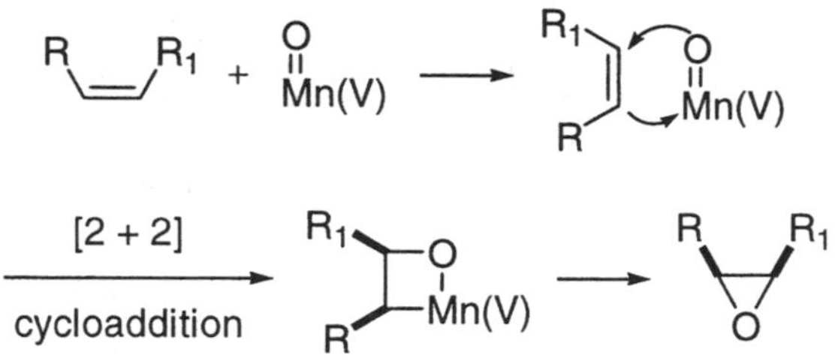

References

1. Zhang, W.; Loebach, J. L.; Wilson, S. R.; Jacobsen, E. N. *J. Am. Chem. Soc.* **1990**, *112*, 2801.
2. Irie, R.; Noda, K.; Ito, Y.; Katsuki, T. *Tetrahedron Lett.* **1991**, *32*, 1055.
3. Zhang, W.; Jacobsen, E. N. *J. Org. Chem.* **1991**, *56*, 2296.
4. Schurig, V.; Betschinger, F. *Chem. Rev.* **1992**, *92*, 873. (Review).
5. Jacobsen, E. N. In *Catalytic Asymmetric Synthesis;* Ojima, I., Ed.; VCH: Weinheim, New York, **1993**, Ch. 4.2. (Review).
6. Palucki, M.; McCormick, G. J.; Jacobsen, E. N. *Tetrahedron Lett.* **1995**, *36*, 5457.
7. Linker, T. *Angew. Chem., Int. Ed. Engl.* **1997**, *36*, 2060.
8. Katsuki, T. In *Catalytic Asymmetric Synthesis;* 2nd ed.; Ojima, I., ed.; Wiley-VCH: New York, **2000**, 287. (Review).
9. El-Bahraoui, J.; Wiest, O.; Feichtinger, D.; Plattner, D. A. *Angew. Chem., Int. Ed.* **2001**, *40*, 2073.
10. Kureshy, R. I.; Khan, N. H.; Abdi, S. H. R.; Patel, S. T.; Iyer, P. K.; Subramanian, P. S.; Jasra, R. V. *J. Catalysis* **2002**, *209*, 99.
11. Katsuki, T. *Synlett* **2003**, 281. (Review).

Jacobsen rearrangement

Treatment of polyalkylbenzenes or polyhalobenzenes with concentrated sulfuric acid to give rearranged and sulfonated polyalkylbenzenes or polyhalobenzenes.

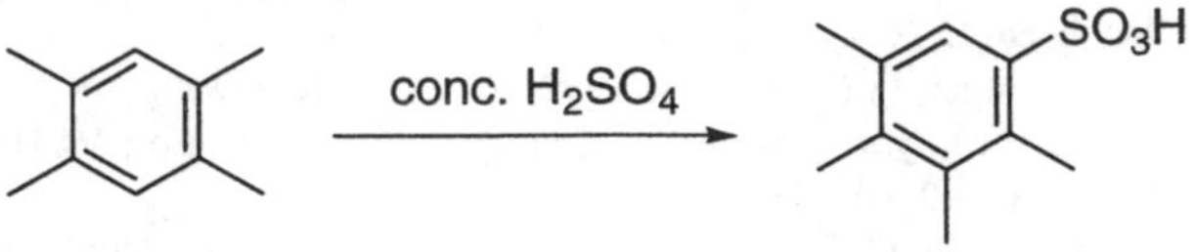

Mechanism 1:

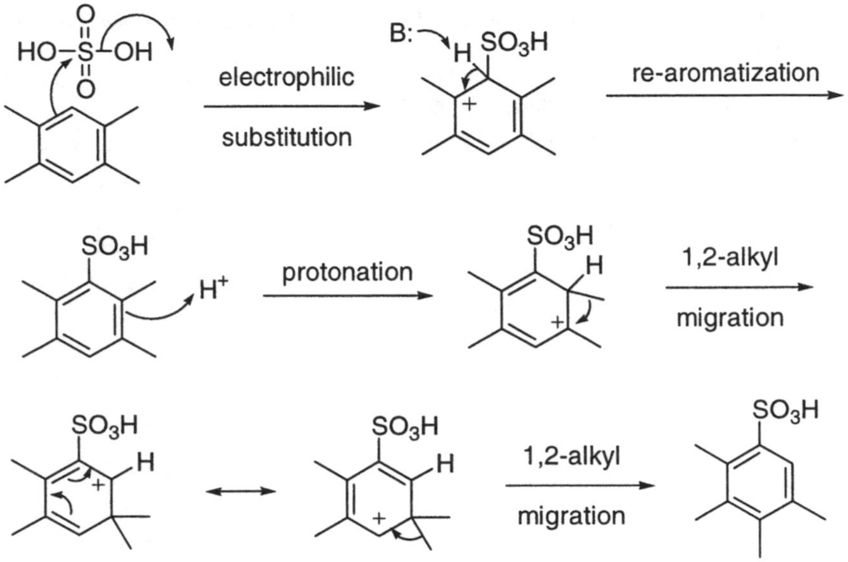

Mechanism 2:

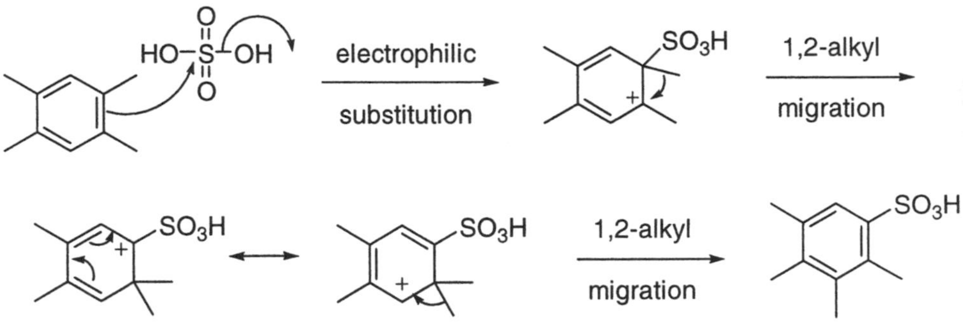

Mechanism 3:

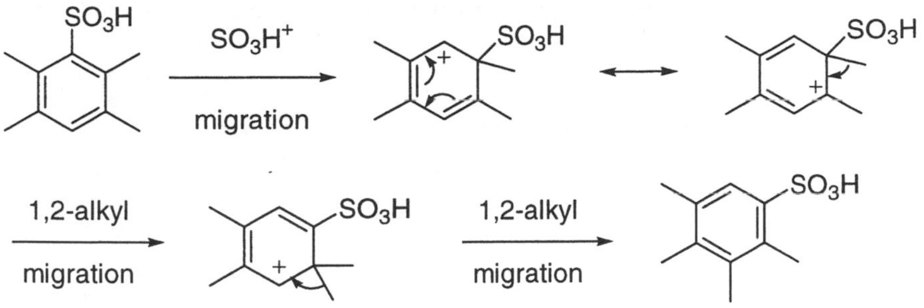

References

1. Jacobsen, O. *Ber. Dtsch. Chem. Ges.* **1952**, *578*, 122.
2. Kilpatrick, M.; Meyer, M. *J. Phys. Chem.* **1961**, *65*, 1312.
3. Marvell, E. N.; Graybill, B. M. *J. Org. Chem.* **1965**, *30*, 4014.
4. Shine, H. J. *Aromatic Rearrangement;* Elsevier: New York, **1967**, pp 23–32, 48–55. (Review).
5. Hart, H.; Janssen, J. F. *J. Org. Chem.* **1970**, *35*, 3637.
6. Suzuki, H.; Sugiyama, T. *Bull. Chem. Soc. Jpn.* **1973**, *46*, 586.
7. Norula, J. L.; Gupta, R. P. *Chem. Era* **1974**, *10*, 7.
8. Solari, E.; Musso, F.; Ferguson, R.; Floriani, C.; Chiesi-Villa, A.; Rizzoli, C. *Angew. Chem., Int. Ed. Engl.* **1995**, *35*, 1510.
9. Dotrong, M.; Lovejoy, S. M.; Wolfe, J. F.; Evers, R. C. *J. Heterocycl. Chem.* **1997**, *34*, 817.

Japp–Klingemann hydrazone synthesis

Hydrazones from α-ketoesters and diazonium salts with the aide of base.

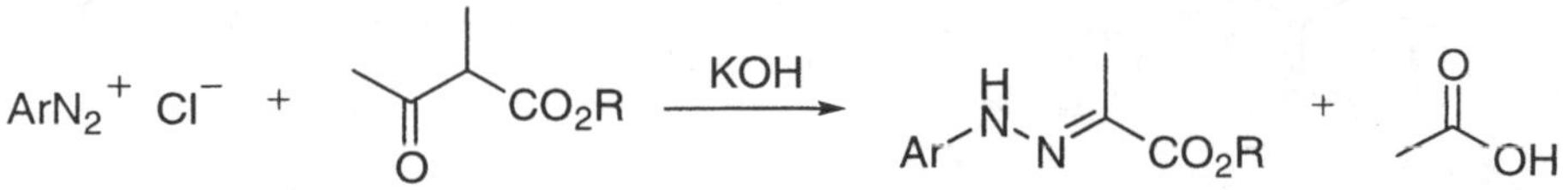

Diazonium salt   α-keto-ester   hydrazone

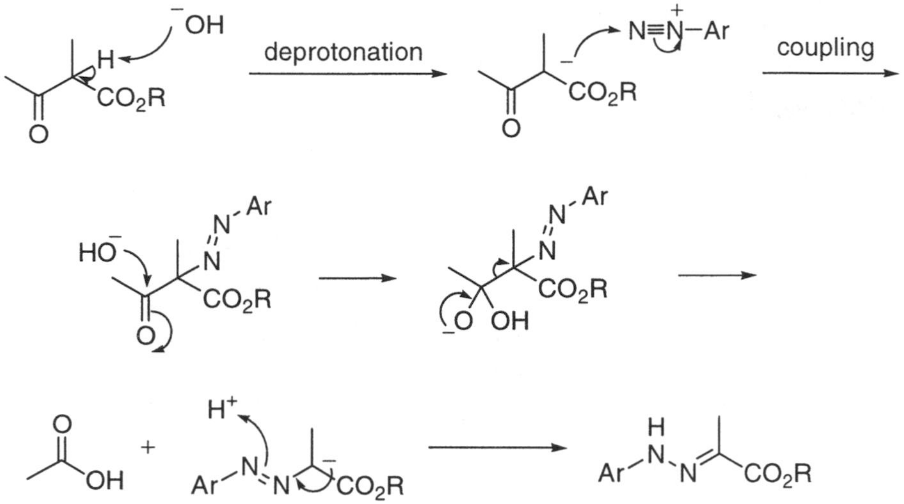

References

1. Japp, F. R.; Klingemann, F. *Justus Liebigs Ann. Chem.* **1888**, *247*, 190.
2. Laduree, D.; Florentin, D.; Robba, M. *J. Heterocycl. Chem.* **1980**, *17*, 1189.
3. Loubinoux, B.; Sinnes, J.-L.; O'Sullivan, A. C.; Winkler, T. *J. Org. Chem.* **1995**, *60*, 953.
4. Saha, C., Miss; Chakraborty, A.; Chowdhury, B. K. *Indian J. Chem.* **1996**, *35B*, 677.
5. Pete, B.; Bitter, I.; Harsanyi, K.; Toke, L. *Heterocycles* **2000**, *53*, 665.
6. Atlan, V.; Kaim, L. E.; Supiot, C. *Chem. Commun.* **2000**, 1385.
7. Shawali, A. S.; Abdallah, M. A.; Mosselhi, M. A. N.; Farghaly, T. A. *Heteroatom Chem.* **2002**, *13*, 136.

Julia–Lythgoe olefination

(E)-Olefins from sulfones and aldehydes.

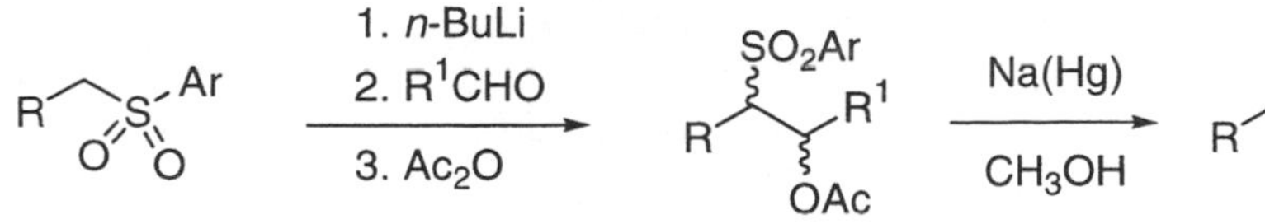

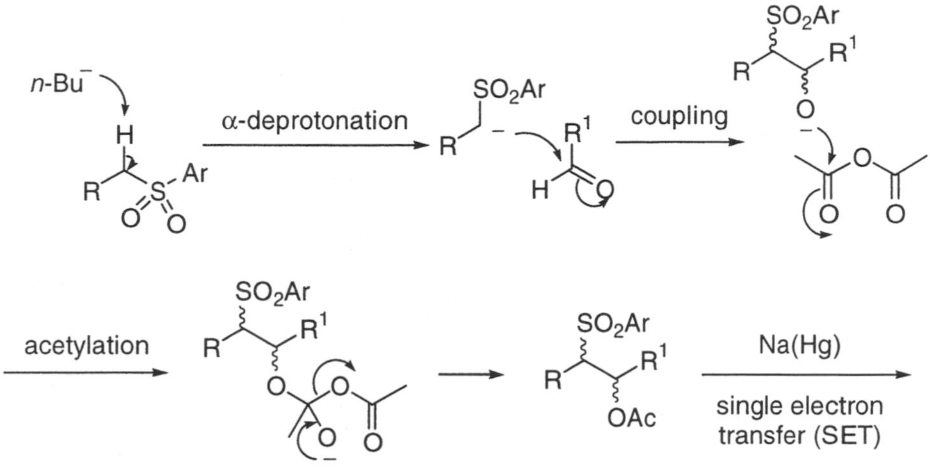

4 possible diastereomers

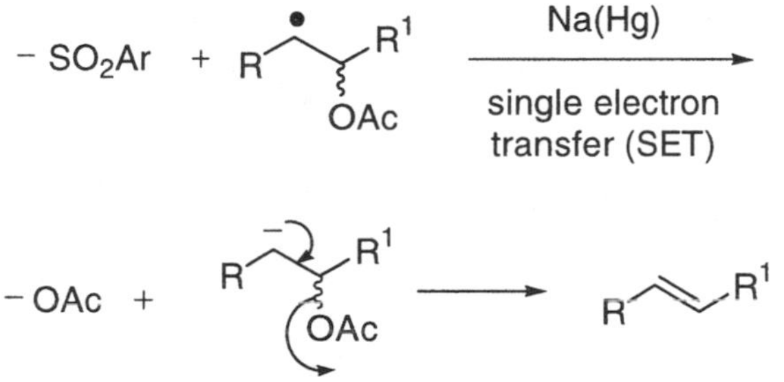

References

1. Julia, M.; Paris, J. M. *Tetrahedron. Lett.* **1973**, 4833.
2. Keck, G. E.; Savin, K. A.; Weglarz, M. A. *J. Org. Chem.* **1995**, *60*, 3194.
3. Marko, I. E.; Murphy, F.; Dolan, S. *Tetrahedron Lett.* **1996**, *37*, 2089.
4. Satoh, T.; Hanaki, N.; Yamada, N.; Asano, T. *Tetrahedron* **2000**, *56*, 6223.
5. Charette, A. B.; Berthelette, C.; St-Martib, D. *Tetrahedron Lett.* **2001**, *42*, 5149.
6. Marko, I. E.; Murphy, F.; Kumps, L.; Ates, A.; Touillaux, R.; Craig, D.; Carballares, S.; Dolan, S. *Tetrahedron* **2001**, *57*, 2609.

7. Breit, B. *Angew. Chem., Int. Ed.* **1998**, 37, 453.
8. Zanoni, G.; Porta, A.; Vidari, G. *J. Org. Chem.* **2002**, *67*, 4346.
9. Marino, J. P.; McClure, M. S.; Holub, D. P.; Comasseto, J. V.; Tucci, F. C. *J. Am. Chem. Soc.* **2002**, *124*, 1664.

Kahne glycosidation

Diastereoselective glycosidation of a sulfoxide at the anomeric center as the glycosyl acceptor. The sulfoxide activation is achieved using Tf_2O.

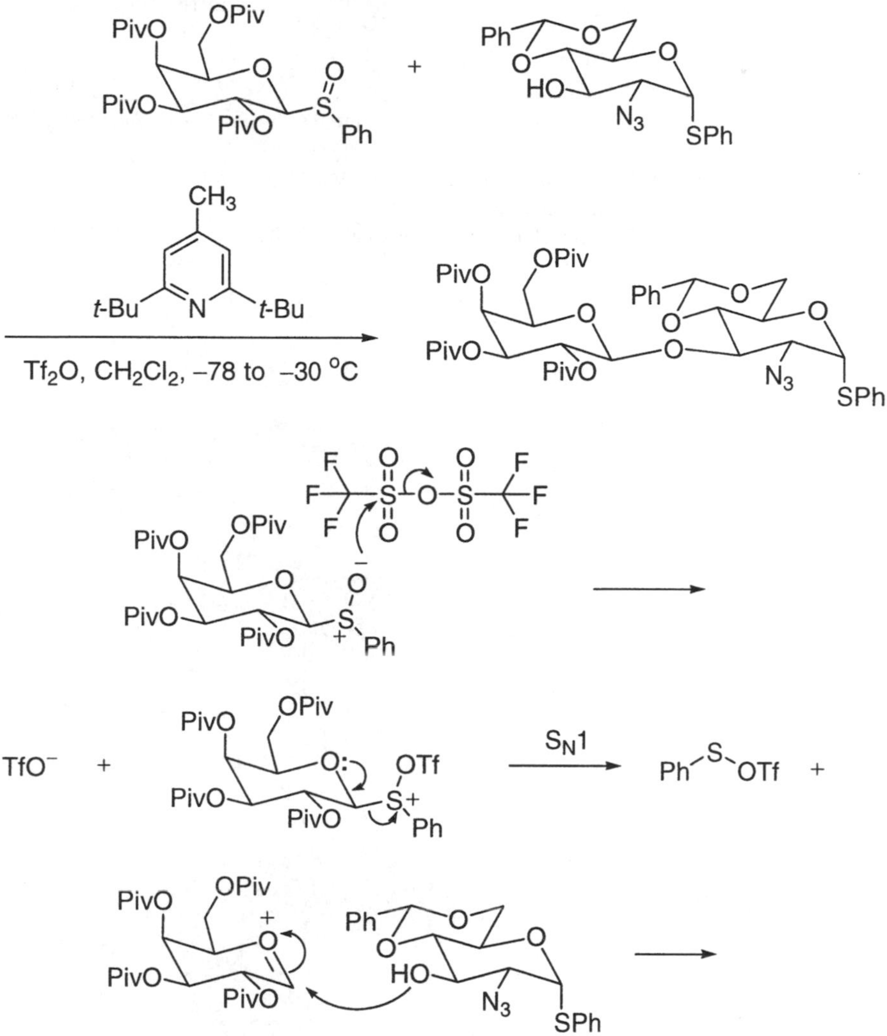

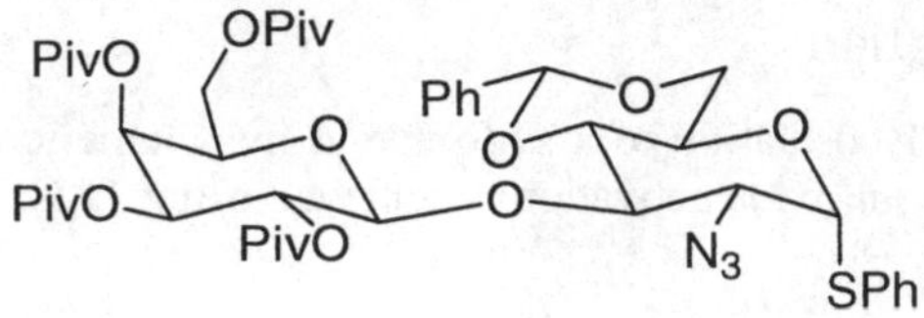

References

1. Yan, L.; Taylor, C. M.; Goodnow, R., Jr.; Kahne, D. *J. Am. Chem. Soc.* **1994**, *116*, 6953.
2. Yan, L.; Kahne, D. *J. Am. Chem. Soc.* **1996**, *118*, 9239.
3. Crich, D.; Li, H. *J. Org. Chem.* **2000**, *65*, 801.
4. Berkowitz, D. B.; Choi, S.; Bhuniya, D.; Shoemaker, R. K. *Org. Lett.* **2000**, *2*, 1149.

Keck stereoselective allylation

Asymmetric allylation of aldehydes with allylstannane in the presence of Lewis acid and chiral BINOL (or other chiral ligands).

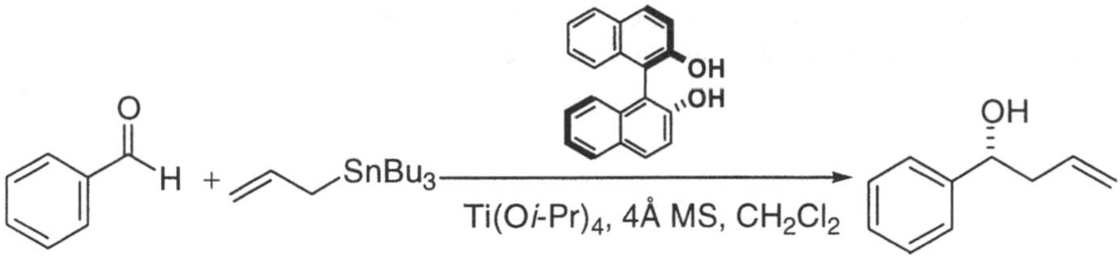

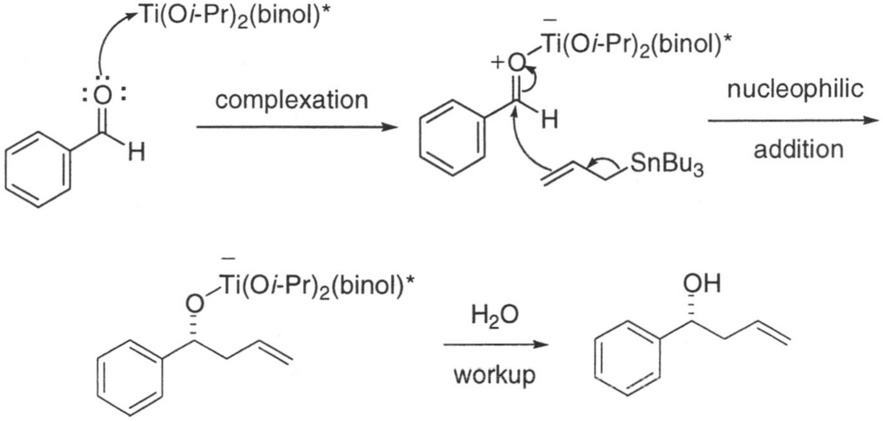

The enantioselectivity is imparted by the steric bias of the chiral ligands which displace *iso*-propoxide of titanium *iso*-propoxide. Therefore, the chiral Lewis acid becomes Ti(O*i*-Pr)$_2$(binol), which is substitutionally labile:

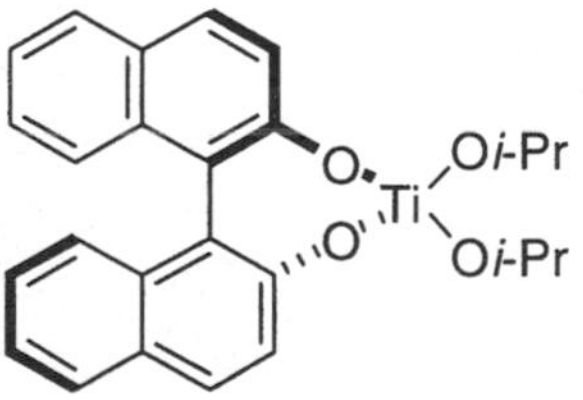

References

1. Keck, G. E.; Tarbet, K. H.; Geraci, L. S. *J. Am. Chem. Soc.* **1993**, *115*, 8467.
2. Keck, G. E.; Geraci, L. S. *Tetrahedron Lett.* **1993**, *34*, 7827.
3. Keck, G. E.; Krishnamurthy, D.; Grier, M. C. *J. Org. Chem.* **1993**, *58*, 6543.

214

4. Roe, B. A.; Boojamra, C. G.; Griggs, J. L.; Bertozzi, C. R. *J. Org. Chem.* **1996**, *61*, 6442.
5. Fürstner, A.; Langemann, K. *J. Am. Chem. Soc.* **1997**, *119*, 9130.
6. Marshall, J. A.; Palovich, M. R. *J. Org. Chem.* **1998**, *63*, 4381.
7. Evans, P. A.; Manangan, T. *J. Org. Chem.* **2000**, *65*, 4523.
8. Keck, G. E.; Wager, C. A.; Wager, T. T.; Savin, K. A.; Covel, J. A.; McLaws, M. D.; Krishnamurthy, D.; Cee, V. J. *Angew. Chem., Int. Ed.* **2001**, *40*, 231.
9. Ginn, J. D.; Padwa, A. *Org. Lett.* **2002**, *4*, 1515.

Keck macrolactonization

Macrolactonization of ω-hydroxyl acids using a combination of DCC, DMAP and DMAP•HCl.

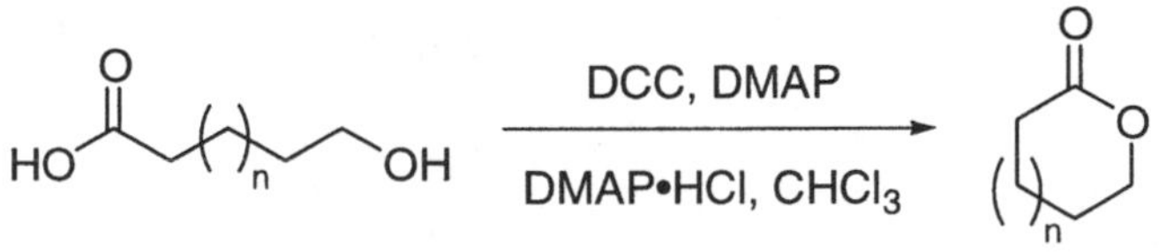
DCC, DMAP
DMAP•HCl, CHCl3

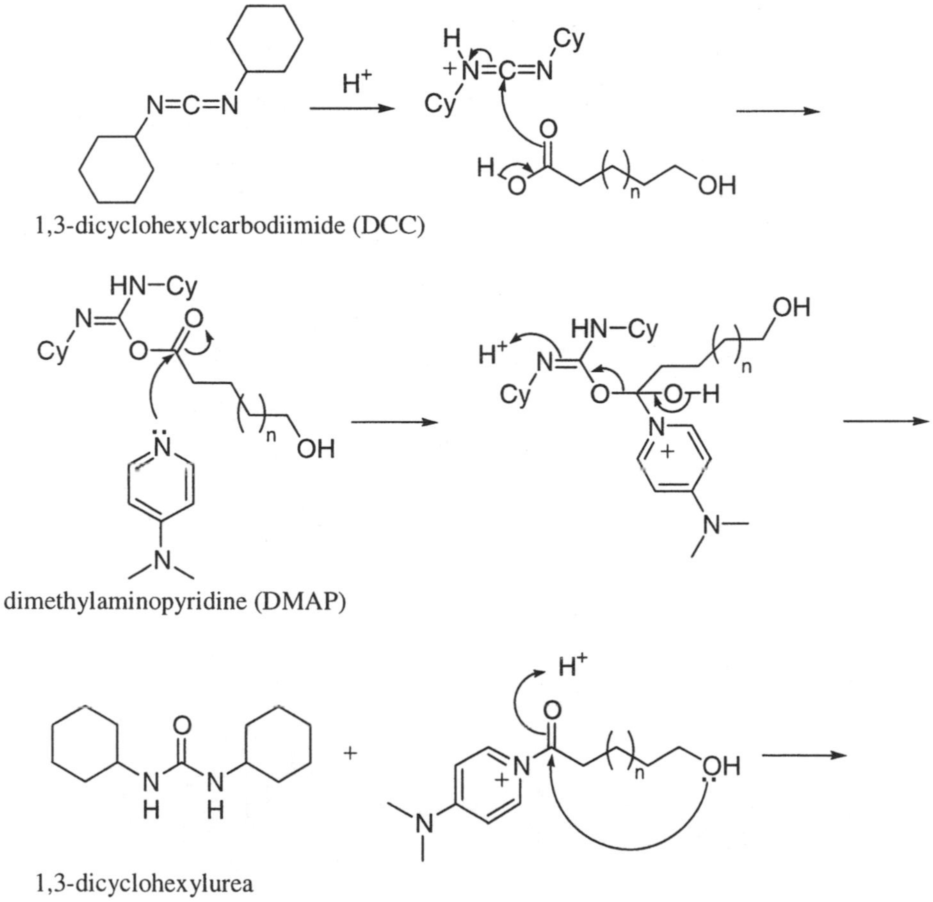
1,3-dicyclohexylcarbodiimide (DCC)
dimethylaminopyridine (DMAP)
1,3-dicyclohexylurea

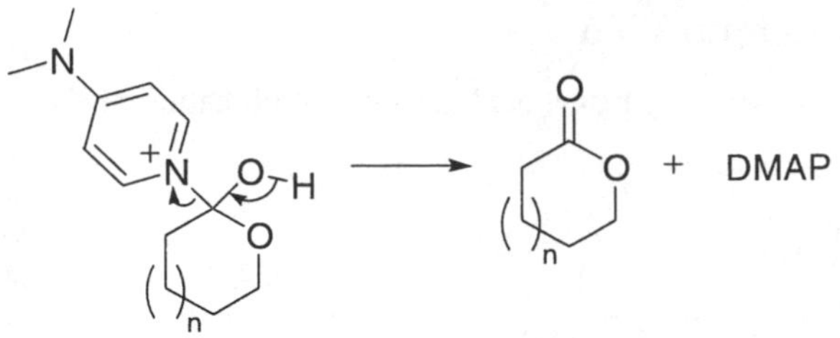

216

References

1. Boden, E. P.; Keck, G. E. *J. Org. Chem.* **1985**, *50*, 2394.
2. Paterson, I.; Yeung, K.-S.; Ward, R. A.; Cumming, J. G.; Smith, J. D. *J. Am. Chem. Soc.* **1994**, *116*, 9391.
3. Keck, G. E.; Sanchez, C.; Wager, C. A. *Tetrahedron Lett.* **2000**, *41*, 8673.
4. Tsai, C.-Y.; Huang, X.; Wong, C.-H. *Tetrahedron Lett.* **2000**, *41*, 9499.
5. Hanessian, S.; Ma, J.; Wang, W. *J. Am. Chem. Soc.* **2001**, *123*, 10200.
6. Lewis, A.; Stefanuti, I.; Swain, S. A.; Smith, S. A.; Taylor, R. J. K. *Org. Biomol. Chem.* **2003**, *1*, 104.

Kemp elimination

Treatment of benzisoxazole with base results in the ring-opening product, salicylonitrile.

benzisoxazole

salicylonitrile

References

1. Casey, M. L.; Kemp, D. S.; Paul, K. G.; Cox, D. D. *J. Org. Chem.* **1973**, *38*, 2294.
2. Kemp, D. S.; Casey, M. L. *J. Am. Chem. Soc.* **1973**, *95*, 6670.
3. Kemp, D. S.; Cox, D. D.; Paul, K. G. *J. Am. Chem. Soc.* **1975**, *97*, 7312.
4. Genre-Grandpierre, A.; Tellier, C.; Loirat, M.-J.; Blanchard, D.; Hodgson, D. R. W.; Hollfelder, F.; Kirby, A. J. *Bioorg. Med. Chem. Lett.* **1997**, *7*, 2497.
5. McCracken, P. G.; Ferguson, C. G.; Vizitiu, D.; Walkinshaw, C. S.; Wang, Y.; Thatcher, G. R. J. *J. Chem. Soc., Perkin Trans. 1* **1999**, 911.
6. Shulman, H.; Keinan, E. *Org. Lett.* **2000**, *2*, 3747.
7. Hollfelder, F.; Kirky, A. J.; Tawfik, D. S. *J. Org. Chem.* **2001**, *66*, 5866.
8. Klijn, J. E.; Engberts, J. B. F. N. *J. Am. Chem. Soc.* **2003**, *125*, 1825.

Kennedy oxidative cyclization

Asymmetric synthesis of tetrahydrofuran by treatment of a γ-hydroxyolefin with Re_2O_7.

trans:cis > 12:1

perrhenate ester

References

1. Kennedy, R. M.; Tang, S. *Tetrahedron Lett.* **1992**, *33*, 3729.
2. Tang, S.; Kennedy, R. M. *Tetrahedron Lett.* **1992**, *33*, 5299.
3. Tang, S.; Kennedy, R. M. *Tetrahedron Lett.* **1992**, *33*, 5303.
4. Tang, S.; Kennedy, R. M. *Tetrahedron Lett.* **1992**, *33*, 7823.
5. Boyce, R. S.; Kennedy, R. M. *Tetrahedron Lett.* **1994**, *35*, 5133.
6. Sinha, S. C.; Sinha, A.; Yazbak, A.; Keinan, E. *J. Org. Chem.* **1996**, *61*, 7640.
7. Sinha, S. C.; Sinha, A.; Santosh, C.; Keinan, E. *J. Am. Chem. Soc.* **1997**, *119*, 12014.
8. Avedissian, H.; Sinha, S. C.; Yazbak, A.; Sinha, A.; Neogi, P.; Sinha, S. C.; Keinan, E. *J. Org. Chem.* **2000**, *65*, 6035.

Kharasch addition reaction

Transition metal-catalyzed radical addition of $CXCl_3$ to olefins.

M = organometallic reagent containing Ru, Re, Mo, W, Fe, Al, B, Cr, Sm, *etc.*

References

1. Kharasch, M. S.; Jensen, E. V.; Urry, W. H. *Science* **1945**, *102*, 2640.
2. Gossage, R. A.; van de Kuil, L. A.; van Koten, G. *Acc. Chem. Res.* **1998**, *31*, 423. (Review).
3. Simal, F.; Wlodarczak, L.; Demonceau, A.; Noels, A. F. *Tetrahedron Lett.* **2000**, *41*, 6071.
4. Tallarico, J. A.; Malnick, L. M.; Snapper, M. L. *J. Org. Chem.* **1999**, *64*, 344.
5. Simal, F.; Wlodarczak, L.; Demonceau, A.; Noels, A. F. *Eur. J. Org. Chem.* **2001**, 2689.
6. De Clercq, B.; Verpoort, F. *Tetrahedron Lett.* **2001**, *42*, 8959.
7. Feng, H.; Kavrakova, I. K.; Pratt, D. A.; Tellinghuisen, J.; Porter, N. A. *J. Org. Chem.* **2002**, *67*, 6050.
8. Van Heerbeek, R.; Kamer, P. C. J.; Van Leeuwen, P. W. N. M.; Reek, J. N. H. *Chem. Rev.* **2002**, *102*, 3717. (Review).
9. De Clercq, B.; Verpoort, F. *Catalysis Lett.* **2002**, *83*, 9.

Knöevenagel condensation

Condensation between carbonyl compounds and activated methylene compounds catalyzed by amines.

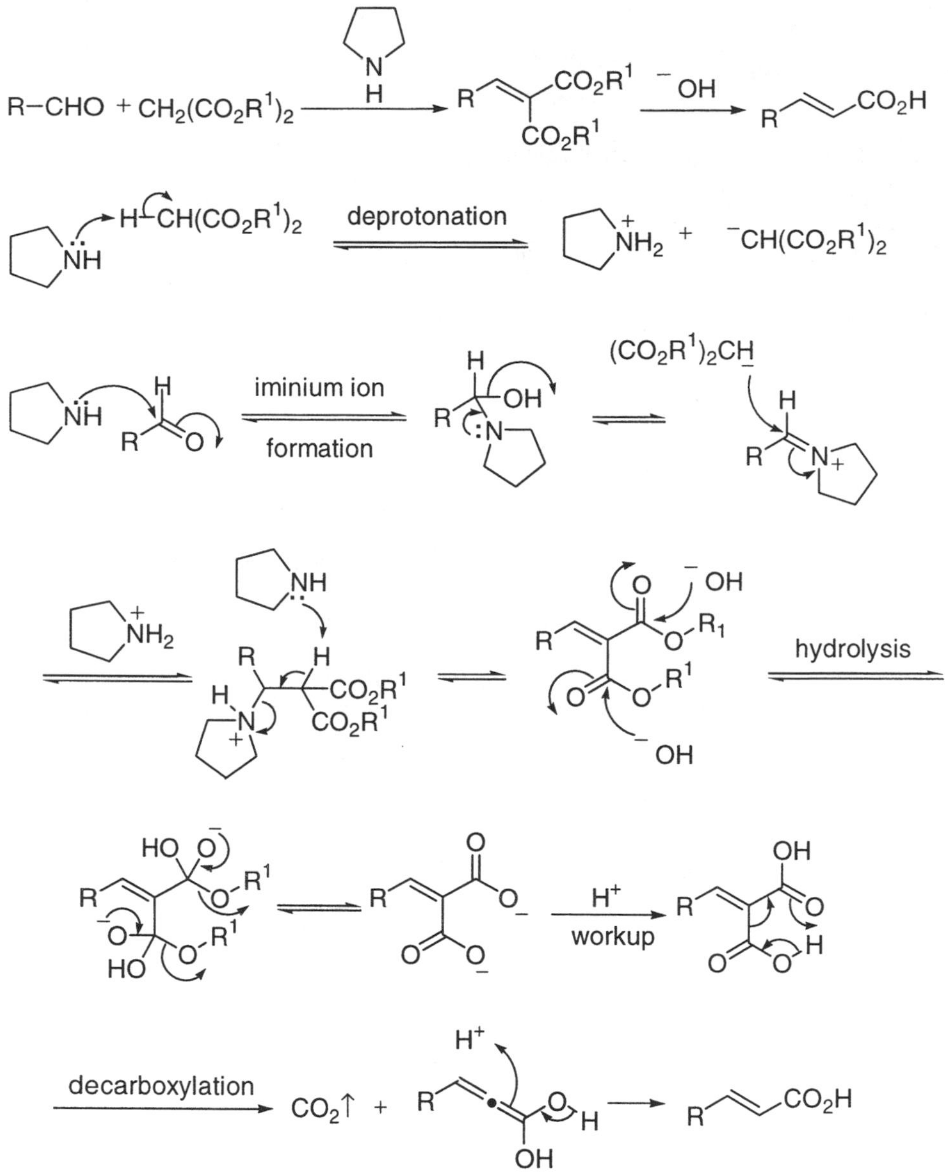

References

1. Knöevenagel, E. *Ber. Dtsch. Chem. Ges.* **1898**, *31*, 2596.
2. Jones, G. *Org. React.* **1967**, *15*, 204. (Review).
3. Van der Baan, J. L.; Bickelhaupt, F. *Tetrahedron* **1974**, *30*, 2088.
4. Green, B.; Khaidem, I. S.; Crane, R. I.; Newaz, S. S. *Tetrahedron* **1975**, *31*, 2997.
5. Angeletti, E.; Canepa, C.; Martinetti, G.; Venturello, P. *J. Chem. Soc., Perkin Trans. 1* **1989**, 105.
6. Paquette, L. A.; Kern, B. E.; Mendez-Andino, J. *Tetrahedron Lett.* **1999**, *40*, 4129.
7. Balalaie, S.; Nemati, N. *Synth. Commun.* **2000**, *30*, 869.
8. Pearson, A. J.; Mesaros, E. F. *Org. Lett.* **2002**, *4*, 2001.
9. Curini, M.; Epifano, F.; Marcotullio, M. C.; Rosati, O.; Tsadjout, A. *Synth. Commun.* **2002**, *32*, 355.
10. Kourouli, T.; Kefalas, P.; Ragoussis, N.; Ragoussis, V. *J. Org. Chem.* **2002**, *67*, 4615.
11. Wada, S.; Suzuki, H. *Tetrahedron Lett.* **2003**, *44*, 399.

222

Knorr pyrrole synthesis

A modification of the Feist–Bénary furan synthesis (p134). Reaction between α-aminoketones, derived fron α-haloketones and ammonia, and β-ketoesters assembles pyrroles.

References

1. Knorr, L. *Ber. Dtsch. Chem. Ges.* **1884**, *17*, 1635.
2. Hort, E. V.; Anderson, L. R. *Kirk-Othmer Encycl. Chem. Technol.;* 3rd Ed.; **1982**, *19*, 499. (Review).
3. Jones, R. A.; Rustidge, D. C.; Cushman, S. M. *Synth. Commun.* **1984**, *14*, 575.
4. Fabiano, E.; Golding, B. T. *J. Chem. Soc., Perkin Trans. 1* **1991**, 3371.
5. Hamby, J. M.; Hodges, J. C. *Heterocycles* **1993**, *35*, 843.
6. Alberola, A.; Ortega, A. G.; Sadaba, M. L.; Sanudo, C. *Tetrahedron* **1999**, *55*, 6555.
7. Braun, R. U.; Zeitler, K.; Mueller, T. J. J. *Org. Lett.* **2001**, *3*, 3297.
8. Elghamry, I. *Synth. Commun.* **2002**, *32*, 897.

Koch carbonylation reaction
(Koch–Haaf carbonylation reaction)

Strong acid-catalyzed tertiary carboxylic acid formation from alcohols or olefins and CO.

the tertiary carbocation is thermodynamically favored

References

1. Koch, H.; Haaf, W. *Justus Liebigs Ann. Chem.* **1958**, *618*, 251.
2. Kell, D. R.; McQuillin, F. J. *J. Chem. Soc., Perkin Trans. 1* **1972**, 2096.
3. Norell, J. R. *J. Org. Chem.* **1972**, *37*, 1971.
4. Booth, B. L.; El-Fekky, T. A. *J. Chem. Soc., Perkin Trans. 1* **1979**, 2441.
5. Langhals, H.; Mergelsberg, I.; Ruechardt, C. *Tetrahedron Lett.* **1981**, *22*, 2365.
6. Farooq, O.; Marcelli, M.; Prakash, G. K. S.; Olah, G. A. *J. Am. Chem. Soc.* **1988**, *110*, 864.
7. Stepanov, A. G.; Luzgin, M. V.; Romannikov, V. N.; Zamaraev, K. I. *J. Am. Chem. Soc.* **1995**, *117*, 3615.
8. Olah, G. A.; Prakash, G. K. S.; Mathew, T.; Marinez, E. R. *Angew. Chem., Int. Ed.* **2000**, *39*, 2547.

224

9. Xu, Q.; Inoue, S.; Tsumori, N.; Mori, H.; Kameda, M.; Tanaka, M.; Fujiwara, M.; Souma, Y. *J. Mol. Catal. A: Chem.* **2001**, *170*, 147.
10. Tsumori, N.; Xu, Q.; Souma, Y.; Mori, H. *J. Mol. Catalysis A: Chem.* **2002**, *179*, 271.

Koenig–Knorr glycosidation

Formation of the β-glycoside from α-halocarbohydrate under the influence of silver salt.

$$+ \quad AgBr \quad + \quad CO_2\uparrow \quad + \quad H_2O$$

oxonium ion

β-anomer is
———————→
favored

β-anomer

References

1. Koenig, W.; Knorr, E. *Ber. Dtsch. Chem. Ges.* **1901**, *34*, 957.
2. Schmidt, R. R. *Angew. Chem.* **1986**, *98*, 213.
3. Greiner, J.; Milius, A.; Riess, J. G. *Tetrahedron Lett.* **1988**, *29*, 2193.
4. Smith, A. B., III; Rivero, R. A.; Hale, K. J.; Vaccaro, H. A. *J. Am. Chem. Soc.* **1991**, *113*, 2092.
5. Li, H.; Li, Q.; Cai, M.-S.; Li, Z.-J. *Carbohydr. Res.* **2000**, *328*, 611.
6. Fürstner, A.; Radkowski, K.; Grabowski, J.; Wirtz, C.; Mynott, R. *J. Org. Chem.* **2000**, *65*, 8758.
7. Josien-Lefebvre, D.; Desmares, G.; Le Drian, C. *Helv. Chim. Acta* **2001**, *84*, 890.
8. Seebacher, W.; Haslinger, E.; Weis, R. *Monatsh. Chem.* **2001**, *132*, 8397.
9. Yashunsky, D. V.; Tsvetkov, Y. E.; Ferguson, M. A. J.; Nikolaev, A. V. *J. Chem. Soc., Perkin Trans. 1* **2002**, 242.
10. Kroger, L.; Thiem, J. *J. Carbohydrate Chem.* **2003**, *22*, 9.

226

Kolbe electrolytic coupling

Electrolysis of carboxylates to afford the coupled products. Homocoupling product is obtained if two carboxylates are the same — decarboxylative dimerization; unsymmetrical product will be produced if the two carboxylates are different.

$$R-CO_2^- \xrightarrow{\text{electrolysis}} R-R + CO_2\uparrow$$

$$R-CO_2^- \xrightarrow{-e} R-CO_2{}^\bullet \longrightarrow CO_2\uparrow + R\bullet \longrightarrow R-R$$

References

1. Kolbe, H. *Justus Liebigs Ann. Chem.* **1849**, *69*, 257.
2. Vijh, A. K.; Conway, B. E. *Chem. Rev.* **1967**, *67*, 623. (Review).
3. Rabjohn, N.; Flasch, G. W., Jr. *J. Org. Chem.* **1981**, *46*, 4082.
4. Feldhuse, M.; Schäfer, H. J. *Tetrahedron* **1985**, *41,* 4213.
5. Becking, L.; Schäfer, H. J. *Tetrahedron Lett.* **1988**, *29,* 2797.
6. Schäfer, H. J. *Comp. Org. Synth.* **1991**, *3*, 633–658. (Review).
7. Seebach, D.; Maestro, M. A.; Sefkow, M.; Adam, G.; Hintermann, S.; Neidlein, A. *Liebigs Ann. Chem.* **1994**, 701.
8. Sugiya, M.; Nohira, H. *Chem. Lett.* **1998**, 479.
9. Hiebl, J.; Blanka, M.; Guttman, A.; Kollmann, H.; Leitner, K.; Mayhofer, G.; Rovenszky, F.; Winkler, K. *Tetrahedron* **1998**, *54,* 2059.
10. Torii, S.; Tanaka, H. in *Organic Electrochemistry* (4th Ed.) Lund, H.; Hammerich, O. Marcel Dekker: New York, N. Y. (**2001**), 499–543. (Review).

Kolbe–Schmitt reaction

Carboxylation of sodium phenoxides with carbon dioxide, mostly at the *ortho* position.

References

1. Kolbe, H. *Justus Liebigs Ann. Chem.* **1860**, *113*, 1125.
2. Schmitt, R. *J. Prakt. Chem.* **1885**, *31*, 397.
3. Lindsey, A. S.; Jeskey, H. *Chem. Rev.* **1957**, *57*, 583. (Review).
4. Kunert, M.; Dinjus, E.; Nauck, M.; Sieler, J. *Ber.* **1997**, *130*, 1461.
5. Kosugi, Y.; Takahashi, K. *Stud. Surf. Sci. Catal.* **1998**, *114*, 487.
6. Kosugi, Y.; Rahim, M. A.; Takahashi, K.; Imaoka, Y.; Kitayama, M. *Appl. Organomet. Chem.* **2000**, *14*, 841.
7. Rahim, M. A.; Matsui, Y.; Kosugi, Y. *Bull. Chem. Soc. Jpn.* **2002**, *75*, 619.

228

Kostanecki reaction

Also known as **Kostanecki–Robinson reaction**. Transformation **1→2** represents an **Allan–Robinson reaction** (see page 3), whereas **1→3** is a **Kostanecki (acylation) reaction**:

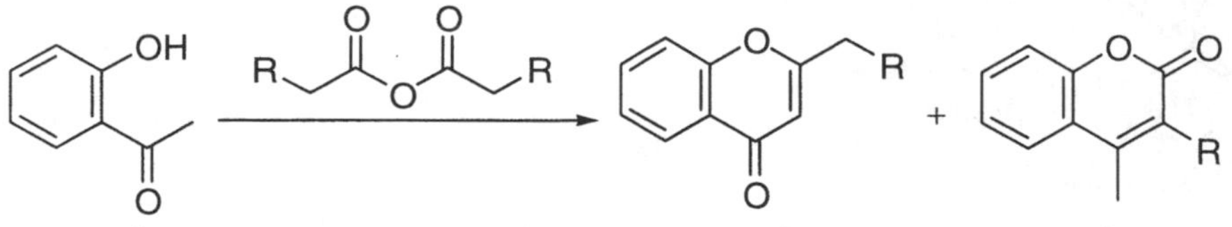

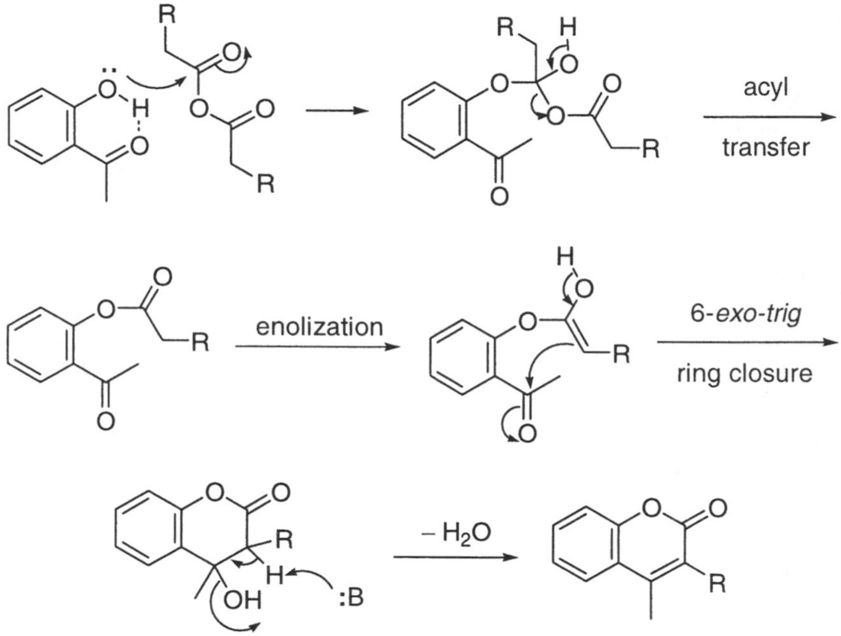

References

1. von Kostanecki, S.; Rozycki, A. *Ber. Dtsch. Chem. Ges.* **1901**, *34*, 102.
2. Cook, D.; McIntyre, J. S. *J. Org. Chem.* **1968**, *33*, 1746.
3. Szell, T.; Dozsai, L.; Zarandy, M.; Menyharth, K. *Tetrahedron* **1969**, *25*, 715.
4. Pardanani, N. H.; Trivedi, K. N. *J. Indian Chem. Soc.* **1972**, *49*, 599.
5. Ahluwalia, V. K. *Indian J. Chem., Sect. B* **1976**, *14B*, 682.
6. Looker, J. H.; McMechan, J. H.; Mader, J. W. *J. Org. Chem.* **1978**, *43*, 2344.

7. Iyer, P. R.; Iyer, C. S. R.; Prasad, K. J. R. *Indian J. Chem., Sect. B* **1983**, *22B*, 1055.
8. Flavin, M. T.; Rizzo, J. D.; Khilevich, A.; Kucherenko, A.; Sheinkman, A. K.; Vilay-chack, V.; Lin, L.; Chen, W.; Mata, E.; Greenwood, E. M.; Pengsuparp, T.; Pezzuto, J. M.; Hughes, S. H.; Flavin, T. M.; Cibulski, M.; Boulanger, W. A.; Shone, R. L.; Xu, Z.-Q. *J. Med. Chem.* **1996**, *39*, 1303.

Krapcho decarboxylation

Nucleophilic decarboxylation of β-ketoesters, malonate esters, α-cyanoesters, or α-sulfonylesters.

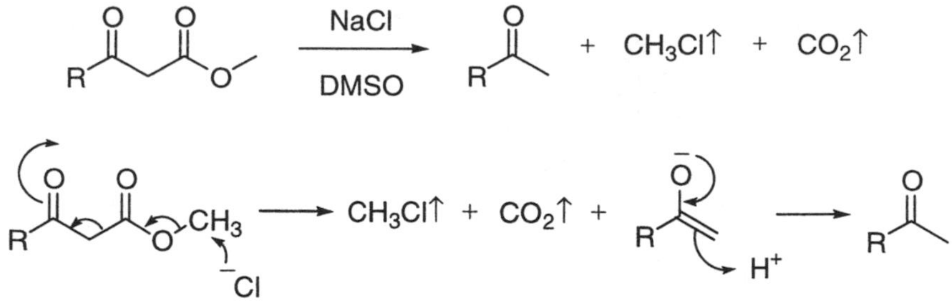

References

1. Krapcho, A. P.; Glynn, G. A.; Grenon, B. J. *Tetrahedron Lett.* **1967**, 215.
2. Flynn, D. L.; Becker, D. P.; Nosal, R.; Zabrowski, D. L. *Tetrahedron Lett.* **1992**, *33*, 7283.
3. Martin, C. J.; Rawson, D. J.; Williams, J. M. J. *Tetrahedron: Asymmetry* **1998**, *9*, 3723.

Kröhnke reaction (pyridine synthesis)

Pyridines from α-pyridinium methyl ketone salts and α,β-unsaturated ketones.

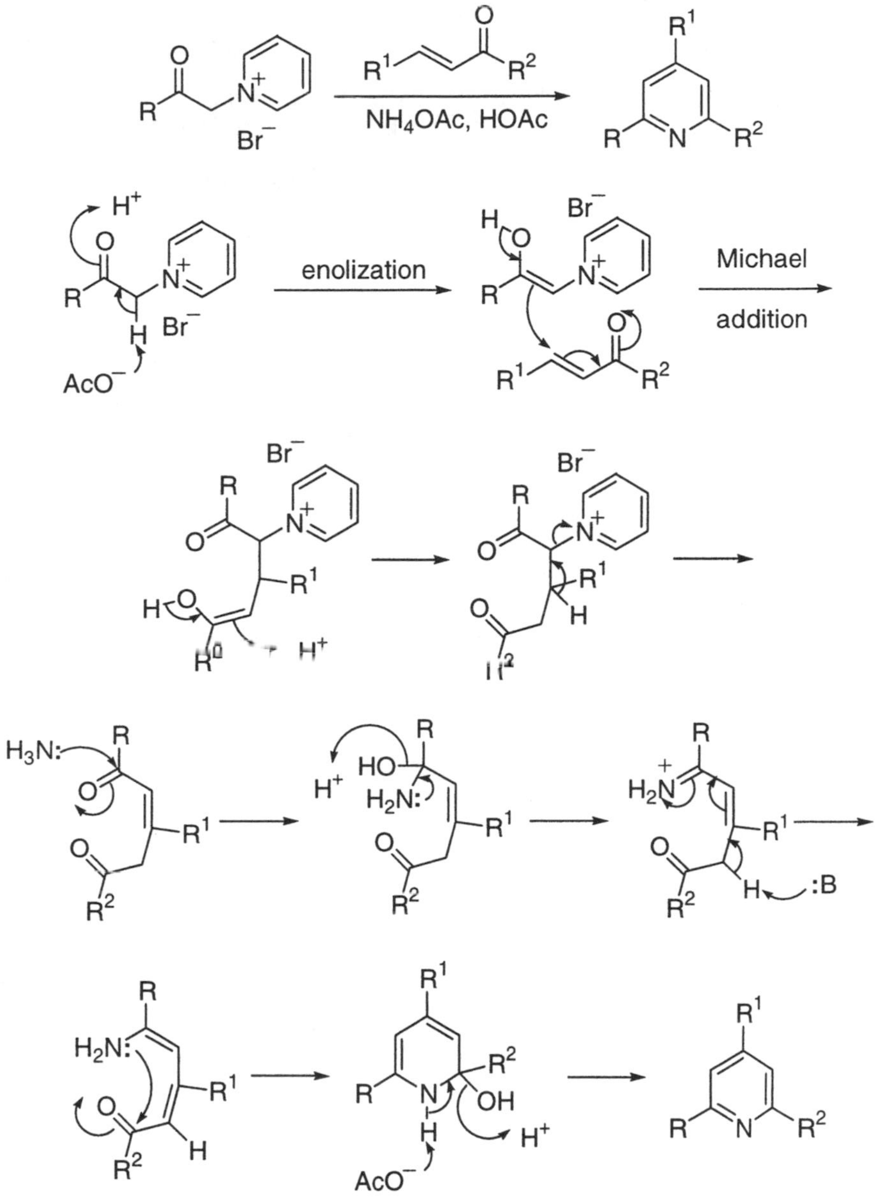

232

References

1. Zecher, W.; Kröhnke, F. *Ber.* **1961**, *94*, 690.
2. Kröhnke, F. *Synthesis* **1976**, 1. (Review).
3. Constable, E. C.; Lewis, J. *Polyhedron* **1982**, *1*, 303.
4. Constable, E. C.; Ward, M. D.; Tocher, D. A. *J. Chem. Soc., Dalton Trans.* **1991**, 1675.
5. Constable, E. C.; Chotalia, R. *J. Chem. Soc., Chem. Commun.* **1992**, 65.
6. Markovac, A.; Ash, A. B.; Stevens, C. L.; Hackley, B. E., Jr.; Steinberg, G. M. *J. Heterocycl. Chem.* **1977**, *14*, 19.
7. Chatterjea, J. N.; Shaw, S. C.; Singh, J. N.; Singh, S. N. *Indian J. Chem., Sect. B* **1977**, *15B*, 430.
8. Kelly, T. R.; Lee, Y.-J.; Mears, R. J. *J. Org. Chem.* **1997**, *62*, 2774.
9. Bark, T.; Von Zelewsky, A. *Chimia* **2000**, *54*, 589.
10. Malkov, A. V.; Bella, M.; Stara, I. G.; Kocovsky, P. *Tetrahedron Lett.* **2001**, *42*, 3045.

Kumada cross-coupling reaction

The Kumada cross-coupling reaction (also occasionally known as the Kharasch cross-coupling reaction) is a nickel- or palladium-catalyzed cross-coupling reaction of a Grignard reagent with an organic halide, triflate, *etc.*

$$R-X \ + \ R^1-MgX \ \xrightarrow{Pd(0)} \ R-R^1 \ + \ MgX_2$$

R−X + L$_2$Pd(0) $\xrightarrow[\text{addition}]{\text{oxidative}}$ $\underset{L}{\overset{R}{\underset{}{Pd}}}\!\!\diagdown X$ (L) $\xrightarrow[\substack{\text{transmetallation} \\ \text{isomerization}}]{R^1-MgX}$

MgX$_2$ + $\underset{R}{\overset{L}{\underset{}{Pd}}}\!\!\diagdown R^1$ (L) $\xrightarrow[\text{elimination}]{\text{reductive}}$ R−R^1 + L$_2$Pd(0)

The Kumada cross-coupling reaction, as well as the Negishi, Stille, Hiyama, and Suzuki cross-coupling reactions, belong to the same category of Pd-catalyzed cross-coupling reactions of organic halides, triflates and other electrophiles with organometallic reagents. These reactions follow a general mechanistic cycle as shown on the next page. There are slight variations for the Hiyama and Suzuki reactions, for which an additional activation step is required for the transmetallation to occur.

The catalytic cycle:

$$L_nPd(II) \ + \ R^1M \ \xrightarrow{\text{transmetallation}} \ L_nPd(II)\underset{R^1}{\overset{R^1}{\diagdown}}$$

$$\xrightarrow[\text{elimination}]{\text{reductive}} \ R^1-R^1 \ + \ L_nPd(0)$$

234

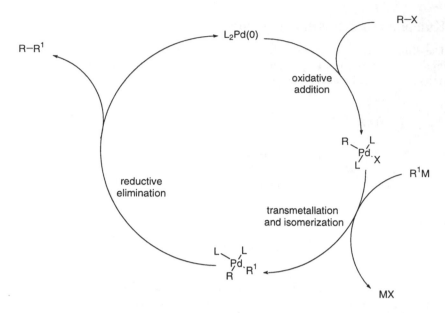

References

1. Tamao, K.; Sumitani, K.; Kiso, Y.; Zembayashi, M.; Fujioka, A.; Kodma, S.-i.; Nakajima, I.; Minato, A.; Kumada, M. *Bull. Chem. Soc. Jpn.* **1976**, *49*, 1958.
2. Kalinin, V. N. *Synthesis* **1992**, 413.
3. Stanforth, S. P. *Tetrahedron* **1998**, *54,* 263.
4. Park, M.; Buck, J. R.; Rizzo, C. J. *Tetrahedron* **1998**, *54*, 12707.
5. Huang, J.; Nolan, S. P. *J. Am. Chem. Soc.* **1999**, *121*, 9889.
6. Uenishi, J.; Matsui, K. *Tetrahedron Lett.* **2001**, *42*, 4353.
7. Li, G. Y. *J. Organomet. Chem.* **2002**, *653*, 63.
8. Anctil, E. J.-G.; Snieckus, V. *J. Organomet. Chem.* **2002**, *653*, 150.
9. Banno, T.; Hayakawa, Y.; Umeno, M. *J. Organomet. Chem.* **2002**, *653*, 288.
10. Tasler, S.; Lipshutz, B. H. *J. Org. Chem.* **2003**, *68*, 1190.

Larock indole synthesis

Indole synthesis using the palladium-catalyzed coupling reaction of an
o-iodoaniline with a propargyl alcohol.

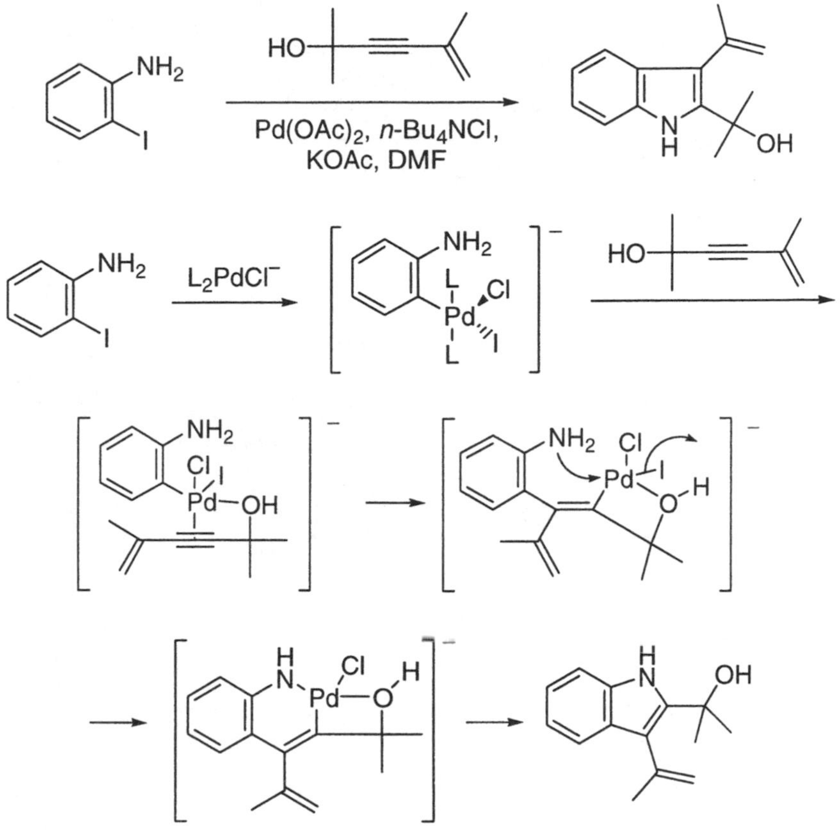

References

1. Larock, R. C.; Yum, E. K. *J. Am. Chem. Soc.* **1991**, *113*, 6689.
2. Larock, R. C.; Yum, E. K.; Refvik, M. D. *J. Org. Chem.* **1998**, *63*, 7652.
3. Larock, R. C. *J. Organomet. Chem.* **1999**, *576*, 111.
4. Walsh, T. F.; Toupence, R. B.; Ujjainwalla, F.; Young, J. R.; Goulet, M. T. *Tetrahedron* **2001**, *57*, 5233.
5. Nishikawa, T.; Wada, K.; Isobe, M. *Biosci. Biotech. Biochem.* **2002**, *66*, 2273.
6. Kalai, T.; Balog, M.; Jeko, J.; Hubbell, W. L.; Hideg, K. *Synthesis* **2002**, 2365.

Lawesson's reagent

2,4-Bis-(4-methoxyphenyl)-[1,3,2,4]dithiadiphosphetane 2,4-disulfide, trans-forms the carbonyl groups of ketones, amides and esters into the corresponding thiocarbonyl compounds.

References

1. Lawesson, S. O.; Perregaad, J.; Scheibye, S.; Meyer, H. J.; Thomsen, I. *Bull. Soc. Chim. Belg.* **1977**, *86*, 679.
2. Navech, J.; Majoral, J. P.; Kraemer, R. *Tetrahedron Lett.* **1983**, *24*, 5885.
3. Cava, M. P.; Levinson, M. I. *Tetrahedron* **1985**, *41*, 5061.
4. Luheshi, A. B. N.; Smalley, R. K.; Kennewell, P. D.; Westwood, R. *Tetrahedron Lett.* **1990**, *31*, 123.
5. Luo, Y.; He, L.; Ding, M.; Yang, G.; Luo, A.; Liu, X.; Wu, T. *Heterocycl. Commun.* **2001**, *7*, 37.
6. He, L.; Luo, Y.; Li, K.; Ding, M.; Lu, A.; Liu, X.; Wu, T.; Cai, F. *Synth. Commun.* **2002**, *32*, 1415.
7. Ishii, A.; Yamashita, R.; Saito, M.; Nakayama, J. *J. Org. Chem.* **2003**, *68*, 1555.

Leuckart–Wallach reaction

Amine synthesis from reductive amination of a carbonyl compound and an amine in the presence of excess formic acid, which serves as the reducing reagent by delivering a hydride.

α-aminoalcohol

iminium ion intermediate

reduction

References

1. Leuckart, R. *Ber. Dtsch. Chem. Ges.* **1885**, *18*, 2341.
2. Wallach, O. *Justus Liebigs Ann. Chem.* **1892**, *272*, 99.
3. Moore, M. L. *Org. React.* **1948**, *5*, 301. (Review).
4. Mechanism, Lukasiewiez, A. *Tetrahedron* **1963**, *19*, 1789.
5. Bach, R. D. *J. Org. Chem.* **1968**, *33*, 1647.
6. Doorenbos, N. J.; Solomons, W. E. *Chem. Ind.* **1970**, 1322.
7. Ito, K.; Oba, H.; Sekiya, M. *Bull. Chem. Soc. Jpn.* **1976**, *49*, 2485.
8. Musumarra, G.; Sergi, C. *Heterocycles* **1994**, *37*, 1033.
9. Kitamura, M.; Lee, D.; Hayashi, S.; Tanaka, S.; Yoshimura, M. *J. Org. Chem.* **2002**, *67*, 8685.

Lieben haloform reaction

Iodoform, a yellow precipitate in water, is often used for detection of methyl ketones.

References

1. Lieben, A. *Justus Liebigs Ann. Chem.* **1870**, *Suppl. 7*, 218.
2. Rothenberg, G.; Sasson, Y. *Tetrahedron* **1996**, *52*, 13641.
3. Tietze, L. F.; Voss, E.; Hartfiel, U. *Org. Synth.* **1990**, *69*, 238.
4. Madler, M. M.; Klucik, J.; Soell, P. S.; Brown, C. W.; Liu, S.; Berlin, K. D.; Benbrook, D. M.; Birckbichler, P. J.; Nelson, E. C. *Org. Prep. Proced. Int.* **1998**, *30*, 230.
5. Connolly, C. J. C.; Hamby, J. M.; Schroeder, M. C.; Barvian, M.; Lu, G. H.; Panek, R. L.; Amar, A.; Shen, C.; Kraker, A. J.; Fry, D. W.; Klohs, W. D.; Doherty, A. M. *Bioorg. Med. Chem. Lett.* **1997**, *7*, 2415.
6. Jablonski, L.; Billard, T.; Langlois, B. R. *Tetrahedron Lett.* **2003**, *44*, 1055.

Liebeskind–Srogl coupling

Palladium-catalyzed cross-coupling between thioesters and organoboronic acids to afford ketones.

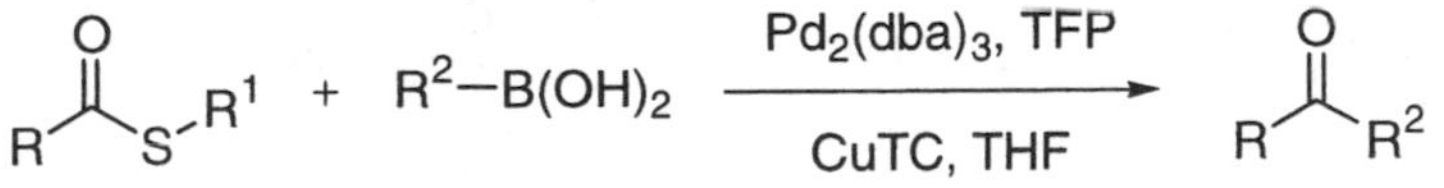

TFP = tris(2-furyl)phosphine, CuTC = copper(I) thiophene-2-carboxylate

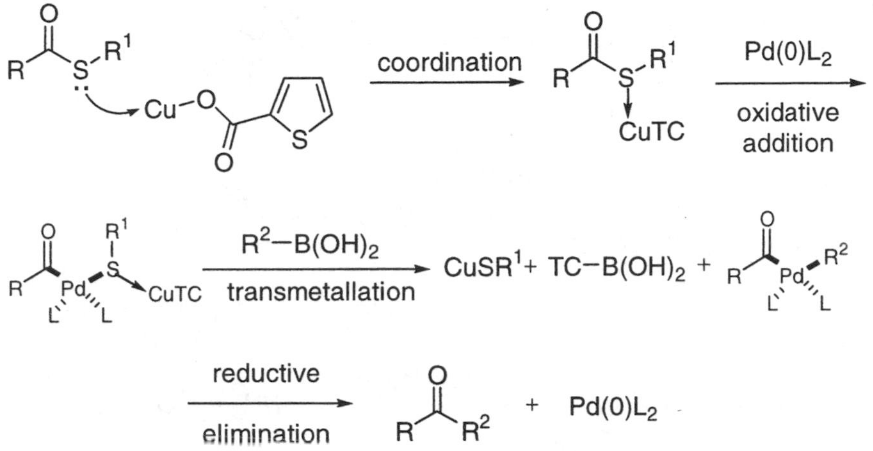

Reference

1. Liebeskind, L. S.; Srogl, J. *J. Am. Chem. Soc.* **2000**, *122*, 11260.
2. Savarin, C.; Srogl, J.; Liebeskind, L. S. *Org. Lett.* **2000**, *2*, 3229.
3. Savarin, C.; Srogl, J.; Liebeskind, L. S. *Org. Lett.* **2001**, *3*, 91.
4. Liebeskind, L. S.; Srogl, J. *Org. Lett.* **2002**, *4*, 979.
5. Liebeskind, L. S.; Srogl, J. *Org. Lett.* **2002**, *4*, 983.

Lossen rearrangement

Treatment of O-acylated hydroxamic acids with base provides isocyanates.

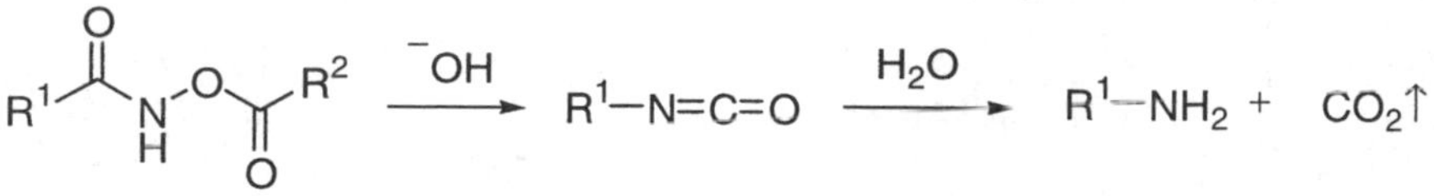

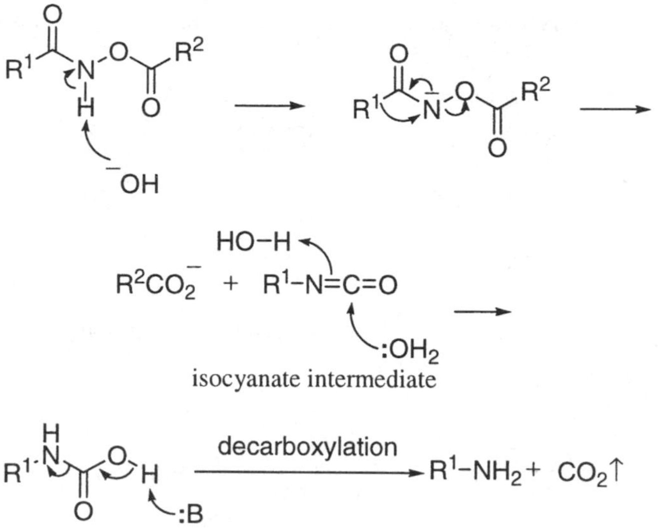

$$R^1-N=C=O \xrightarrow{H_2O} R^1-NH_2 + CO_2\uparrow$$

isocyanate intermediate

$$R^2CO_2^- + R^1-N=C=O$$

$$R^1-NH_2 + CO_2\uparrow$$

decarboxylation

References

1. Lossen, W. *Ann.* **1872**, *161*, 347.
2. Bauer, L.; Exner, O. *Angew. Chem.* **1974**, *86*, 419.
3. Lipczynska-Kochany, E. *Wiad. Chem.* **1982**, *36*, 735.
4. Casteel, D. A.; Gephart, R. S.; Morgan, T. *Heterocycles* **1993**, *36*, 485.
5. Zalipsky, S. *Chem. Commun.* **1998**, 69.
6. Anilkumar, R.; Chandrasekhar, S.; Sridhar, M. *Tetrahedron Lett.* **2000**, *41*, 5291.
7. Needs, P. W.; Rigby, N. M.; Ring, S. G.; MacDougall, A. *Carbohydr. Res.* **2001**, *333*, 47.

Luche reduction

1,2-Reduction of enones using $NaBH_4$–$CeCl_3$.

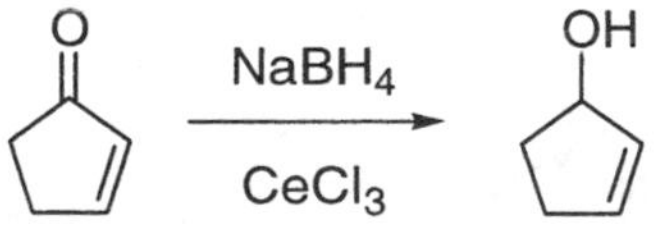

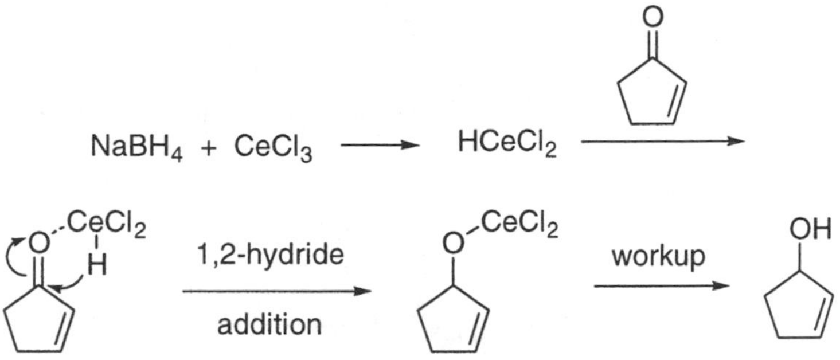

References

1. Luche, J.-L. *J. Am. Chem. Soc.* **1978**, *100*, 2226.
2. Li, K.; Hamann, L. G.; Koreeda, M. *Tetrahedron Lett.* **1992**, *33*, 6569.
3. Cook, G. P.; Greenberg, M. M. *J. Org. Chem.* **1994**, *59*, 4704.
4. Hutton, C; Jolliff, T.; Mitchell, H.; Warren, S. *Tetrahedron Lett.* **1995**, *36*, 7905.
5. Moreno-Dorado, F. J.; Guerra, F. M.; Aladro, F. J., Bustamante, J M ; Jurec, Z. D.; Massanet, G. M. *Tetrahedron* **1999**, *55*, 6997.
6. Barluenga, J.; Fananas, F. J.; Sanz, R.; Garcia, F.; Garcia, N. *Tetrahedron Lett.* **1999**, *40*, 4735.
7. Haukaas, M. H.; O'Doherty, G. A. *Org. Lett.* **2001**, *3*, 401.
8. Uttaro, J.-P.; Audran, G.; Galano, J.-M.; Monti, H. *Tetrahedron Lett.* **2002**, *43*, 2757.

McFadyen–Stevens reduction

Treatment of acylbenzenesulfonylhydrazines with base delivers the corresponding aldehydes.

References

1. McFadyen, J. S.; Stevens, T. S. *J. Chem. Soc.* **1936,** 584.
2. Newman, M. S.; Caflish, E. G., Jr. *J. Am. Chem. Soc.* **1958,** *80,* 862.
3. Sprecher, M.; Feldkimel, M.; Wilchek, M. *J. Org. Chem.* **1961,** *26,* 3664.
4. Jensen, K. A.; Holm, A. *Acta. Chem. Scand.* **1961,** *15,* 1787.
5. Babad, H.; Herbert, W.; Stiles, A. W. *Tetrahedron Lett.* **1966,** 2927.
6. Graboyes, H.; Anderson, E. L.; Levinson, S. H.; Resnick, T. M. *J. Heterocycl. Chem.* **1975,** *12,* 1225.
7. Eichler, E.; Rooney, C. S.; Williams, H. W. R. *J. Heterocycl. Chem.* **1976,** *13,* 841.
8. Nair, M.; Shechter, H. *J. Chem. Soc., Chem. Commun.* **1978,** 793.
9. Dudman, C. C.; Grice, P.; Reese, C. B. *Tetrahedron Lett.* **1980,** *21,* 4645.
10. Manna, R. K.; Jaisankar, P.; Giri, V. S. *Synth. Commun.* **1998,** *28,* 9.
11. Jaisankar, P.; Pal, B.; Giri, V. S. *Synth. Commun.* **2002,** *32,* 2569.

McLafferty fragmentation

Intramolecular fragmentation of carbonyls in mass spectroscopy.

References

1. McLafferty, F. W. *Anal. Chem.* **1956**, *28*, 306.
2. Gilpin, J. A.; McLafferty, F. W. *Anal. Chem.* **1957**, *29*, 990.
3. Zollinger, M.; Seibl, J. *Org. Mass Spectrom.* **1985**, *20*, 649.
4. Kingston, D. G. I.; Bursey, J. T.; Bursey, M. M. *Chem. Rev.* **1974**, *74*, 215.
5. Budzikiewicz, H.; Bold, P. *Org. Mass Spectrom.* **1991**, *26*, 709.
6. Stringer, M. B.; Underwood, D. J.; Bowie, J. H.; Allison, C. E.; Donchi, K. F.; Derrick, P. J. *Org. Mass Spectrom.* **1991**, *27*, 270.
7. Lightner, D. A.; Steinberg, F. S.; Huestis, L. D. *J. Mass Spectrom. Soc. Jpn.* **1998**, *46*, 11.
8. Alvarez, R. M.; Fernandez, A. H.; Chioua, M.; Perez, P. R.; Vilchez, N. V.; Torres, F. G. *Rapid Commun. Mass Spectrom.* **1999**, *13*, 2480.
9. Rychlik, M. *J. Mass Spectrom.* **2001**, *36*, 555.
10. Jiang, N.; Wang, J.-B.; He, M.-Y. *Chin. J. Chem.* **2002**, *20*, 789.

McMurry coupling

Olefination of carbonyls with low-valent titanium such as Ti(0) derived from TiCl$_3$/LiAlH$_4$. Single-electron process.

$$R^1 \underset{R^2}{\overset{}{>}}=O \xrightarrow[\text{2. H}_2\text{O}]{\text{1. TiCl}_3, \text{ LiAlH}_4} \underset{R^2\ \ R^2}{\overset{R^1\ \ R^1}{>\!\!=\!\!<}} + \text{ TiO}$$

$$\text{Ti(III)Cl}_3 + \text{ LiAlH}_4 \longrightarrow \text{ Ti(0)}$$

$$R^1 \underset{R^2}{\overset{}{>}}=O \quad :\text{Ti(0)} \xrightarrow[\text{transfer}]{\text{single electron}} \text{radical anion intermediate} \xrightarrow{\text{homocoupling}}$$

radical anion intermediate

oxide-coated titanium surface

References

1. McMurry, J. E.; Fleming, M. P. *J. Am. Chem. Soc.* **1974**, *96*, 4708.
2. McMurry, J. E. *Chem. Rev.* **1989**, *89*, 1513. (Review).
3. Ephritikhine, M. *Chem. Commun.* **1998**, 2549.
4. Hirao, T. *Synlett* **1999**, 175.
5. Yamato, T.; Fujita, K.; Tsuzuki, H. *J. Chem. Soc., Perkin Trans. 1* **2001**, 2089.
6. Sabelle, S.; Hydrio, J.; Leclerc, E.; Mioskowski, C.; Renard, P.-Y. *Tetrahedron Lett.* **2002**, *43*, 3645.
7. Williams, D. R.; Heidebrecht, R. W., Jr. *J. Am. Chem. Soc.* **2003**, *125*, 1843.
8. Kowalski, K.; Vessieres, A.; Top, S.; Jaouen, G.; Zakrzewski, J. *Tetrahedron Lett.* **2003**, *44*, 2749.
9. Honda, T.; Namiki, H.; Nagase, H.; Mizutani, H. *Tetrahedron Lett.* **2003**, *44*, 3035.

Madelung indole synthesis

Indoles from the intramolecular cyclization of 2-(acylamino)-toluenes using strong bases.

References

1. Madelung, W. *Ber. Dtsch. Chem. Ges.* **1912**, *45*, 1128.
2. Houlihan, W. J.; Parrino, V. A.; Uike, Y. *J. Org. Chem.* **1981**, *46*, 4511.
3. Houlihan, W. J.; Uike, Y.; Parrino, V. A. *J. Org. Chem.* **1981**, *46*, 4515.
4. Orlemans, E. O. M.; Schreuder, A. H.; Conti, P. G. M.; Verboom, W.; Reinhoudt, D. N. *Tetrahedron* **1987**, *43*, 3817.
5. Smith, A. B., III; Haseltine, J. N.; Visnick, M. *Tetrahedron* **1989**, *45*, 2431.
6. Wacker, D. A.; Kasireddy, P. *Tetrahedron Lett.* **2002**, *43*, 5189.

Mannich reaction

Three-component aminomethylation from amine, formaldehyde and a compound with an acidic methylene moiety.

When R = H, the $^+NH_2=CH_2$ salt is known as **Eschenmoser's salt**

The Mannich reaction can also operate under basic conditions:

Mannich Base

References

1. Mannich, C.; Krosche, W. *Arch. Pharm.* **1912**, *250*, 647.
2. Bordunov, A. V.; Bradshaw, J. S.; Pastushok, V. N.; Izatt, R. M. *Synlett* **1996**, 933.
3. Arend, M.; Westermann, B.; Risch, N. *Angew. Chem., Int. Ed.* **1998**, *37*, 1045.
4. Padwa, A.; Waterson, A. G. *J. Org. Chem.* **2000**, *65*, 235.
5. List, B. *J. Am. Chem. Soc.* **2000**, *122*, 9336.
6. Schlienger, N.; Bryce, M. R.; Hansen, T. K. *Tetrahedron* **2000**, *56*, 10023.
7. Bur, S. K.; Martin, S. F. *Tetrahedron* **2001**, *57*, 3221. (Review).
8. McReynolds, M. D.; Hanson, P. R. *Chemtracts* **2001**, *14*, 796. (Review).

9.   Vicario, J. L.; Badía, D.; Carrillo, L. *Org. Lett.* **2001**, *3*, 773.
10. Ranu, B. C.; Samanta, S.; Guchhait, S. K. *Tetrahedron* **2002**, *58*, 983.
11. Martin, S. F. *Acc. Chem. Res.* **2002**, *35*, 895–904. (Review).
12. Padwa, A.; Bur, S. K.; Danca, D. M.; Ginn, J. D.; Lynch, S. M. *Synlett* **2002**, 851–862. (Review).
13. Yang, X.-F.; Wang, M.; Varma, R. S.; Li, C.-J. *Org. Lett.* **2003**, *5*, 657.

248

Marshall boronate fragmentation

Cf. Grob fragmentation (p166). In fact, Marshall boronate fragmentation belongs to the Grob fragmentation category.

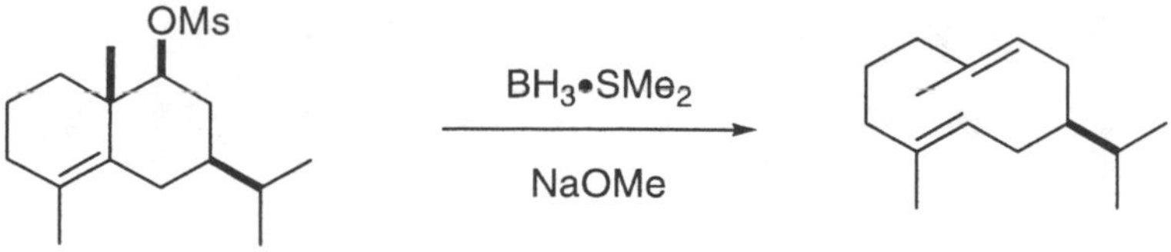

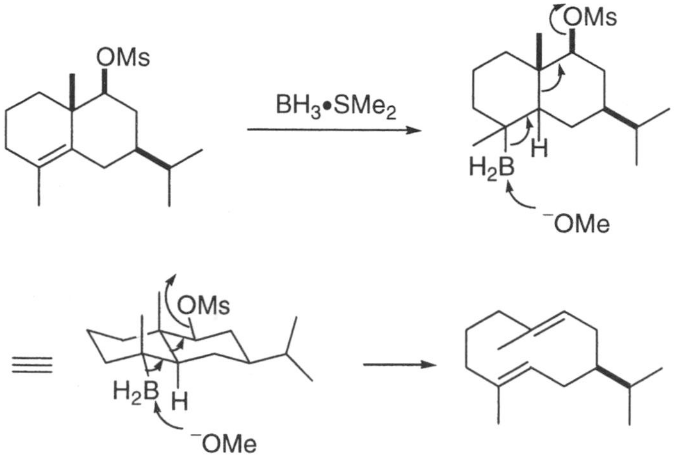

References

1. Marshall, J. A.; Huffman, W. F. *J. Am. Chem. Soc.* **1970**, *92*, 6358.
2. Marshall, J. A. *Synthesis* **1971**, 229.
3. Wharton, P. S.; Sundin, C. E.; Johnson, D. W.; Kluende, H. C. *J. Org. Chem.* **1972**, *37*, 34.
4. Minnard, A. J.; Wijinberg, J. B. P. A.; de Groot, A. *Tetrahedron* **1994**, *50*, 4755.
5. Zhabinskii, V.; Minnard, A. J.; Wijinberg, J. B. P. A.; de Groot, A. *J. Org. Chem.* **1996**, *61*, 4022.
6. Minnard, A. J.; Stork, G. A.; Wijinberg, J. B. P. A.; de Groot, A. *J. Org. Chem.* **1997**, *62*, 2344.

Martin's sulfurane dehydrating reagent

Dehydrates secondary and tertiary alcohols to give olefins, but forms ethers with primary alcohols. *Cf.* Burgess dehydrating reagent.

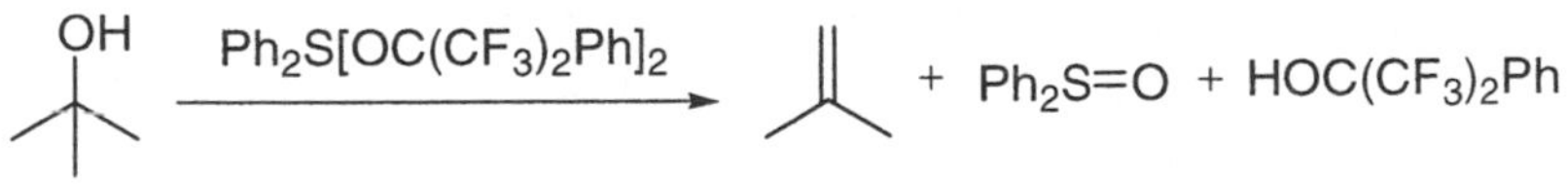

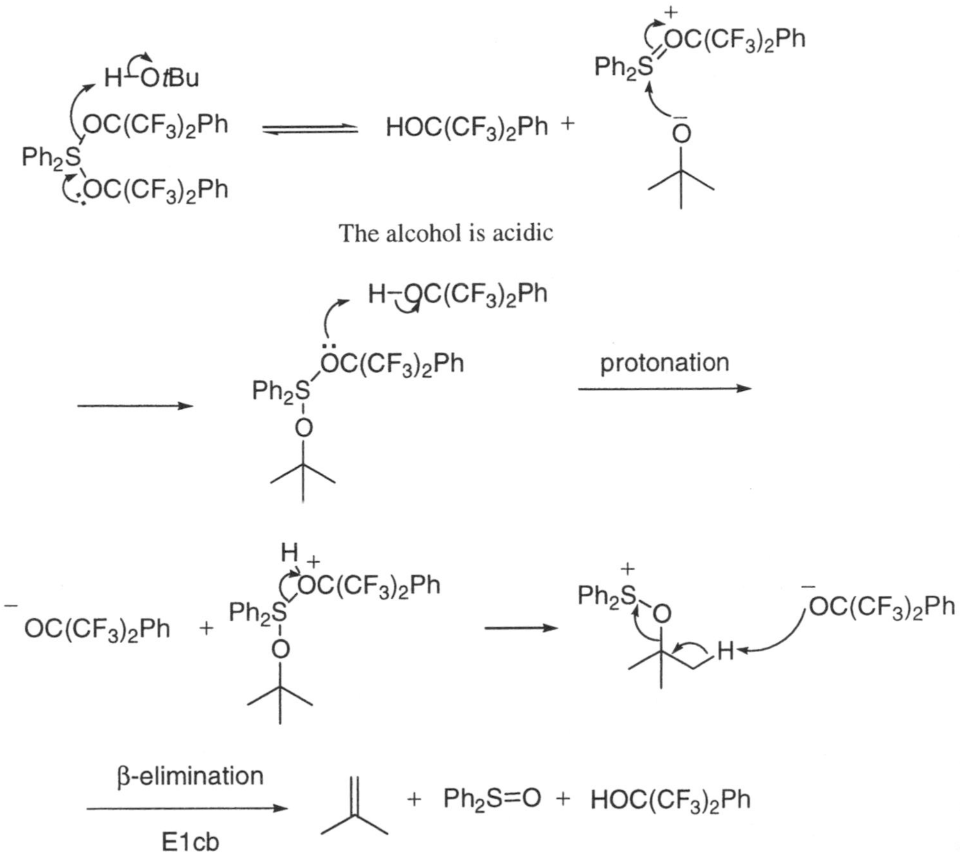

The alcohol is acidic

protonation

β-elimination

E1cb

References

1. Martin, J. C.; Arhart, R. J. *J. Am. Chem. Soc.* **1971**, *93*, 2339, 2341.
2. Martin, J. C.; Arhart, R. J. *J. Am. Chem. Soc.* **1971**, *93*, 4327.
3. Martin, J. C.; Arhart, R. J.; Franz, J. A.; Perozzi, E. F.; Kaplan, L. *J. Org. Synth.* **1973**, *53*, 1850.
4. Bartlett, P. D.; Aida, T.; Chu, H.-K.; Fang, T.-S. *J. Am. Chem. Soc.* **1980**, *102*, 3515.
5. Tse, B.; Kishi, Y. *J. Org. Chem.* **1994**, *59*, 7807.

250

6. Winkler, J. D.; Stelmach, J. E.; Axten, J. *Tetrahedron Lett.* **1996**, *37*, 4317.
7. Gais, H. J.; Schmiedl, G.; Ossenkamp, R. K. L. *Liebigs Ann.* **1997**, 2419.
8. Box, J. M.; Harwood, L. M.; Humphreys, J. L.; Morris, G. A.; Redon, P. M.; White-head, R. C. *Synlett* **2002**, 358.
9. Myers, A. G.; Glatthar, R.; Hammond, M.; Harrington, P. M.; Kuo, E. Y.; Liang, J.; Schaus, S. E.; Wu, Y.; Xiang, J.-N. *J. Am. Chem. Soc.* **2002**, *124*, 5380.

Masamune–Roush conditions

Applicable to base-sensitive aldehydes and phosphonates for the Horner–Wadsworth–Emmons reaction

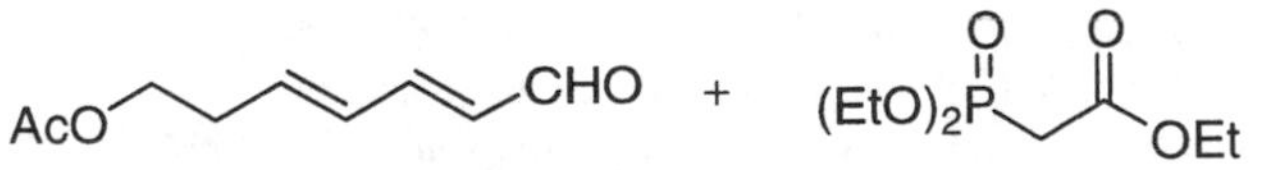

α-keto or α-alkoxycarbonyl phosphonate required

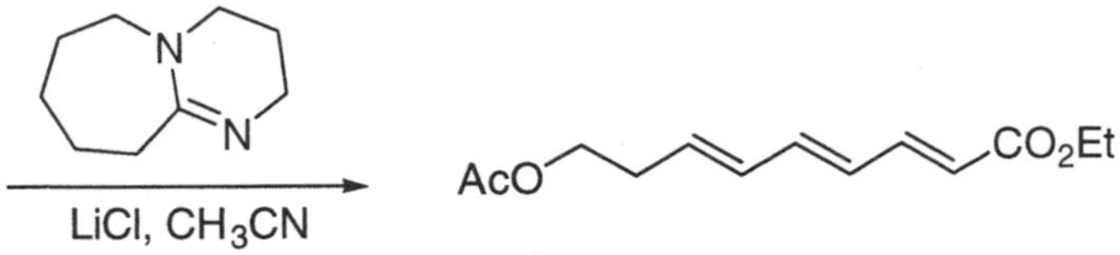

LiCl, CH₃CN

1,8-diazabicyclo[5.4.0]undec-7-ene (DBU)

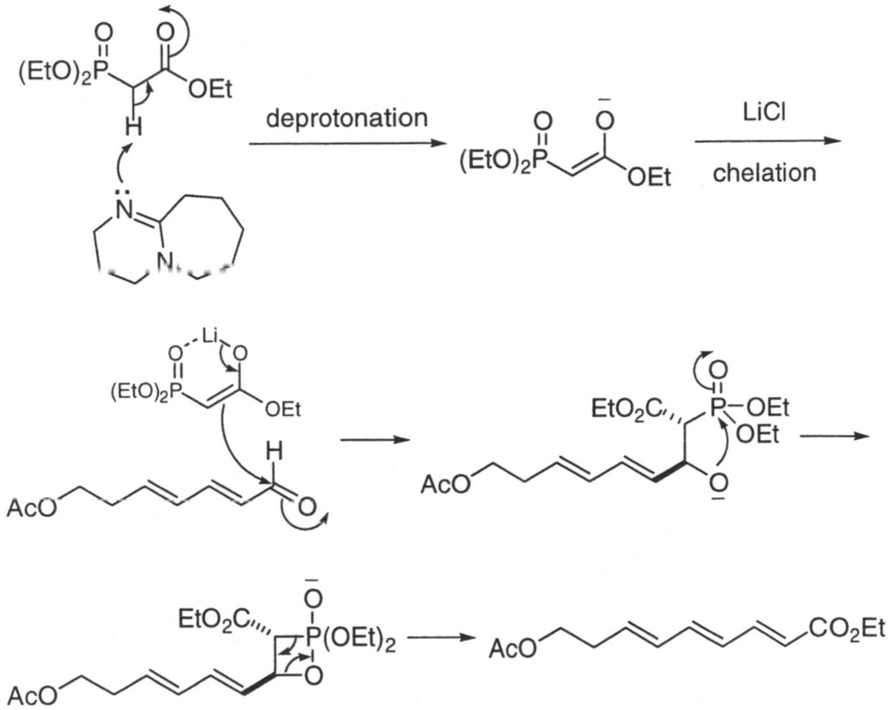

deprotonation

LiCl

chelation

References

1. Blanchette, M. A.; Choy, W.; Davis, J. T.; Essenfeld, A. P.; Masamune, S.; Roush, W. R.; Sakai, T. *Tetrahedron Lett.* **1984**, *25*, 2183.
2. Marshall, J. A.; DuBay, W. J. *J. Org. Chem.* **1994**, *59*, 1703.
3. Tius, M. A.; Fauq, A. H. *J. Am. Chem. Soc.* **1986**, *108*, 1035.
4. Tius, M. A.; Fauq, A. H. *J. Am. Chem. Soc.* **1986**, *108*, 6389.
5. Rychnovsky, S. D.; Khire, U. R.; Yang, G. *J. Am. Chem. Soc.* **1997**, *119*, 2058.
6. Dixon, D. J.; Foster, A. C.; Ley, S. V. *Org. Lett.* **2000**, *2*, 123.

Meerwein arylation

Radical arylation of unsaturated compounds by diazonium salts.

$$ArN_2^+ \ Cl^- \ + \ \underset{R^1}{\overset{H}{R}}{=}Z \ \xrightarrow{CuCl_2} \ \underset{R^1}{\overset{Ar}{R}}{=}Z$$

$$Z = Ar, C{=}C, C{=}O, CN, H$$

$$ArN_2^+ \ Cl^- \ \xrightarrow{CuCl_2} \ Ar\bullet \ + \ N_2\uparrow \ + \ CuCl \ + \ Cl_2\uparrow$$

$$Ar\bullet \ \underset{R^1}{\overset{H}{R}}{=}Z \ \xrightarrow[\text{addition}]{\text{radical}} \ \underset{R^1}{\overset{Ar}{R}}{\bullet}Z \ \xrightarrow{CuCl_2} \ \underset{R^1}{\overset{Ar}{R}}{=}Z + CuCl$$

References

1. Meerwein, H.; Buchner, E.; van Emster, K. *J. Prakt. Chem.* **1939**, *152*, 237.
2. Rondestvedt, C. S., Jr. *Org. React.* **1976**, *24*, 225.
3. Raucher, S.; Koolpe, G. A. *J. Org. Chem.* **1983**, *48*, 2066.
4. Sutter, P.; Weis, C. D. *J. Heterocycl. Chem.* **1987**, *24*, 69.
5. Schmidt, A. H.; Bohmin, G. Diedrich, H. *Synthesis* **1990**, 579.
6. Nock, H.; Schottenberger, H. *J. Org. Chem.* **1993**, *58*, 7045.
7. Takahashi, I.; Muramatsu, O.; Fukuhara, J.; Hosokawa, Y.; Takeyama, N.; Morita, T.; Kitajima, H. *Chem. Lett.* **1994**, 465.
8. Brunner, H.; Bluchel, C.; Doyle, M. P. *J. Organomet. Chem.* **1997**, *541*, 89.
9. Mella, M.; Coppo, P.; Guizzardi, B.; Fagnoni, M.; Freccero, M.; Albini, A. *J. Org. Chem.* **2001**, *66*, 6344.
10. Milanesi, S.; Fagnoni, M.; Albini, A. *Chem. Commun.* **2003**, 216.

Meerwein–Ponndorf–Verley reduction

Reduction of ketones to the corresponding alcohols using Al(O*i*-Pr)$_3$ in isopropanol.

cyclic transition state

References

1. Meerwein, H.; Schmidt, R. *Justus Liebigs Ann. Chem.* **1925**, *444*, 221.
2. Ashby, E. C. *Acc. Chem. Res.* **1988**, *21*, 414. (Review).
3. de Graauw, C. F.; Peters, J. A.; van Bekkum, H.; Huskens, J. *Synthesis* **1994**, 1007.
4. Aremo, N.; Hase, T. *Org. React.* **2001**, *42*, 3637. (Review).
5. Campbell, E. J.; Zhou, H.; Nguyen, S. T. *Angew. Chem., Int. Ed. Engl.* **2002**, *41*, 1020.
6. Faller, J. W.; Lavoie, A. R. *Organometallics* **2002**, *21*, 2010.
7. Nishide, K.; Node, M. *Chirality* **2002**, *14*, 759.
8. Jerome, J. E.; Sergent, R. H. *Chem. Ind.* **2003**, *89*, 97.

Meinwald rearrangement

Treatment of bicyclic epoxides with acid affords rearranged aldehydes.

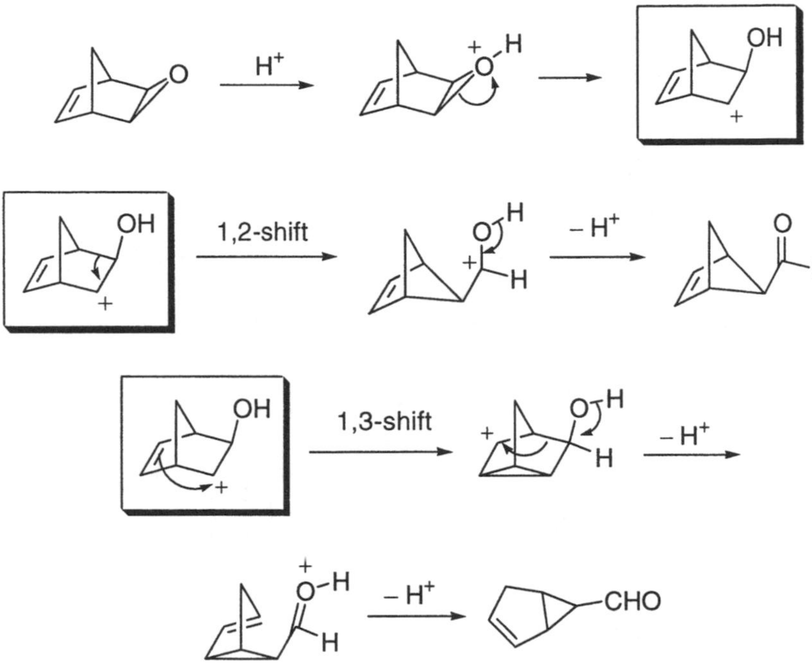

References

1. Meinwald, J.; Labana, S. S.; Chadha, M. S. *J. Am. Chem. Soc.* **1962**, *85*, 582.
2. Meinwald, J.; Labana, S. S.; Labana, L. L.; Wahl, G. H., Jr. *Tetrahedron Lett.* **1965**, *23*, 1789.
3. Niwayama, S.; Noguchi, H.; Ohno, M.; Kobayashi, S. *Tetrahedron Lett.* **1993**, *34*, 665.
4. Niwayama, S.; Kobayashi, S.; Ohno, M. *J. Am. Chem. Soc.* **1994**, *116*, 3290.
5. Kim, W.; Kim, H.; Rhee, H. *Heterocycles* **2000**, *53*, 219.
6. Rhee, H.; Yoon, D.-O.; Jung, M. E. *Nucleosides, Nucleotides Nucleic Acids* **2000**, *19*, 619.
7. Sun, H.; Yang, J.; Amaral, K. E.; Horenstein, B. A. *Tetrahedron Lett.* **2001**, *42*, 2451.

Meisenheimer complex

Also known as Meisenheimer–Jackson salt, the stable intermediate for certain S$_N$Ar reactions.

Sanger's reagent

ipso attack

ipso substitution

Meisenheimer complex (**Meisenheimer–Jackson salt**)

The reaction using Sanger's reagent is faster than using the corresponding chloro-, bromo-, and iodo-dinitrobenzene — the fluoro-Meisenheimer complex is the most stabilized because F is the most electron-withdrawing. The reaction rate does not depend upon the leaving ability of the leaving group.

References

1. Meisenheimer, J. *Justus Liebigs Ann. Chem.* **1902**, *323*, 205.
2. Strauss, M. J. *Acc. Chem. Res.* **1974**, *7*, 181. (Review).
3. Bernasconi, C. F. *Acc. Chem. Res.* **1978**, *11*, 147. (Review).
4. Terrier, F. *Chem. Rev.* **1982**, *82*, 77. (Review).
5. Buncel, E.; Dust, J. M.; Manderville, R. A. *J. Am. Chem. Soc.* **1996**, *118*, 6072.
6. Sepulcri, P.; Goumont, R.; Halle, J.-C.; Buncel, E.; Terrier, F. *Chem. Commun.* **1997**, 789.

7.   Weiss, R.; Schwab, O.; Hampel, F. *Chem.— Eur. J.* **1999**, *5*, 968.
8.   Hoshino, K.; Ozawa, N.; Kokado, H.; Seki, H.; Tokunaga, T.; Ishikawa, T. *J. Org. Chem.* **1999**, *64*, 4572.
9.   Adam, W.; Makosza, M.; Zhao, C.-G.; Surowiec, M. *J. Org. Chem.* **2000**, *65*, 1099.
10. Gallardo, I.; Guirado, G.; Marquet, J. *J. Org. Chem.* **2002**, *67*, 2548.
11. Kim, H.-Y.; Song, H.-G. *Appl. Microbiol. Biotech.* **2003**, *61*, 150.

258

Meisenheimer rearrangement

[1,2]-Sigmatropic rearrangement:

$$R_1\text{-}N\text{-}R \xrightarrow{\Delta} R_2\text{-}N\text{-}O\text{-}R$$

[2,3]-Sigmatropic rearrangement:

$$\xrightarrow{\Delta}$$

References

1. Meisenheimer, J. *Ber. Dtsch. Chem. Ges.* **1919**, *52*, 1667.
2. [1,2]-Sigmatropic rearrangement, Castagnoli, N. Jr.; Craig, J. C.; Melikian, A. P.; Roy, S. K. *Tetrahedron* **1970**, *26*, 4319.
3. [2,3]-Sigmatropic rearrangement, Yamamoto, Y.; Oda, J.; Inouye, Y. *J. Org. Chem.* **1976**, *41*, 303.
4. Johnston, R. A. W. *Mech. Mol. Migr.* **1969**, *2*, 249. (Review).
5. Kurihara, T.; Sakamoto, Y.; Matsumoto, H.; Kawabata, N.; Harusawa, S.; Yoneda, R. *Chem. Pharm. Bull.* **1994**, *42*, 475.
6. Molina, J. M.; El-Bergmi, R.; Dobado, J. A.; Portal, D. *J. Org. Chem.* **2000**, *65*, 8574.
7. Blanchet, J.; Bonin, M.; Micouin, L.; Husson, H.-P. *Tetrahedron Lett.* **2000**, *41*, 8279.

Meyer–Schuster rearrangement

The isomerization of secondary and tertiary α-acetylenic alcohols to α,β-unsaturated carbonyl compounds *via* 1,3-shift. When the acetylenic group is terminal, the products are aldehydes, whereas the internal acetylenes give ketones. *Cf.* Rupe rearrangement.

References

1. Swaminathan, S.; Narayanan, K. V. *Chem. Rev.* **1971**, *71*, 429. (Review).
2. Edens, M.; Boerner, D.; Chase, C. R.; Nass, D.; Schiavelli, M. D. *J. Org. Chem.* **1977**, *42*, 3403.
3. Cachia, P.; Darby, N.; Mak, T. C. W.; Money, T.; Trotter, J. *Can. J. Chem.* **1980**, *58*, 1172.
4. Andres, J.; Cardenas, R.; Silla, E.; Tapia, O. *J. Am. Chem. Soc.* **1988**, *110*, 666.
5. Tapia, O.; Lluch, J. M.; Cardenas, R.; Andres, J. *J. Am. Chem. Soc.* **1989**, *111*, 829.
6. Omar, E. A.; Tu, C.; Wigal, C. T.; Braun, L. L. *J. Heterocycl. Chem.* **1992**, *29*, 947.
7. Yoshimatsu, M.; Naito, M.; Kawahigashi, M.; Shimizu, H.; Kataoka, T. *J. Org. Chem.* **1995**, *60*, 4798.
8. Lorber, C. Y.; Osborn, J. A. *Tetrahedron Lett.* **1996**, *37*, 853.
9. Chihab-Eddine, A.; Daich, A.; Jilale, A.; Decroix, B. *J. Heterocycl. Chem.* **2000**, *37*, 1543.

Michael addition

Conjugate addition of a carbon-nucleophile to an α,β-unsaturated system.

e.g.:

e.g.:

References

1. Michael, A. *J. Prakt. Chem.* **1887**, *35*, 349.
2. Hunt, D. A. *Org. Prep. Proced. Int.* **1989**, *21*, 705.
3. D'Angelo, J.; Desmaele, D.; Dumas, F.; Guingant, Ae. *Tetrahedron: Asymmetry* **1992**, *3*, 459.
4. Hoz, S. *Acc. Chem. Res.* **1993**, *26*, 69. (Review).
5. Ihara, M.; Fukumoto, K. *Angew. Chem., Int. Ed. Engl.* **1993**, *32*, 1010.
6. Itoh, T.; Shirakami, S. *Heterocycles* **2001**, *55*, 37.
7. Cai, C.; Soloshonok, V. A.; Hruby, V. J. *J. Org. Chem.* **2001**, *66*, 1339.
8. Sundararajan, G.; Prabagaran, N. *Org. Lett.* **2001**, *3*, 389.
9. Amstrong, A.; Critchley, T. J.; Gourdel-Martin, M.-E.; Kelsey, R. D.; Mortlock, A. A. *Perkin 1* **2002**, 1344.
10. Bolm, C.; Kasyan, A.; Heider, P.; Saladin, S.; Drauz, K.; Günther, K.; Wagner, C. *Org. Lett.* **2002**, *4*, 2265.
11. Eilitz, U.; Leßmann, F.; Seidelmann, O.; Wendisch, V. *Tetrahedron: Asymmetry* **2003**, *14*, 189.

Michaelis–Arbuzov phosphonate synthesis

Phosphonate synthesis from the reaction of alkyl halides with phosphites.

General scheme:

$$(R^1O)_3P \quad + \quad R_2-X \quad \xrightarrow{\Delta} \quad R_2-\overset{\overset{\displaystyle O}{\|}}{\underset{\underset{\displaystyle OR^1}{|}}{P}}-OR^1 \quad + \quad R^1-X$$

R^1 = alkyl, *etc.*; R_2 = alkyl, acyl, *etc.*; X = Cl, Br, I

e.g.:

References

1. Michaelis, A.; Kaehne, R. *Ber.* **31**, 1048 (1898).
2. Arbuzov, A. E. *J. Russ. Phys. Chem. Soc.* **1906**, *38*, 687.
3. Swaminathan, S.; Narayanan, K. V. *Chem. Rev.* **1971**, *71*, 429. (Review).
4. Gellespie, P.; Ramirez, F.; Ugi, I.; Marquarding, D. *Angew. Chem., Int. Ed. Engl.* **1973**, *12*, 91.
5. Bhattacharya, A. K.; Thyagarajan, G. *Chem. Rev.* **1981**, *81*, 415. (Review).
6. Waschbüsch, R.; Carran, J.; Marinetti, A.; Savignac, P. *Synthesis* **1997**, 672.
7. Kato, T.; Tejima, M.; Ebiike, H.; Achiwa, K. *Chem. Pharm. Bull.* **1996**, *44*, 1132.
8. Griffith, J. A.; McCauley, D. J.; Barrans, R. E., Jr.; Herlinger, A. W. *Synth. Commun.* **1998**, *28*, 4317.
9. Kiddle, J. J.; Gurley, A. F. *Phosphorus, Sulfur Silicon Relat. Elem.* **2000**, *160*, 195.
10. Bhattacharya, A. K.; Stolz, F.; Schmidt, R. R. *Tetrahedron Lett.* **2001**, *42*, 5393.
11. Nifantiev, E. E.; Khrebtova, S. B.; Kulikova, Y. V.; Predvoditelev, D. A.; Kukhareva, T. S.; Petrovskii, P. V.; Rose, M.; Meier, C. *Phosphorus, Sulfur Silicon Relat. Elem.* **2002**, *177*, 251.
12. Battaggia, S.; Vyle, J. S. *Tetrahedron Lett.* **2003**, *44*, 861.

Midland reduction

Asymmetric reduction of ketones using Alpine-borane®.
Alpine-borane® = B-isopinocampheyl-9-borabicyclo[3.3.1]nonane.

(1R)-(+)-α-pinene   9-BBN   (R)-Alpine-borane

9-BBN = 9-borabicyclo[3.3.1]nonane

References

1. Midland, M. M.; Tramontano, A.; Zederic, S. A. *J. Am. Chem. Soc.* **1979**, *101*, 2352.
2. Midland, M. M.; McDowell, D. C.; Hatch, R. L.; Tramontano, A. *J. Am. Chem. Soc.* **1980**, *102*, 867.
3. Brown, H. C.; Pai, G. G.; Jadhav, P. K. *J. Am. Chem. Soc.* **1984**, *106*, 1531.
4. Brown, H. C.; Pai, G. G. *J. Org. Chem.* **1982**, *47*, 1606.
5. Singh, V. K. *Synthesis* **1992**, 605.
6. Williams, D. R.; Fromhold, M. G.; Earley, J. D. *Org. Lett.* **2001**, *3*, 2721.

Miller–Loudon–Snyder nitrile synthesis

Conversion of aldehydes to nitriles using hydroxyamine and *p*-nitrobenzonitrile sequentially.

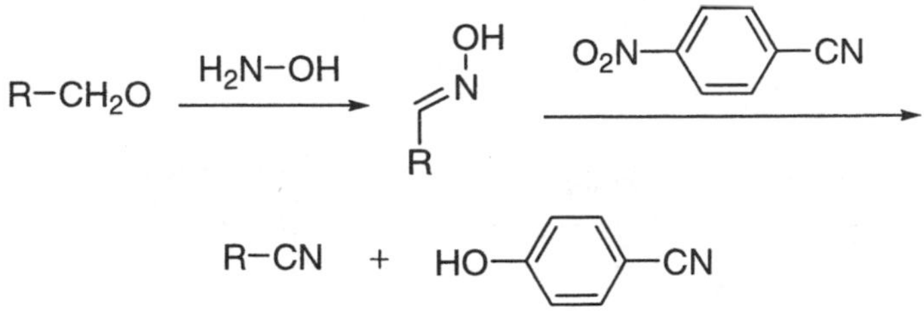

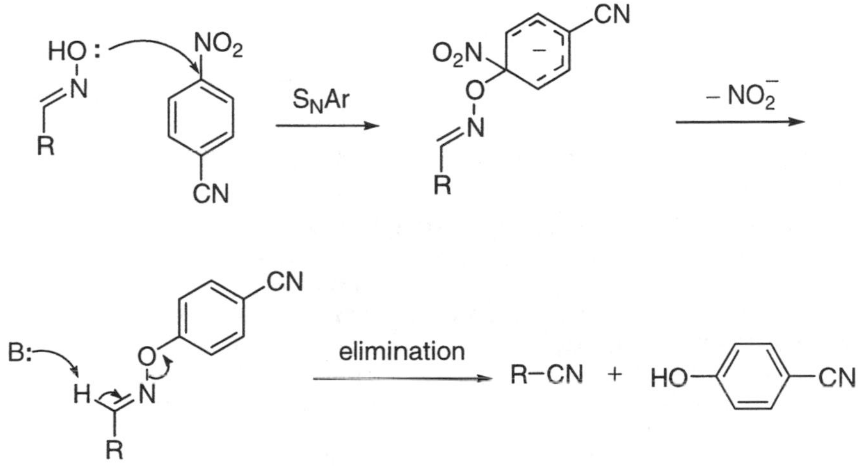

References

1. Snyder, M. R. *J. Org. Chem.* **1974**, *39*, 3343.
2. Miller, M. J.; Loudon, G. M. *J. Org. Chem.* **1975**, *40*, 126.
3. Snyder, M. R. *J. Org. Chem.* **1975**, *40*, 2879.
4. Kumar, H. M. S.; Reddy, B. V. S.; Reddy, P. T.; Yadav, J. S. *Synthesis* **1999**, 586.
5. Chakraborti, A. K.; Kaur, G. *Tetrahedron* **1999**, *55*, 13265.
6. Paraskar, A. S.; Jagtap, H. S.; Sudalai, A. *J. Chem. Res. (S)*, **2000**, 30.
7. Das, B.; Ramesh, C.; Madhusudhan, P. *Synlett* **2000**, 1599.
8. Srinivas, K. V. N. S.; Reddy, E. B.; Das, B. *Synlett* **2002**, 625.

Mislow–Evans rearrangement

[2,3]-Sigmatropic rearrangement of allylic sulfide.

References

1. Tang, R.; Mislow, K. *J. Am. Chem. Soc.* **1970**, *92*, 2100.
2. Evans, D. A.; Andrews, G. C.; Sims, C. L. *J. Am. Chem. Soc.* **1971**, *93*, 4956.
3. Evans, D. A.; Andrews, G. C. *J. Am. Chem. Soc.* **1972**, *94*, 3672.
4. Evans, D. A.; Andrews, G. C. *Acc. Chem. Res.* **1974**, *7*, 147. (Review).
5. Masaki, Y.; Sakuma, K.; Kaji, K. *Chem. Pharm. Bull.* **1985**, *33*, 2531.
6. Jones-Hertzog, D. K.; Jorgensen, W. L. *J. Am. Chem. Soc.* **1995**, *117*, 9077.
7. Jones-Hertzog, D. K.; Jorgensen, W. L. *J. Org. Chem.* **1995**, *60*, 6682.
8. Mapp, A. K.; Heathcock, C. H. *J. Org. Chem.* **1999**, *64*, 23.
9. Zhou, Z. S.; Flohr, A.; Hilvert, D. *J. Org. Chem.* **1999**, *64*, 8334.
10. Shinada, T.; Fuji, T.; Ohtani, Y.; Yoshida, Y.; Ohfune, Y. *Synlett* **2002**, 1341.

Mitsunobu reaction

S_N2 inversion of an alcohol by a nucleophile using diethyl azodicarboxylate (DEAD) and triphenylphosphine.

Diethyl azodicarboxylate (DEAD)

References

1. Mitsunobu, O.; Yamada, M. *Bull. Chem. Soc. Jpn.* **1967**, *40*, 2380.
2. Mitsunobu, O. *Synthesis* **1981**, 1. (Review).
3. Hughes, D. L. *Org. React.* **1992**, *42*, 335–656. (Review).
4. Hughes, D. L. *Org. Prep. Proc. Int.* **1996**, *28*, 127. (Review).
5. Barrett, A. G. M.; Roberts, R. S.; Schroeder, J. *Org. Lett.* **2000**, *2*, 2999.
6. Racero, J. C.; Macias-Sanchez, A. J.; *et al. J. Org. Chem.* **2000**, *65*, 7786.
7. Langlois, N.; Calvez, O. *Tetrahedron Lett.* **2000**, *41*, 8285.
8. Charette, A. B.; Janes, M. K.; Boezio, A. A. *J. Org. Chem.* **2001**, *66*, 2178.
9. Ahn, C.; Correia, R.; DeShong, P. *J. Org. Chem.* **2002**, *67*, 1751.
10. Dandapani, S.; Curran, D. P. *Tetrahedron* **2002**, *58*, 3855.
11. Bitter, I.; Csokai, V. *Tetrahedron Lett.* **2003**, *44*, 2261.

Miyaura boration reaction

Palladium-catalyzed reaction of aryl halides with diboron reagent to produce aryl-boronates.

References

1. Ishiyama, T.; Murata, M.; Miyaura, N. *J. Org. Chem.* **1995**, *60*, 7508.
2. Miyaura, N.; Suzuki, A. *Chem. Rev.* **1995**, *95*, 2457. (Review).
3. Suzuki, A. *J. Organomet. Chem.* **1995**, *576*, 147. (Review).
4. Carbonnelle, A.-C.; Zhu, J. *Org. Lett.* **2000**, *2*, 3477.
5. Willis, D. M.; Strongin, R. M. *Tetrahedron Lett.* **2000**, *41*, 8683.
6. Takahashi, K.; Takagi, J.; Ishiyama, T.; Miyaura, N. *Chem. Lett.* **2000**, 126.
7. Todd, M. H.; Abell, C. *J. Comb. Chem.* **2001**, *3*, 319.
8. Giroux, A. *Tetrahedron Lett.* **2003**, *44*, 233.

Moffatt oxidation

Oxidation of alcohols using DCC and DMSO, aka "Pfitzner-Moffatt oxidation".

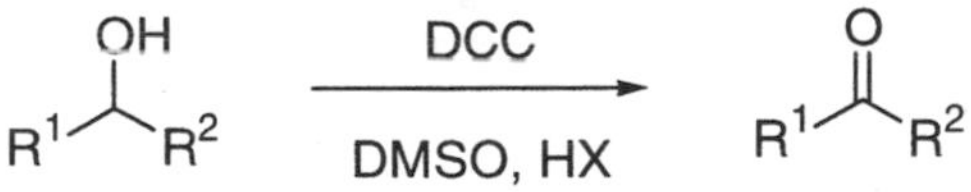

DCC, 1,3-dicyclohexylcarbodiimide

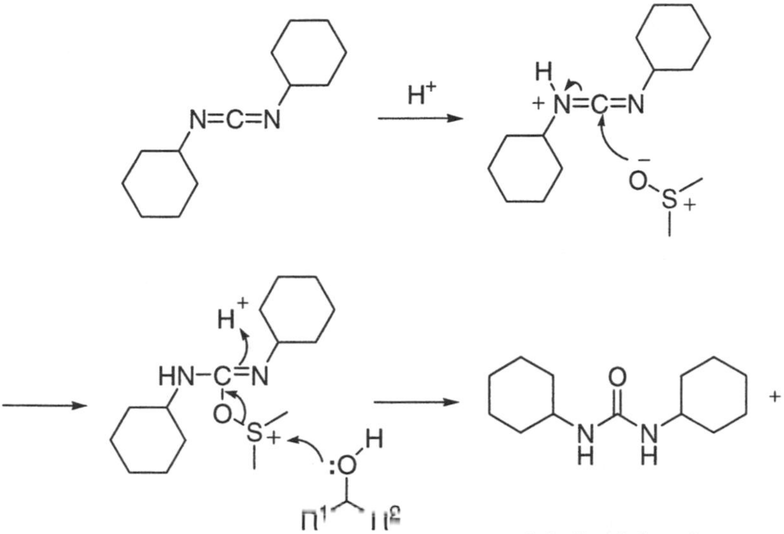

1,3-dicyclohexylurea

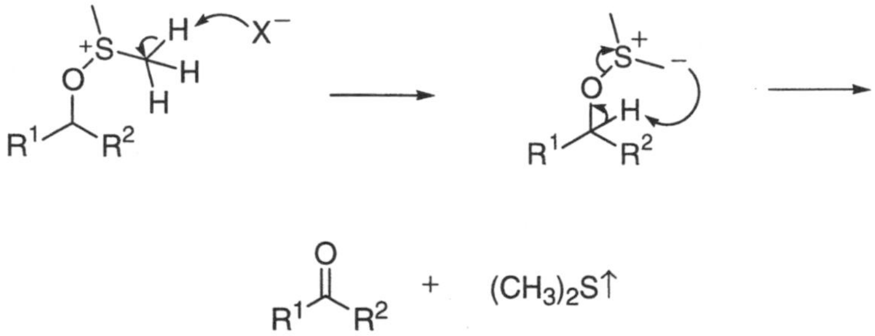

$$R^1 \overset{O}{\underset{}{\mathord{\parallel}}} R^2 \quad + \quad (CH_3)_2S\uparrow$$

References

1. Pfitzner, K. E.; Moffatt, J. G. *J. Am. Chem. Soc.* **1963**, *85*, 3027.
2. Schobert, R. *Synthesis* **1987**, 741.
3. Liu, H. J.; Nyangulu, J. M. *Tetrahedron Lett.* **1988**, *29*, 3167.
4. Tidwell, T. T. *Org. React.* **1990**, *39*, 297. (Review).

5.   Gordon, J. F.; Hanson, J. R.; Jarvis, A. G.; Ratcliffe, A. H. *J. Chem. Soc., Perkin Trans. 1*, **1992**, 3019.

6.   Krysan, D. J.; Haight, A. R.; Lallaman, J. E.; *et al. Org. Prep. Proced. Int.* **1993**, *25*, 437.

7.   Wnuk, S. F.; Ro, B.-O.; Valdez, C. A.; *et al. J. Med. Chem.* **2002**, *45*, 2651.

Morgan–Walls reaction (Pictet–Hubert reaction)

Morgan–Walls reaction

Pictet–Hubert reaction

References

1. Pictet, A.; Hubert, A. *Ber. Dtsch. Chem. Ges.* **1896**, *29*, 1182.
2. Morgan, C. T.; Walls, L. P. *J. Chem. Soc.* **1931**, 2447.
3. Morgan, C. T.; Walls, L. P. *J. Chem. Soc.* **1932**, 2225.
4. Gilman, H.; Eisch, J. *J. Am. Chem. Soc.* **1957**, *79*, 4423.
5. Hollingsworth, B. L.; Petrow, V. *J. Chem. Soc.* **1961**, 3664.
6. Nagarajan, K.; Shah, R. K. *Indian J. Chem.* **1972**, *10*, 450.
7. Sivasubramanian, S.; Muthusubramanian, S.; Ramasamy, S.; Arumugam, N. *Indian J. Chem., Sect. B* **1981**, *20B*, 552.
8. Atwell, G. J.; Baguley, B. C.; Denny, W. A. *J. Med. Chem.* **1988**, *31*, 774.
9. Peytou, V.; Condom, R.; Patino, N.; Guedj, R.; Aubertin, A.-M.; Gelus, N.; Bailly, C.; Terreux, R.; Cabrol-Bass, D. *J. Med. Chem.* **1999**, *42*, 4042.

Mori–Ban indole synthesis

Intramolecular Heck reaction of *o*-halo-aniline with pendant olefin to prepare indole.

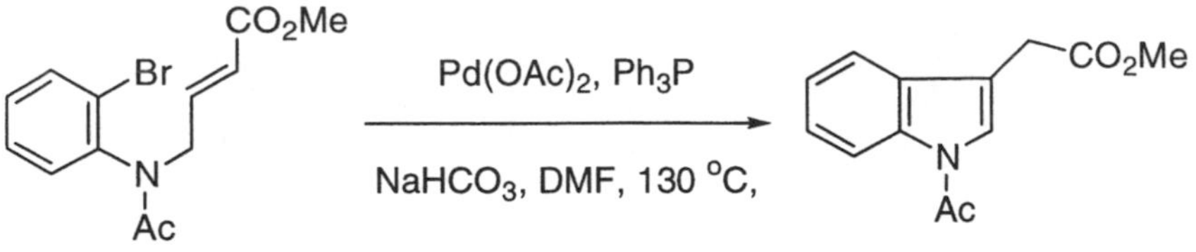

Reduction of $Pd(OAc)_2$ to $Pd(0)$ using Ph_3P:

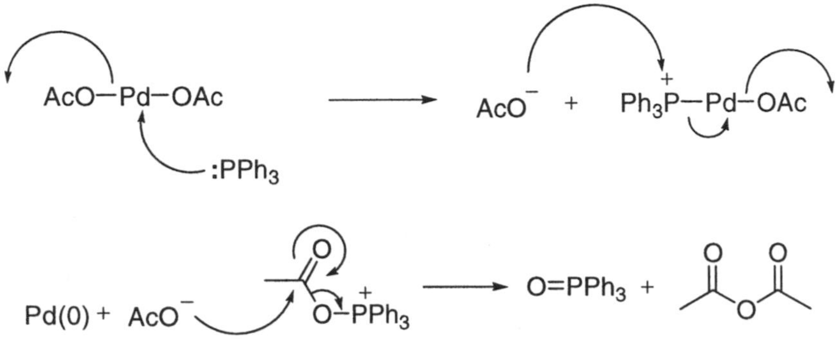

Mori–Ban indole synthesis:

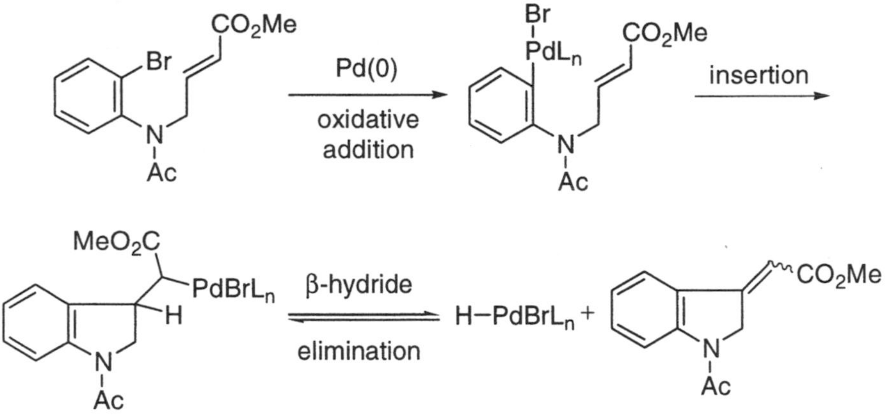

Regeneration of Pd(0):

$$H-PdBrL_n + NaHCO_3 \longrightarrow Pd(0) + NaBr + H_2O + CO_2\uparrow$$

References

1. Mori–Ban indole synthesis, (a) Mori, M.; Chiba, K.; Ban, Y. *Tetrahedron Lett.* **1977**, *12*, 1037; (b) Ban, Y.; Wakamatsu, T.; Mori, M. *Heterocycles* **1977**, *6*, 1711.
2. Reduction of Pd(OAc)$_2$ to Pd(0), (a) Amatore C.; Carre, E.; Jutand, A.; M'Barki, M. A.; Meyer, G. *Organometallics* **1995**, *14*, 5605; (b) Amatore C.; Carre, E.; M'Barki, M. A. *Organometallics* **1995**, *14*, 1818; (c) Amatore C.; Jutand, A.; M'Barki, M. A. *Organometallics* **1992**, *11*, 3009; (d) Amatore C.; Azzabi, M; Jutand, A. *J. Am. Chem. Soc.* **1991**, *113*, 8375.
3. Li, J. J. *J. Org. Chem.* **1999**, *64*, 8425.
4. Gelpke, A. E. S.; Veerman, J. J. N.; Goedheijt, M. S.; Kamer, P. C. J.; Van Leuwen, P. W. N. M.; Hiemstra, H. *Tetrahedron* **1999**, *55*, 6657.
5. Sparks, S. M.; Shea, K. J. *Tetrahedron Lett.* **2000**, *41*, 6721.
6. Bosch, J.; Roca, T.; Armengol, M.; Fernandez-Forner, D. *Tetrahedron* **2001**, *57*, 1041.

272

Morin rearrangement

Acid-catalyzed conversion of penicillin sulfoxides to cephalosporins.   The rearrangement seems to be general for a variety of other heterocyclic sulfoxides as well.

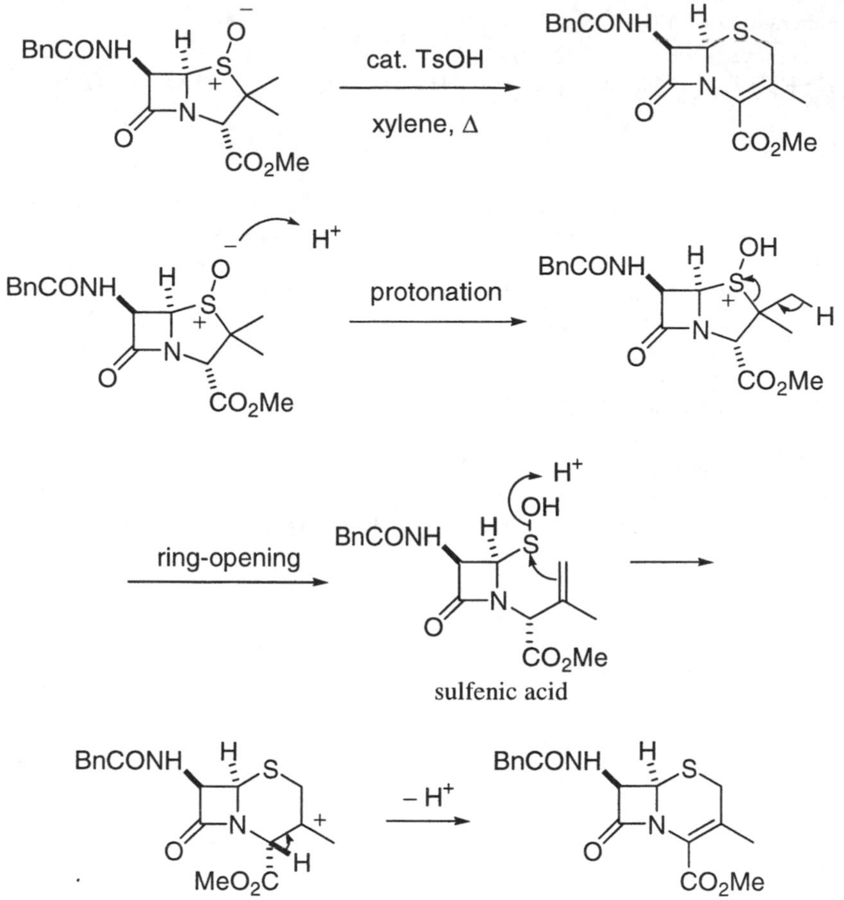

sulfenic acid

References

1.   Morin, R. B.; Jackson, B. G.; Mueller, R. A.; Lavagnino, E. R.; Scanlon, W. B.; Andrews, S. L. *J. Am. Chem. Soc.* **1963**, *85*, 1896.
2.   Morin, R. B.; Jackson, B. G.; Mueller, R. A.; Lavagnino, E. R.; Scanlon, W. B.; Andrews, S. L. *J. Am. Chem. Soc.* **1969**, *91*, 1401.
3.   Morin, R. B.; Spry, D. O. *J. Chem. Soc., Chem. Commun.* **1970**, 335.
4.   Gottstein, W. J.; Misco, P. F.; Cheney, L. C. *J. Org. Chem.* **1972**, *37*, 2765.

5. Chen, C. H. *Tetrahedron Lett.* **1976**, 17, 25.
6. Mah, H.; Nam, K. D.; Hahn, H.-G. *J. Heterocycl. Chem.* **1989**, *26,* 1447.
7. Farina, V.; Kant, J. *Synlett* **1994**, 565.
8. Hart, D. J.; Magomedov, N. A. *J. Org. Chem.* **1999**, *64*, 2990.
9. Freed, J. D.; Hart, D. J.; Magomedov, N. A. *J. Org. Chem.* **2001**, *66*, 839.

Mukaiyama aldol reaction

Lewis acid-catalyzed aldol condensation of aldehyde and silyl enol ether.

$$R-CHO + R^1\underset{OSiMe_3}{\overset{R^2}{\diagdown}} \xrightarrow[\text{acid}]{\text{Lewis}} R\overset{OH}{\diagdown}\overset{O}{\diagdown}R^2$$

Mukaiyama Michael addition

Lewis acid-catalyzed Michael addition of silyl enol ether to α,β-unsaturated system.

References

1. Mukaiyama, T.; Narasaka, K.; Banno, K. *Chem. Lett.* **1973**, 1011.
2. Mukaiyama, T.; Narasaka, K.; Banno, K. *J. Am. Chem. Soc.* **1974**, *96*, 7503.
3. Langer, P.; Koehler, V. *Org. Lett.* **2000**, *2*, 1597.
4. Matsukawa, S.; Okano, N.; Imamoto, T. *Tetrahedron Lett.* **2000**, *41*, 103.
5. Delas, C.; Blacque, O.; Moise, C. *Tetrahedron Lett.* **2000**, *41*, 8269.
6. Ishihara, K.; Kondo, S.; Yamamoto, H. *J. Org. Chem.* **2000**, *65*, 9125.
7. Kumareswaran, R.; Reddy, B. G.; Vankar, Y. D. *Tetrahedron Lett.* **2001**, *42*, 7493.
8. Armstrong, A.; Critchley, T. J.; Gourdel-Martin, M.-E.; Kelsey, R. D.; Mortlock, A. A. *J. Chem. Soc., Perkin Trans. 1* **2002**, 1344.
9. Clézio, I. L.; Escudier, J.-M.; Vigroux, A. *Org. Lett.* **2003**, *5*, 161.
10. Muñoz-Muñiz, O.; Quintanar-Audelo, M.; Juaristi, E. *J. Org. Chem.* **2003**, *68*, 1622.

Mukaiyama esterification

Esterification using Mukaiyama reagent such as 2-chloro-1-methyl-pyridinium iodide (Mukaiyama reagent).

General scheme:

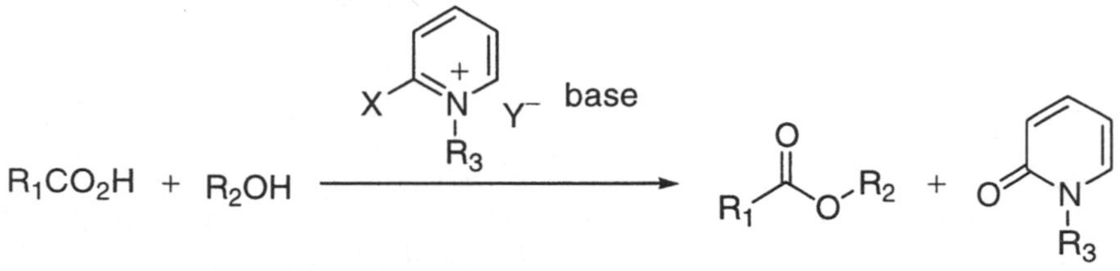

$$R_1CO_2H \ + \ R_2OH \ \xrightarrow{\text{base}} \ R_1\text{-CO-O-}R_2 \ + \ \text{pyridinone}$$

X = F, Cl, Br

e.g.

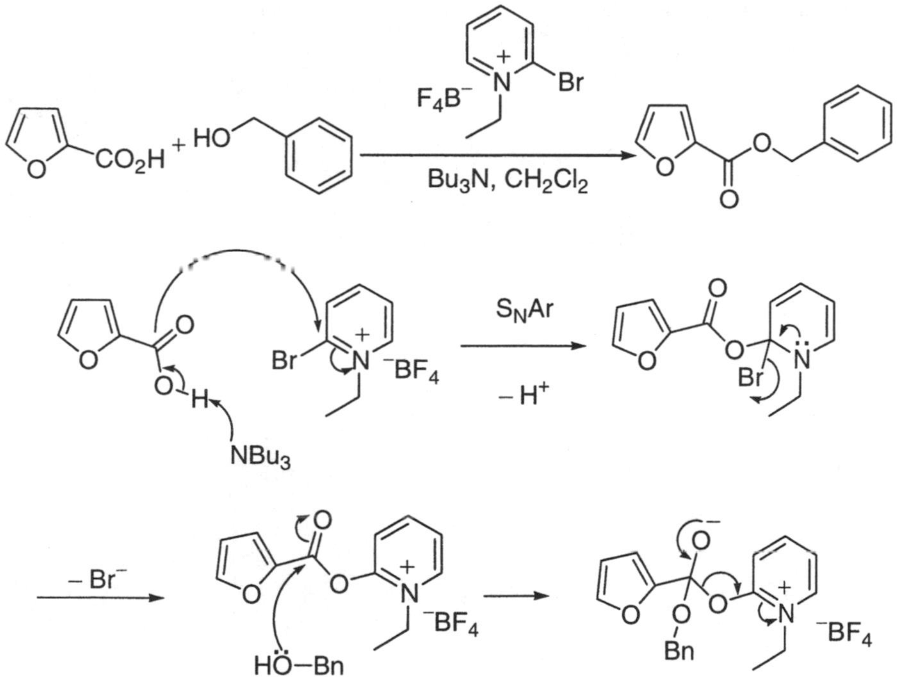

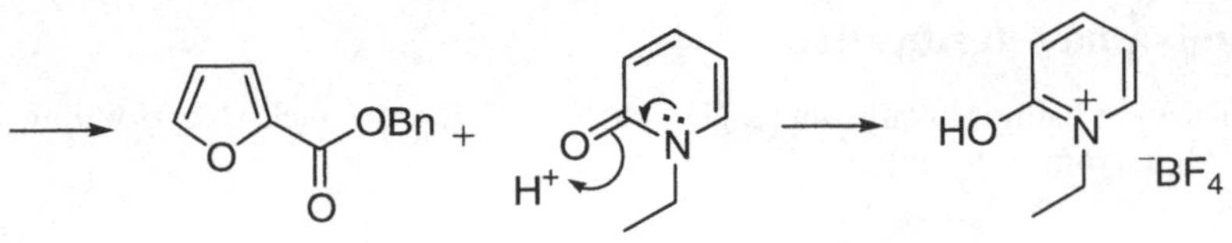

276

Amide formation using the Mukaiyama reagent follows a similar mechanistic pathway [4].

References

1. Mukaiyama, T.; Usui, M.; Shimada, E.; Saigo, K. *Chem. Lett.* **1975**, 1045.
2. Hojo, K.; Kobayashi, S.; Soai, K.; Ikeda, S.; Mukaiyama, T. *Chem. Lett.* **1977**, 635.
3. Mukaiyama, T. *Angew. Chem., Int. Ed. Engl.* **1979**, *18*, 707.
4. For amide formation, see: Huang, H.; Iwasawa, N.; Mukaiyama, T. *Chem. Lett.* **1984**, 1465.
5. Nicolaou, K. C.; Bunnage, M. E.; Koide, K. *J. Am. Chem. Soc.* **1994**, *116*, 8402.
6. Yong, Y. F.; Kowalski, J. A.; Lipton, M. A. *J. Org. Chem.* **1997**, *62*, 1540.
7. Folmer, J. J.; Acero, C.; Thai, D. L.; Rapoport, H. *J. Org. Chem.* **1998**, *63*, 8170.

Myers–Saito cyclization

Cf. Bergman cyclization and Schmittel cyclization.

allenyl enyne   diradical

References

1. Myers, A. G.; Proteau, P. J.; Handel, T. M. *J. Am. Chem. Soc.* **1988**, *110*, 7212.
2. Saito, K.; Watanabe, T.; Takahashi, K. *Chem. Lett.* **1989**, 2099.
3. Saito, I.; Nagata, R.; Yamanaka, H.; Murahashi, E. *Tetrahedron Lett.* **1990**, *31* 2907.
4. Myers, A. G.; Dragovich, P. S.; Kuo, E. Y. *J. Am. Chem. Soc.* **1992**, *114*, 9369.
5. Schmittel, M.; Strittmatter, M.; Kiau, S. *Tetrahedron Lett.* **1995**, *36*, 4975.
6. Engels, B.; Lennartz, C.; Hanrath, M.; Schmittel, M.; Strittmatter, M. *Angew. Chem., Int. Ed.* **1998**, *37*, 1960.
7. Ferri, F.; Bruckner, R.; Herges, R. *New J. Chem.* **1998**, *22*, 531.
8. Bruckner, R; Suffert, J. *Synlett* **1999**, 657–679. (Review).
9. Kim, C.-S.; Diez, C.; Russell, K. C. *Chem. — Eur. J.* **2000**, *6*, 1555.
10. Cramer, C. J.; Kormos, B. L.; Seierstad, M.; Sherer, E. C.; Winget, P. *Org. Lett.* **2001**, *3*, 1881.

11. Stahl, F.; Moran, D.; Schleyer, P. von R.; Prall, M.; Schreiner, P. R. *J. Org. Chem.* **2002**, *67*, 1453.
12. Musch, P. W; Remenyi, C.; Helten, H.; Engels, B. *J. Am. Chem. Soc.* **2002**, *124*, 1823.

Nametkin rearrangement (Retropinacol rearrangement)

Rearrangement of chlorocamphene involving a methyl migration.

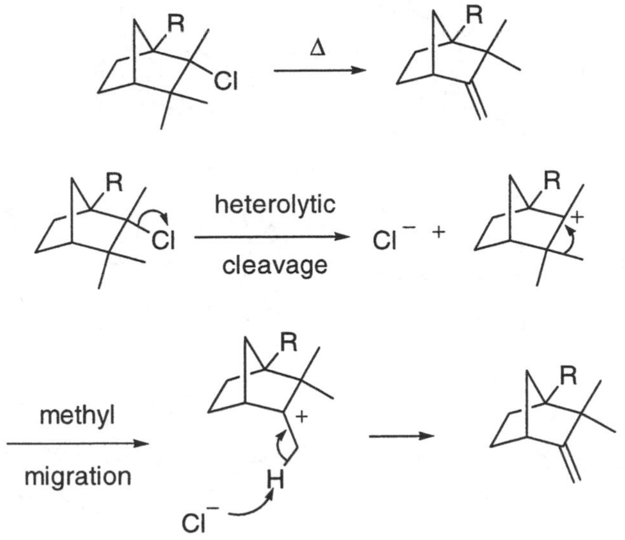

References

1. Nametkin, S. S. *Justus Liebigs Ann. Chem.* **1923**, *432*, 207.
2. Bernstein, D. *Tetrahedron Lett.* **1967**, 2281.
3. Kossanyi, J.; Furth, B.; Morizur, J. P. *Tetrahedron* **1970**, *26*, 395.
4. Moews, P. C.; Knox, J. R.; Vaughan, W. R. *J. Am. Chem. Soc.* **1978**, *100*, 260.
5. Starling, S. M.; Vonwiller, S. C.; Reek, J. N. H. *J. Org. Chem.* **1998**, *63*, 2262.
6. Martinez, A. G.; Vilar, E. T.; Fraile, A. G.; Fernandez, A. H.; De La Moya, C. S. *Tetrahedron* **1998**, *54*, 4607.

Nazarov cyclization

Acid-catalyzed electrocyclic formation of cyclopentenone from divinyl ketone.

References

1. Nazarov, I. N. Torgov, I. B.; Terekhova, L. N. *Bull. Acad. Sci. (USSR)* **1942**, 200.
2. Habermas, K. L.; Denmark, S. E.; Jones, T. K. *Org. React.* **1994**, *45*, 1. (Review).
3. Kuroda, C.; Koshio, H.; Koito, A.; Sumiya, H.; Murase, A.; Hitono, Y. *Tetrahedron* **2000**, *56*, 6441.
4. Giese, S.; Kastrup, L.; Stiens, D.; West, F. G. *Angew. Chem., Int. Ed.* **2000**, *39*, 1970.
5. Kim, S.-H.; Cha, J. K. *Synthesis* **2000**, 2113.
6. Giese, S.; West, F. G. *Tetrahedron* **2000**, *56*, 10221.
7. Fernández M., A.; Martin de la Nava, E. M.; González, R. R. *Tetrahedron* **2001**, *57*, 1049.
8. Harmata, M.; Lee, D. R. *J. Am. Chem. Soc.* **2002**, *124*, 14328.
9. Leclerc, E.; Tius, M. A. *Org. Lett.* **2003**, *5*, 1171.

Neber rearrangement

α-Aminoketone from tosyl ketoxime and base.

$$R^1 \overset{N^{-OTs}}{\underset{R^2}{\bigg|}} \quad \xrightarrow[\text{2. H}_2\text{O}]{\text{1. KOEt}} \quad R^1 \overset{NH_2}{\underset{O}{\bigg|}} R^2 \quad + \quad TsOH$$

ketoxime α-aminoketone

deprotonation cyclization

TsOH +

azirine intermediate

hydrolysis

References

1. Neber, P. W.; v. Friedolsheim, A. *Justus Liebigs Ann. Chem.* **1926**, *449*, 109.
2. O'Brien, C. *Chem. Rev.* **1964**, *64*, 81. (Review).
3. Kakehi, A.; Ito, S.; Manabe, T.; Maeda, T.; Imai, K. *J. Org. Chem.* **1977**, *42*, 2514.
4. Friis, P.; Larsen, P. O.; Olsen, C. E. *J. Chem. Soc., Perkin Trans. 1* **1977**, 661.
5. Corkins, H. G.; Storace, L.; Osgood, E. *J. Org. Chem.* **1980**, *45*, 3156.
6. Parcell, R. F.; Sanchez, J. P. *J. Org. Chem.* **1981**, *46*, 5229.
7. Verstappen, M. M. H.; Ariaans, G. J. A.; Zwanenburg, B. *J. Am. Chem. Soc.* **1996**, *118*, 8491.
8. Mphahlele, M. J. *Phosphorus, Sulfur Silicon Relat. Elem.* **1999**, *144–146*, 351.
9. Banert, K.; Hagedorn, M.; Liedtke, C.; Melzer, A.; Schoffler, C. *Eur. J. Org. Chem.* **2000**, 257.
10. Palacios, F.; Ochoa de Retana, A. M.; Gil, J. I. *Tetrahedron Lett.* **2002**, *41*, 5363.
11. Ooi, T.; Takahashi, M.; Doda, K.; Maruoka, K. *J. Am. Chem. Soc.* **2002**, *124*, 7640.

Nef reaction

Conversion of a primary or secondary nitroalkane into the corresponding carbonyl compound.

nitronate

nitronic acid

References

1. Nef, J. U. *Justus Liebigs Ann. Chem.* **1894**, *280*, 263.
2. Pinnick, H. W. *Org. React.* **1990**, *38*, 655. (Review).
3. Hwu, J. R.; Gilbert, B. A. *J. Am. Chem. Soc.* **1991**, *113*, 5917.
4. Ceccherelli, P.; Curini, M.; Marcotullio, M. C.; Epifano, F.; Rosati, O. *Synth. Commun.* **1998**, *28*, 3057.
5. Adam, W.; Makosza, M.; Saha-Moeller, C. R.; Zhao, C.-G. *Synlett* **1998**, 1335.
6. Shahi, S. P.; Vankar, Y. D. *Synth. Commun.* **1999**, *29*, 4321.
7. Capecchi, T.; de Koning, C. B.; Michael, J. P. *Perkin 1* **2000**, 2681.
8. Ballini, R.; Bosica, G.; Fiorini, D.; Petrini, M. *Tetrahedron Lett.* **2002**, *43*, 5233.
9. Petrus, L.; Petrusova, M.; Pham-Huu, D.-P.; Lattova, E.; Pribulova, B.; Turjan, J. *Monatsh. Chem.* **2002**, *133*, 383.

Negishi cross-coupling reaction

Palladium-catalyzed cross-coupling reaction of organozinc reagents with organic halides, triflates, *etc.* For the catalytic cycle, see the Kumada coupling on page 234.

$$R-X \ + \ R^1-ZnX \ \xrightarrow{\ Pd(0)\ } \ R-R^1 \ + \ ZnX_2$$

$$R-X + L_2Pd(0) \ \xrightarrow[\text{addition}]{\text{oxidative}} \ R-\underset{L}{\overset{L}{Pd}}-X \ \xrightarrow[\substack{\text{transmetallation}\\\text{isomerization}}]{R^1-ZnX}$$

$$ZnX_2 \ + \ R-\underset{R}{\overset{L}{Pd}}-R^1 \ \xrightarrow[\text{elimination}]{\text{reductive}} \ R-R^1 \ + \ L_2Pd(0)$$

References

1. Negishi, E.-I.; Baba, S. *J. Chem. Soc., Chem. Commun.* **1976**, 596.
2. Negishi, E.-I.; *et al. J. Org. Chem.* **1977**, *42*, 1821.
3. Negishi, E.-I. *Acc. Chem. Res.* **1982**, *15*, 340. (Review).
4. Erdik, E. *Tetrahedron* **1992**, *48*, 9577. (Review).
5. Negishi, E.-I.; Liu, F. In *Metal-Catalyzed Cross-Coupling Reactions;* **1998**, Diederich, F.; Stang, P. J. eds.; Wiley–VCH Verlag GmbH: Weinheim, Germany, pp 1–47. (Review).
6. Yus, M.; Gomis, J. *Tetrahedron Lett.* **2001**, *42*, 5721.
7. Lutzen, A.; Hapke, M. *Eur. J. Org. Chem.* **2002**, 2292.
8. Fang, Y.-Q.; Polson, M. I. J.; Hanan, G. S. *Inorg. Chem.* **2003**, *42*, 5.
9. Arvanitis, A. G.; Arnold, C. R.; Fitzgerald, L. W.; Frietze, W. E.; Olson, R. E.; Gilligan, P. J.; Robertson, D. W. *Bioorg. Med. Chem. Lett.* **2003**, *13*, 289.
10. Ma, S.; Ren, H.; Wei, Q. *J. Am. Chem. Soc.* **2003**, *125*, 4817.

Nenitzescu indole synthesis

5-Hydroxylindole from condensation of *p*-benzoquinone and β-aminocrotonate.

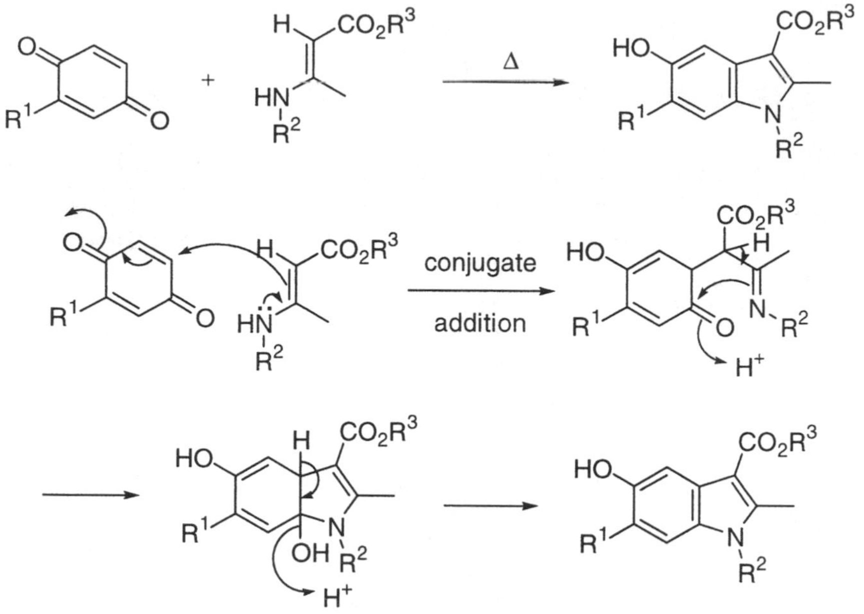

Alternatively:

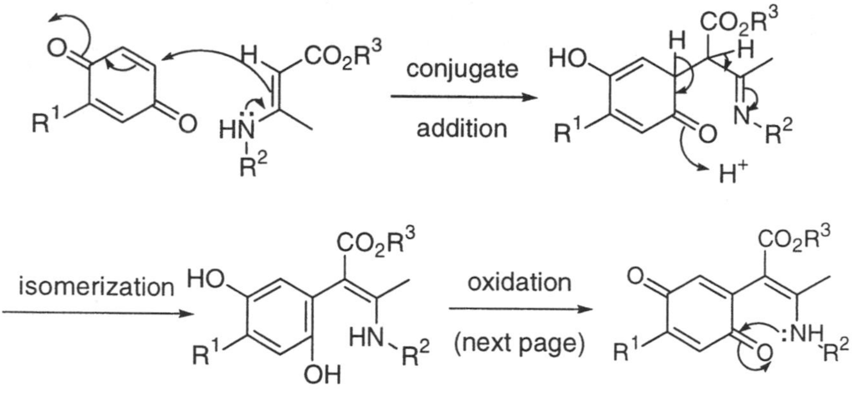

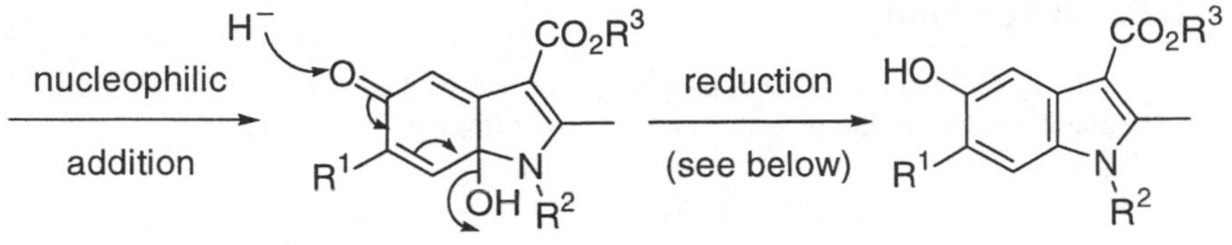

The internal oxidation-reduction process might involve a bimolecular face-to-face electronic transfer complex (in nitromethane) [3]:

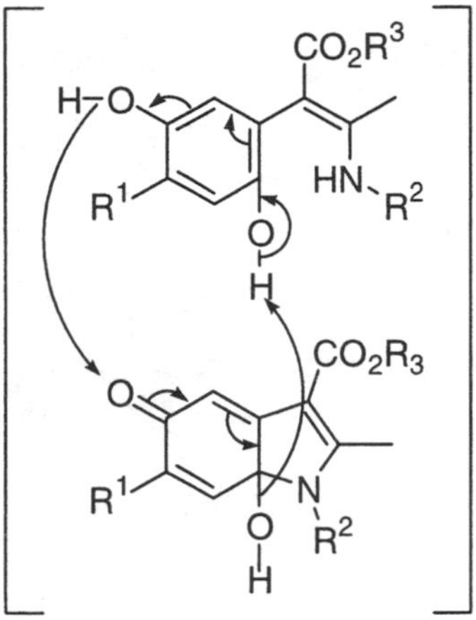

References

1. Nenitzescu, C. D. *Bull. Soc. Chim. Romania* **1929**, *11*, 37.
2. Allen, Jr. G. R. *Org. React.* **1973**, *20*, 337. (Review).
3. Bernier, J. L.; Henichart, J. P. *J. Org. Chem.* **1981**, *46*, 4197.
4. Kinugawa, M.; Arai, H.; Nishikawa, H.; Sakaguchi, A.; Ogasa, T.; Tomioka, S.; Kasai, M. *J. Chem. Soc., Perkin Trans. 1* **1995**, 2677.
5. Mukhanova, T. I.; Panisheva, E. K.; Lyubchanskaya, V. M.; Alekseeva, L. M.; Sheinker, Y. N.; Granik, V. G. *Tetrahedron* **1997**, *53*, 177.
6. Ketcha, D. M.; Wilson, L. J.; Portlock, D. E. *Tetrahedron Lett.* **2000**, *41*, 6253.
7. Brase, S.; Gil, C.; Knepper, K. *Bioorg. Med. Chem. Lett.* **2002**, *10*, 2415.

Nicholas reaction

Hexacarbonyldicobalt-stabilized propargyl cation is captured by a nucleophile. Subsequent oxidative demetallation then gives propargylated product.

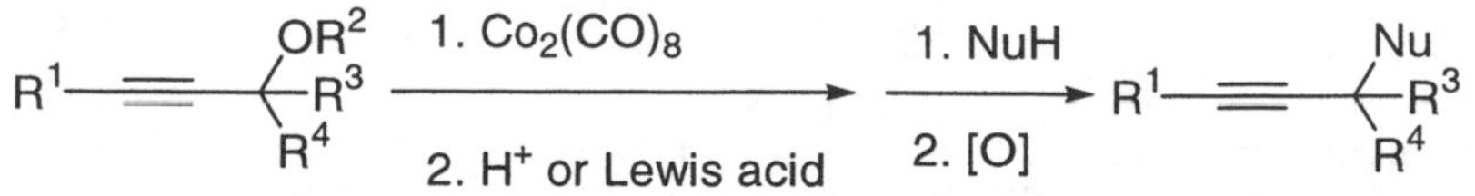

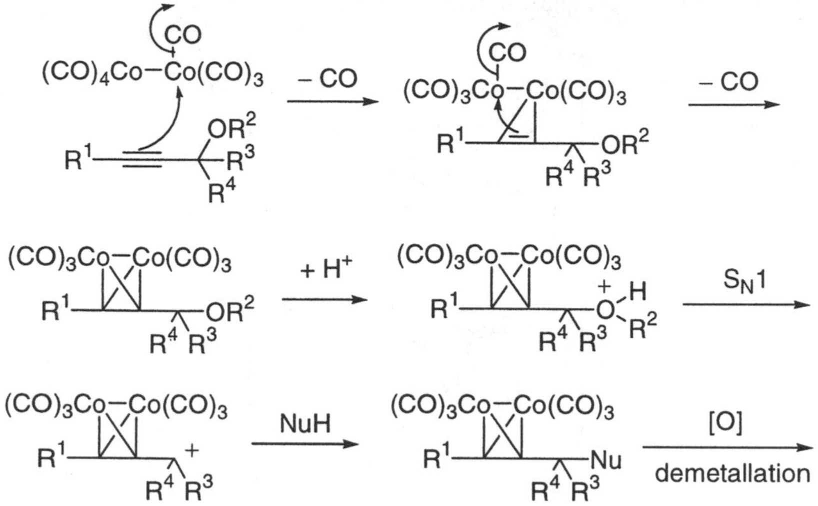

propargyl cation intermediate (stabilized by the hexacarbonyldicobalt complex).

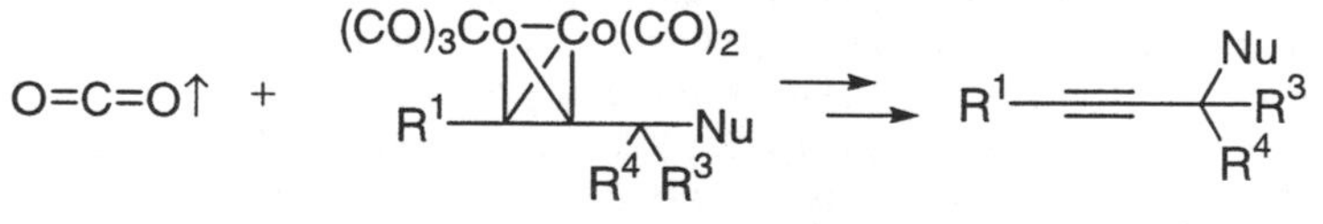

References

1. Nicholas, K. M. *J. Organomet. Chem* **1972**, *C21*, 44.
2. Lockwood, R. F.; Nicholas, K. M. *Tetrahedron Lett.* **1977**, 4163.
3. Nicholas, K. M. *Acc. Chem. Res.* **1987**, *20*, 207. (Review).
4. Roth, K. D. *Synlett* **1992**, 435.
5. Diaz, D.; Martin, V. S. *Tetrahedron Lett.* **2000**, *41*, 743.
6. Guo, R.; Green, J. R. *Synlett* **2000**, 746.
7. Green, J. R. *Curr. Org. Chem.* **2001**, *5*, 809.
8. Teobald, B. J. *Tetrahedron* **2002**, *58*, 4133-4170. (Review).
9. Takase, M.; Morikawa, T.; Abe, H.; Inouye, M. *Org. Lett.* **2003**, *5*, 625.

Noyori asymmetric hydrogenation

Asymmetric reduction of carbonyl *via* hydrogenation catalyzed by ruthenium(II) BINAP complex.

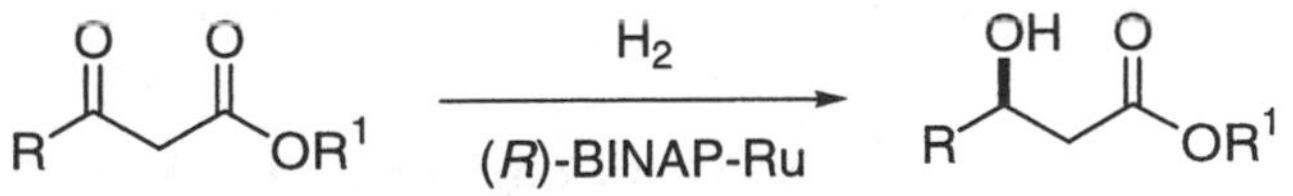

(*R*)-BINAP-Ru =

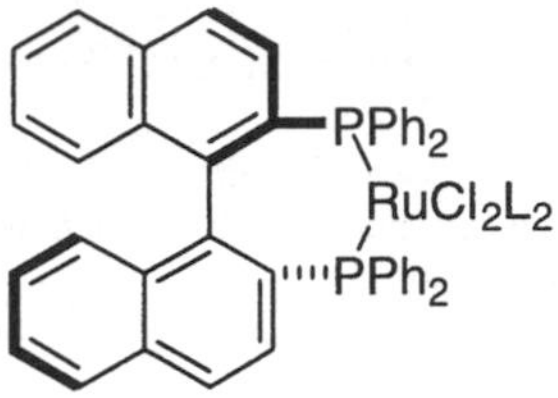

$$[RuCl_2(binap)(solv)_2] \xrightarrow[-HCl]{H_2} [RuHCl(binap)(solv)_2]$$

The catalytic cycle:

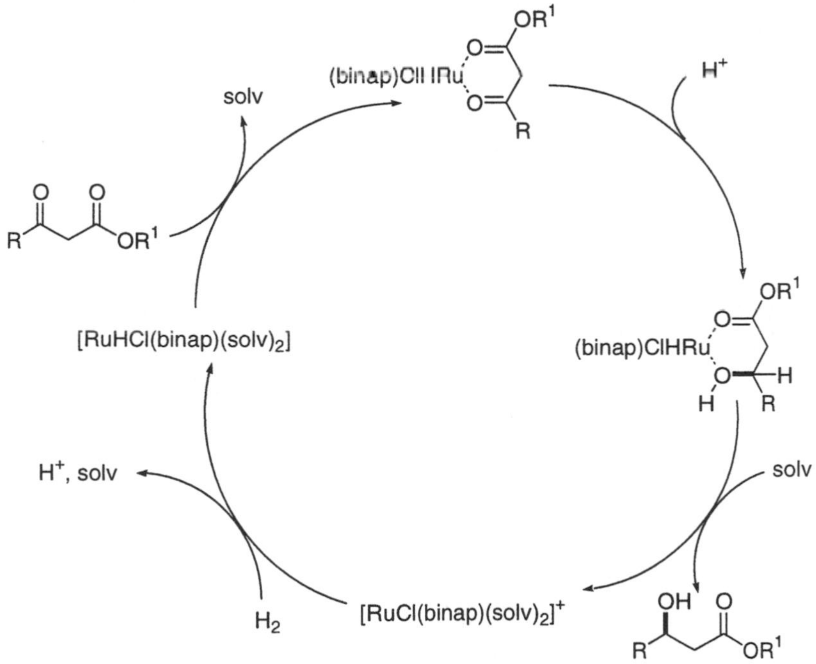

288

References

1. Noyori, R.; *et al. J. Am. Chem. Soc.* **1986**, *108*, 7117.
2. Noyori, R.; Ohkuma, T.; Kitamura, H.; Takaya, H.; Sayo, H.; Kumobayashi, S.; Akutagawa, S. *J. Am. Chem. Soc.* **1987**, *109*, 5856.
3. Case-Green, S. C.; Davies, S. G.; Hedgecock, C. J. R. *Synlett* **1991**, 781.
4. King, S. A.; Thompson, A. S.; King, A. O.; Verhoeven, T. R. *J. Org. Chem.* **1992**, *57*, 6689.
5. Noyori, R. In *Asymmetric Catalysis in Organic Synthesis;* Ojima, I., ed.; Wiley: New York, **1994**, chapter 2. (Review).
6. Chung, J. Y. L.; Zhao, D.; Hughes, D. L.; Mcnamara, J. M.; Grabowski, E. J. J.; Reider, P. J. *Tetrahedron Lett.* **1995**, *36*, 7379.
7. Bayston, D. J.; Travers, C. B.; Polywka, M. E. C. *Tetrahedron: Asymmetry* **1998**, *9*, 2015.
8. Noyori, R.; Ohkuma, T. *Angew. Chem., Int. Ed.* **2001**, *40*, 40.
9. Noyori, R. *Angew. Chem., Int. Ed.* **2002**, *41*, 2008. (Review, Nobel Prize Address).
10. Berkessel, A.; Schubert, T. J. S.; Mueller, T. N. *J. Am. Chem. Soc.* **2002**, *124*, 8693.
11. Fujii, K.; Maki, K.; Kanai, M.; Shibasaki, M. *Org. Lett.* **2003**, *5*, 733.

Nozaki–Hiyama–Kishi reaction

Cr–Ni bimetallic catalyst-promoted redox addition of vinyl halides to aldehydes.

Transmetallation and then reduction by Me_2S

References

1. Jin, H.; Uenishi, J.; Christ, W. J.; Kishi, Y. *J. Am. Chem. Soc.* **1986**, *108*, 5644.
2. Takai, K.; Tagahira, M.; Kuroda, T.; Oshima, K.; Utimoto, K.; Nozaki, H. *J. Am. Chem. Soc.* **1986**, *108*, 6048.
3. Wessjohann, L. A.; Scheid, G. *Synthesis* **1991**, 1. (Review).
4. Kress, M. H.; Ruel, R.; Miller, L. W. H.; Kishi, Y. *Tetrahedron Lett.* **1993**, *34*, 5999.
5. Boeckman, R. K., Jr.; Hudack, R. A., Jr. *J. Org. Chem.* **1998**, *63*, 3524.
6. Fürstner, A. *Chem. Rev.* **1999**, *99*, 991. (Review).
7. Kuroboshi, M.; Tanaka, M.; Kishimoto, S. Goto, K.; Mochizuki, M.; Tanaka, H. *Tetrahedron Lett.* **2000**, *41*, 81.
8. Dai, W.-M.; Wu, A.; Hamaguchi, W. *Tetrahedron Lett.* **2001**, *42*, 4211.
9. Schrekker, H. S.; de Bolster, M. W. G.; Orru, R. V. A.; Wessjohann, L. A. *J. Org. Chem.* **2002**, *67*, 1975.
10. Wan, Z.-K.; Choi, H.-W.; Kang, F.-A.; Nakajima, K.; Demeke, D.; Kishi, Y. *Org. Lett.* **2002**, *4*, 4431.
11. Choi, H.-W.; Nakajima, K.; Demeke, D.; Kang, F.-A.; Wan, Z.-K.; Jun, H.-S.; Kishi, Y. *Org. Lett.* **2002**, *4*, 4435.
12. Berkessel, A.; Menche, D.; Sklorz, C. A.; Schroder, M.; Paterson, I. *Angew. Chem., Int. Ed.* **2003**, *42*, 1032.

290

Oppenauer oxidation

Alkoxide-catalyzed oxidation of secondary alcohols.

cyclic transition state

References

1. Oppenauer, R. V. *Rec. Trav. Chim.* **1937**, *56*, 137.
2. de Graauw, C. F.; Peters, J. A.; van Bekkum, H.; Huskens, J. *Synthesis* **1994**, 1007.
3. Almeida, M. L. S.; Kocovsky, P.; Bäckvall, J.-E. *J. Org. Chem.* **1996**, *61*, 6587.
4. Akamanchi, K. G.; Chaudhari, B. A. *Tetrahedron Lett.* **1997**, *38*, 6925.
5. Raja, T.; Jyothi, T. M.; Sreekumar, K.; Talawar, M. B.; Santhanalakshmi, J.; Rao, B. S. *Bull. Chem. Soc. Jpn.* **1999**, *72*, 2117.
6. Nait Ajjou, A. *Tetrahedron Lett.* **2001**, *42*, 13.
7. Ooi, T.; Otsuka, H.; Miura, T.; Ichikawa, H.; Maruoka, K. *Org. Lett.* **2002**, *4*, 2669.
8. Suzuki, T.; Morita, K.; Tsuchida, M.; Hiroi, K. *J. Org. Chem.* **2003**, *68*, 1601.
9. Auge, J.; Lubin-Germain, N.; Seghrouchni, L. *Tetrahedron Lett.* **2003**, *44*, 819.

Orton rearrangement

Transformation of *N*-chloroanilides to the corresponding *para*-chloroanilides.
Cf. Fischer–Hepp rearrangement.

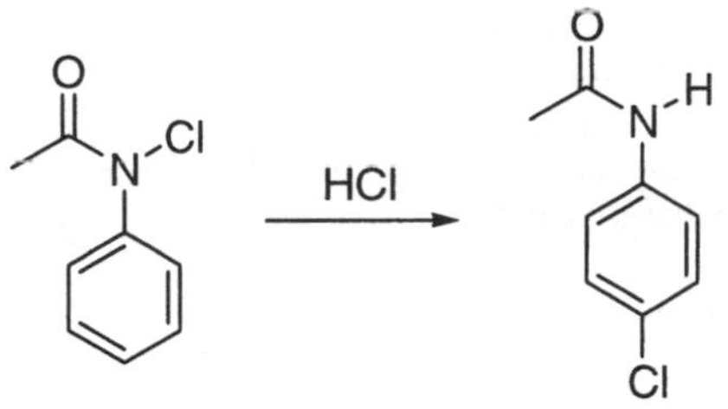

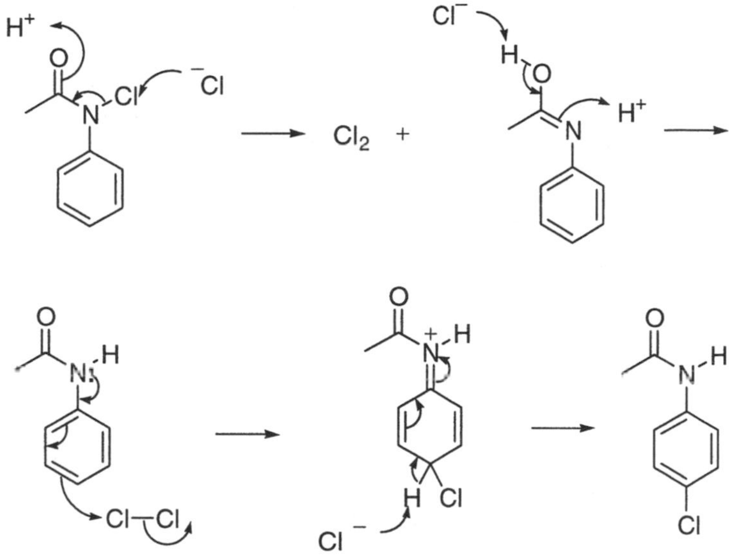

Alternatively:

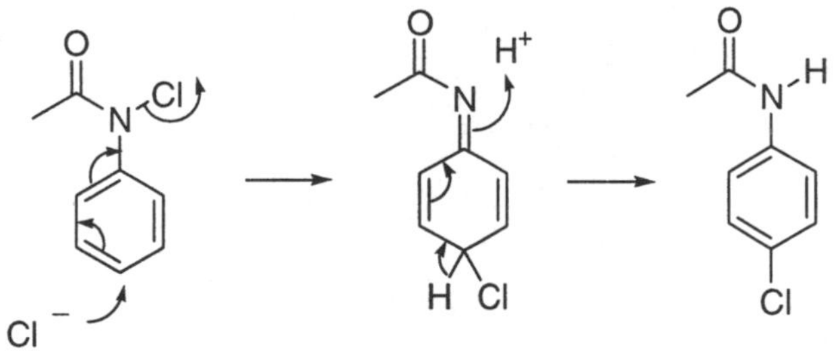

References

1. Verma, S. M.; Srivastava, R. C. *Indian J. Chem.* **1965**, *3*, 266.
2. Scott, J. M. W.; Martin, J. G. *Can. J. Chem.* **1965**, *43*, 732.
3. Scott, J. M. W.; Martin, J. G. *Can. J. Chem.* **1966**, *44*, 2901.
4. Shine, H. J. *Aromatic Rearrangement;* Elsevier: New York, **1967**, *221*, 362. (Review).
5. Golding, P. D.; Reddy, S.; Scott, J. M. W.; White, V. A.; Winter, J. G. *Can. J. Chem.* **1981**, *59*, 839.
6. Yamamoto, J.; Matsumoto, H. *Chem. Express* **1988**, *3*, 419.
7. Kannan, P.; Venkatachalaphathy, C.; Pitchumani, K. *Indian J. Chem., Sect. B* **1999**, *38B*, 384.
8. Ghosh, S.; Baul, S. *Synth. Commun.* **2001**, *31*, 2783.

Overman rearrangement

Stereoselective transformation of allylic alcohol to allylic trichloroacetamide *via* trichloroacetimidate intermediate.

trichloroacetimidate

Δ, [3,3]-sigmatropic rearrangement

References

1. Overman, L. E. *Acc. Chem. Res.* **1971**, *4*, 49. (Review).
2. Overman, L. E. *J. Am. Chem. Soc.* **1974**, *96*, 597.
3. Overman, L. E. *J. Am. Chem. Soc.* **1976**, *98*, 2901.
4. Isobe, M.; Fukuda, Y.; Nishikawa, T.; Chabert, P.; Kawai, T.; Goto, T. *Tetrahedron Lett.* **1990**, *31*, 3327.
5. Eguchi, T.; Koudate, T.; Kakinuma, K. *Tetrahedron* **1993**, *49*, 4527.
6. Toshio, N.; Masanori, A.; Norio, O.; Minoru, I. *J. Org. Chem.* **1998**, *63*, 188.
7. Cho, C.-G.; Lim, Y.-K.; Lee, K.-S.; Jung, I.-H.; Yoon, M.-Y. *Synth. Commun.* **2000**, *30*, 1643.
8. Martin, C.; Prunck, W.; Bortolussi, M.; Bloch, R. *Tetrahedron: Asymmetry* **2000**, *11*, 1585.
9. Demay, S.; Kotschy, A.; Knochel, P. *Synthesis* **2001**, 863.
10. Oishi, T.; Ando, K.; Inomiya, K.; Sato, H.; Iida, M.; Chida, N. *Org. Lett.* **2002**, *4*, 151.
11. Reilly, M.; Anthony, D. R.; Gallagher, C. *Tetrahedron Lett.* **2003**, *44*, 2927.
12. O'Brien, P.; Pilgram, C. D. *Org. Biomol. Chem.* **2003**, *1*, 523.

Paal–Knorr furan synthesis

Acid-catalyzed cyclization of 1,4-ketones to furans.

References

1. Haley, J. F., Jr.; Keehn, P. M. *Tetrahedron Lett.* **1973**, 4017.
2. Amarnath, V.; Amarnath, K. *J. Org. Chem.* **1995**, *60*, 301.
3. Truel, I.; Mohamed-Hachi, A.; About-Jaudet, E.; Collignon, N. *Synth. Commun.* **1997**, *27*, 1165.
4. Friedrichsen, W. In *Comprehensive Heterocyclic Chemistry II;* Katritzky, A. R.; Rees, C. W.; Scrivan, E. F. V. eds.; Pergamon: Oxford, **1996**, *Vol. 2*, p352. (Review).
5. Truel, I.; Mohamed-Hachi, A.; About-Jaudet, E.; Collignon, N. *Synth. Commun.* **1997**, *27*, 1165.
6. Stauffer, F.; Neier, R. *Org. Lett.* **2000**, *2*, 3535.

Paal–Knorr pyrrole synthesis

Reaction between 1,4-ketones and primary amines (or ammonia) to give pyrroles.

References

1. Paal, C. *Ber. Dtsch. Chem. Ges.* **1885**, *18*, 367.
2. Hori, I.; Igarashi, M. *Bull. Chem. Soc. Jpn.* **1971**, *44*, 2856.
3. Chiu, P. K.; Lui, K. H.; Maini, P. N.; Sammes, M. P. *J. Chem. Soc., Chem. Commun.* **1987**, 109.
4. Chiu, P. K.; Sammes, M. P. *Tetrahedron* **1988**, *44*, 3531.
5. Chiu, P. K.; Sammes, M. P. *Tetrahedron* **1990**, *46*, 3439.
6. Yu, S.-X.; Le Quesne, P. W. *Tetrahedron Lett.* **1995**, *36*, 6205.
7. Robertson, J.; Hatley, R. J. D.; Watkin, D. J. *Perkin 1* **2000**, 3389.
8. Braun, R. U.; Zeitler, K.; Mueller, T. J. J. *Org. Lett.* **2001**, *3*, 3297.
9. Gorlitzer, K.; Fabian, J.; Frohberg, P.; Drutkowski, G. *Pharmazie* **2002**, *57*, 243.
10. Quiclet-Sire, B.; Quintero, L.; Sanchez-Jimenez, G.; Zard, Z. *Synlett* **2003**, 75.

Parham cyclization

Annulation of aryl halides with *ortho* side chains with pendant *ortho* electrophilic moiety *via* the treatment with organolithium reagent, involving halogen-metal exchange and subsequent nucleophilic cyclization to form 4 to 7-membered rings.

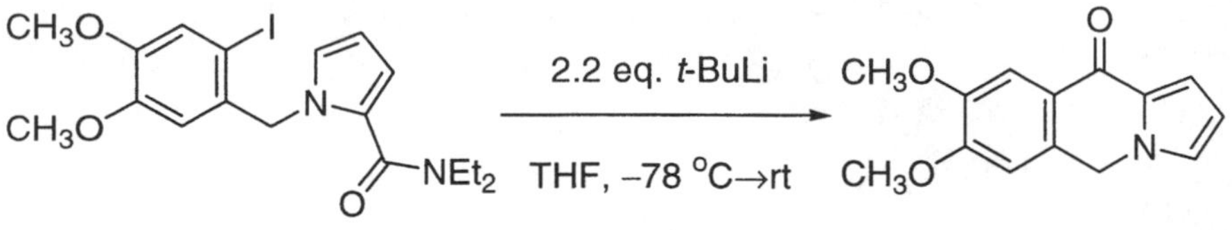

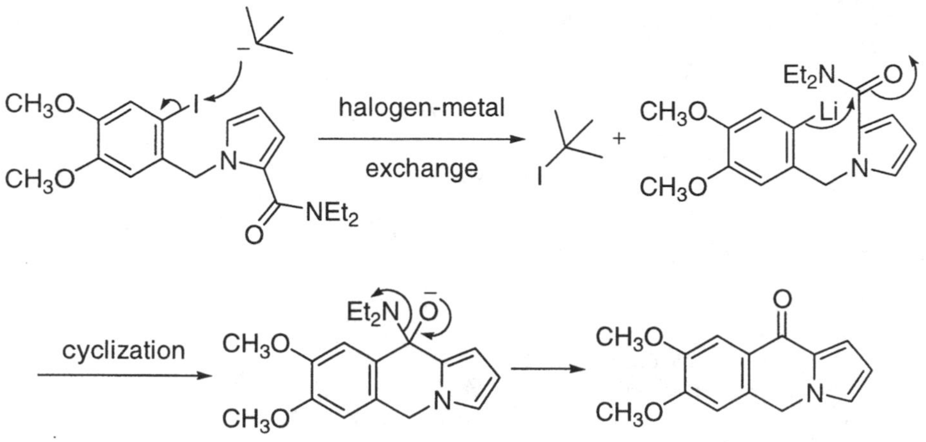

The fate of the second equivalent of *t*-BuLi:

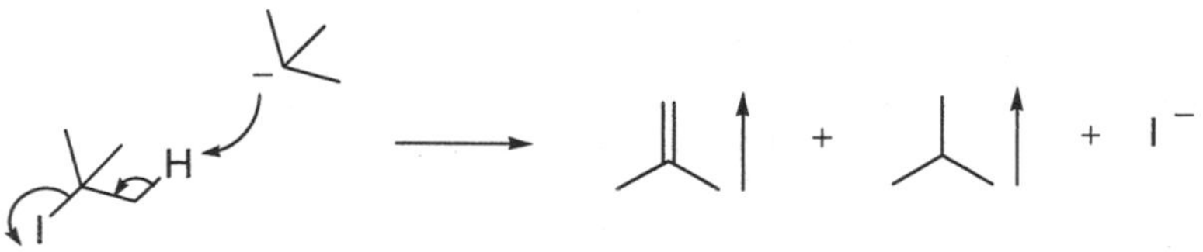

References

1. Parham, W. E.; Jones, L. D. *J. Org. Chem.* **1975**, *40*, 2394.
2. Parham, W. E.; Jones, L. D. *J. Org. Chem.* **1976**, *41*, 1184.
3. Bradsher, C. K.; Hunt, D. A. *Org. Prep. Proced. Int.* **1978**, *10*, 267.
4. Bradsher, C. K.; Hunt, D. A. *J. Org. Chem.* **1981**, *46*, 4608.
5. Parham, W. E.; Bradsher, C. K. *Acc. Chem. Res.* **1982**, *15*, 305. (Review).
6. Quallich, G. J.; Fox, D. E.; Friedmann, R. C.; Murtiashaw, C. W. *J. Org. Chem.* **1992**, *57*, 761.

7. Couture, A.; Deniau, E.; Grandclaudon, P. *J. Chem. Soc., Chem. Commun.* **1994**, 1329.

8. Gray, M.; Tinkl, M.; Snieckus, V. In *Comprehensive Organometallic Chemistry II*; Abel, E. W., Stone, F. G. A., Wilkinson, G., Eds.; Pergamon: Exeter, 1995; Vol. 11; p 66. (Review).

9. Collado, M. I.; Manteca, I.; Sotomayor, N.; Villa, M.-J.; Lete, E. *J. Org. Chem.* **1997**, *62*, 2080.

10. Osante, I.; Collado, M. I.; Lete, E.; Sotomayor, N. *Synlett* **2000**, 101.

11. Ardeo, A.; Lete, E.; Sotomayor, N. *Tetrahedron Lett.* **2000**, *41*, 5211.

12. Osante, I.; Collado, M. I.; Lete, E.; Sotomayor, N. *Eur. J. Org. Chem.* **2001**, 1267.

13. Ardeo, A.; Collado, M. I.; Osante, I.; Ruiz, J.; Sotomayor, N.; Lete, E. In *Targets in Heterocyclic Systems Vol. 5*; Atanassi, O., Spinelli, D., Eds.; Italian Society of Chemistry: Rome, 2001; p 393. (Review).

14. Mealy, M. M.; Bailey, W. F. *J. Organomet. Chem.* **2002**, 649, 59.

15. Sotomayor, N.; Lete, E. *Current Org. Chem.* **2003**, *7*, 275. (Review).

16. Ruiz, J.; Sotomayor, N.; Lete, E. *Org. Lett.* **2003**, *5*, 1115.

Passerini reaction

Three-component condensation (3CC) of carboxylic acids, *C*-isocyanides, and oxo compounds to afford α-acyloxycarboxamides. *Cf.* Ugi reaction.

R^1—N$\overset{+}{\equiv}$C$^-$ + R^2$\overset{O}{\underset{}{\Large\triangle}}$R^3 + R^4—CO$_2$H $\longrightarrow$ R^1-N$\overset{H}{\underset{O}{}}$...R^{2}R^3 O ...O...R$_4$

isocyanide

R^2$\overset{O}{\underset{}{}}$R^3 + R^4—CO$_2$H $\longrightarrow$

$\xrightarrow[\text{transfer}]{\text{acyl}}$

$\longrightarrow$ R^1-N$\overset{H}{\underset{O}{}}$...R^{2}R^3 O ...O...R^4

References

1. Passerini, M. *Gazz. Chim. Ital.* **1921**, *51*, 126, 181.
2. Ferosie, I. *Aldrichimica Acta* **1971**, *4*, 21.
3. Ugi, I.; Lohberger, S.; Karl, R. In *Comprehensive Organic Synthesis*; Trost, B. M.; Fleming, I., Eds.; Pergamon: Oxford, **1991**, *Vol. 2*, p.1083. (Review).
4. Ziegler, T.; Kaisers, H.-J.; Schlomer, R.; Koch, C. *Tetrahedron* **1999**, *55*, 8397.
5. Banfi, L.; Guanti, G.; Riva, R. *Chem. Commun.* **2000**, 985.
6. Semple, J. E.; Owens, T. D.; Nguyen, K.; Levy, O. E. *Org. Lett.* **2000**, *2*, 2769.
7. Owens, T. D.; Semple, J. E. *Org. Lett.* **2001**, *3*, 3301.
8. Xia, Q.; Ganem, B. *Org. Lett.* **2002**, *4*, 1631.
9. Basso, A.; Banfi, L.; Riva, R.; Piaggio, P.; Guanti, G. *Tetrahedron Lett.* **2003**, *44*, 2367.

Paterno–Büchi reaction

Photo-induced oxetane formation from a ketone and an olefin.

oxetane

n, π* triplet

triplet diradical   singlet diradical

References

1. Paterno, E.; Chieffi, G. *Gazz. Chim. Ital.* **1909**, *39*, 341.
2. Büchi, G.; Inman, C. G.; Lipinsky, E. S. *J. Am. Chem. Soc.* **1954**, *76*, 4327.
3. Porco, J. A., Jr.; Schreiber, S. L. In *Comprehensive Organic Synthesis* Trost, B. M.; Fleming, I., Eds.; Pergamon: Oxford, **1991**, *Vol. 5*, 151–192.
4. Fleming, S. A.; Gao, J. J. *Tetrahedron Lett.* **1997**, *38*, 5407.
5. Hubig, S. M.; Sun, D.; Kochi, J. K. *J. Chem. Soc., Perkin Trans. 2* **1999**, 781.
6. D'Auria, M.; Racioppi, R.; Romaniello, G. *Eur. J. Org. Chem.* **2000**, 3265.
7. Bach, T.; Brummerhop, H.; Harms, K. *Chem.--Eur. J.* **2000**, *6*, 3838.
8. Bach, T. *Synlett* **2000**, 1699.
9. Abe, M.; Tachibana, K.; Fujimoto, K.; Nojima, M. *Synthesis* **2001**, 1243.
10. D'Auria, M.; Emanuele, L.; Poggi, G.; Racioppi, R.; Romaniello, G. *Tetrahedron* **2002**, *58*, 5045.
11. Griesbeck, A. G. *Synlett* **2003**, 451.

Pauson–Khand cyclopentenone synthesis

Formal [2 + 2 +1] cycloaddition of an alkene, alkyne, and carbon monoxide mediated by octacarbonyl dicobalt.

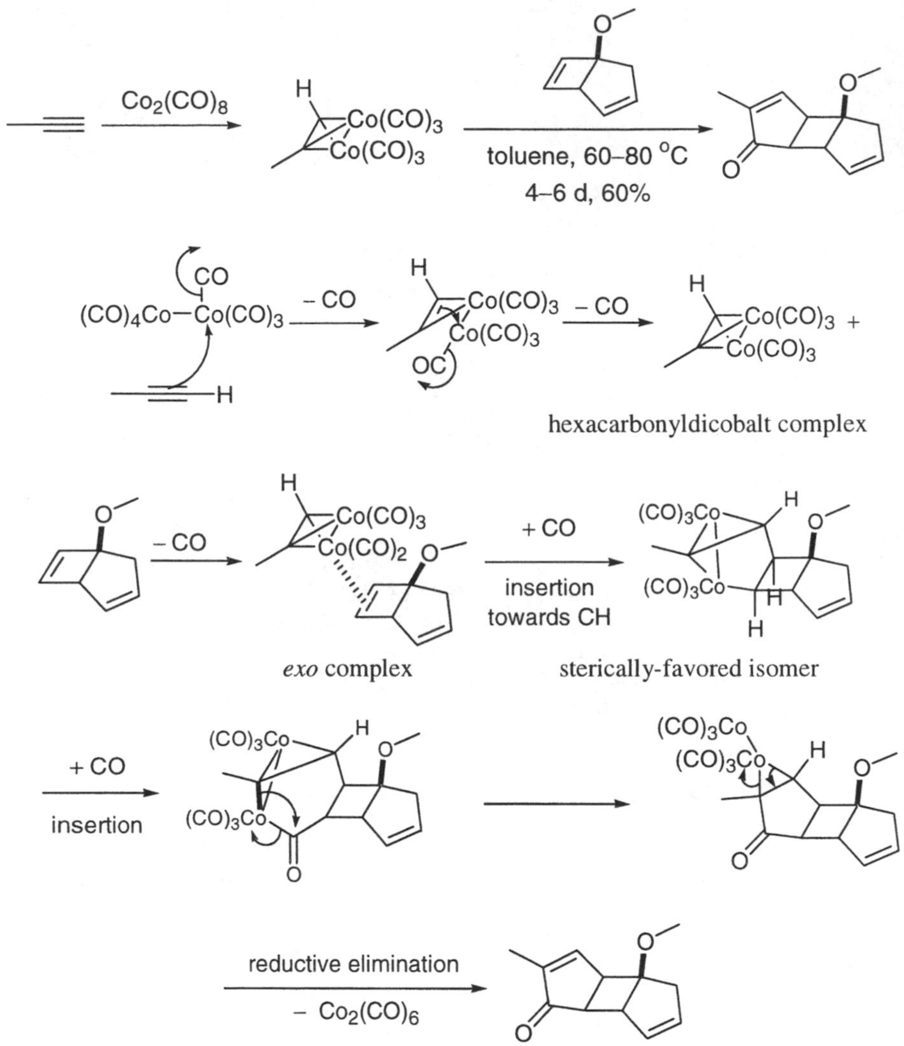

hexacarbonyldicobalt complex

exo complex

sterically-favored isomer

References

1. Bladon, P.; Khand, M. J.; Pauson, P. L. *J. Chem. Res. (M)*, **1977**, 153.
2. Pauson, P. L. *Tetrahedron* **1985**, *41*, 5855.
3. Schore, N. E. *Chem. Rev.* **1988**, *88*, 1081. (Review).

4. Schore, N. E. In *Comprehensive Organic Synthesis*; Paquette, L. A.; Fleming, I.; Trost, B. M., Eds.; Pergamon: Oxford, **1991**, *Vol. 5*, p.1037. (Review).

5. Schore, N. E. *Org. React.* **1991**, *Vol. 40*, pp 1–90. (Review).

6. Brummond, K. M.; Kent, J. L. *Tetrahedron* **2000**, *56*, 3263.

7. Son, S. U.; Lee, S. I.; Chung, Y. K. *Angew. Chem., Int. Ed.* **2000**, *39*, 4158.

8. Kraft, M. E.; Fu, Z.; Boñaga, L. V. R. *Tetrahedron Lett.* **2001**, *42*, 1427.

9. Muto, R.; Ogasawara, K. *Tetrahedron Lett.* **2001**, *42*, 4143.

10. Areces, P.; Durán, M. Á.; Plumet, J.; Hursthouse, M. B.; Light, M. E. *J. Org. Chem.* **2002**, *67*, 3506.

11. Mukai, C.; Nomura, I.; Kitagaki, S. *J. Org. Chem.* **2003**, *68*, 1376.

Payne rearrangement

Base-promoted isomerization of 2,3-epoxy alcohols.

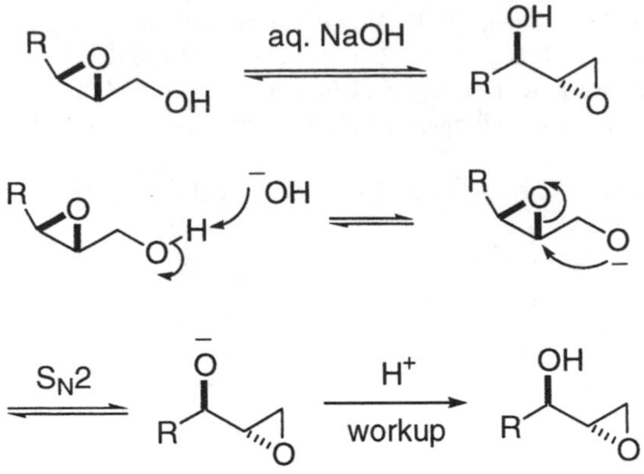

References

1. Payne, G. B. *J. Org. Chem.* **1962**, *27*, 3819.
2. Page, P. C. B.; Rayner, C. M.; Sutherland, I. O. *J. Chem. Soc., Perkin Trans. 1*, **1990**, 1375.
3. Konosu, T.; Miyaoka, T.; Tajima, Y.; Oida, S. *Chem. Pharm. Bull.* **1992**, *40*, 562.
4. Dols, P. P. M. A.; Arnouts, E. G.; Rohaan, J.; Klunder, A. J. H.; Zwanenburg, B. *Tetrahedron* **1994**, *50*, 3473.
5. Ibuka, T. *Chem. Soc. Rev.* **1998**, *27*, 145. (Review).
6. Bickley, J. F.; Gillmore, A. T.; Roberts, S. M.; Skidmore, J.; Steiner, A. *J. Chem. Soc., Perkin Trans. 1* **2001**, 1109.
7. Tamamura, H.; Hori, T.; Otaka, A.; Fujii, N. *J. Chem. Soc., Perkin Trans. 1* **2002**, 577.

Pechmann condensation (coumarin synthesis)

Lewis acid-mediated condensation of phenol with β-ketoester to produce coumarin.

References

1. v. Pechmann, H.; Duisberg, C. *Ber. Dtsch. Chem. Ges.* **1883**, *16*, 2119.
2. Hirata, T.; Suga, T. *Bull. Chem. Soc. Jpn.* **1974**, *47*, 244.
3. Chaudhari, D. D. *Chem. Ind.* **1983**, 568.
4. Holden, M. S.; Crouch, R. D. *J. Chem. Educ.* **1998**, *75*, 1631.
5. Corrie, J. E. T. *J. Chem. Soc., Perkin Trans. 1* **1990**, 2151.
6. Hua, D. H.; Saha, S.; Roche, D.; Maeng, J. C.; Iguchi, S.; Baldwin, C. *J. Org. Chem.* **1992**, *57*, 399.
7. Biswas, G. K.; Basu, K.; Barua, A. K.; Bhattacharyya, P. *Indian J. Chem., Sect. B* **1992**, *31B*, 628.
8. Li, T.-S.; Zhang, Z.-H.; Yang, F.; Fu, C.-G. *J. Chem. Res., (S)* **1998**, 38.
9. Sugino, T.; Tanaka, K. *Chem. Lett.* **2001**, 110.
10. Khandekar, A. C.; Khandekar, B. M. *Synlett.* **2002**, 152.
11. Shockravi, A.; Heravi, M. M.; Valizadeh, H. *Phosphorus, Sulfur Silicon Relat. Elem.* **2003**, *178*, 143.

Pechmann pyrazole synthesis

Pyrazoles from the 1,3-dipolar cycloaddition of diazo compounds and alkynes.

$$H\text{---}\equiv\text{---}H \ + \ H_2C\overset{+}{=}N\overset{-}{=}N \longrightarrow$$

acetylene   diazomethane

Reference

1. v. Pechmann, H.; Duisberg, C. *Ber. Dtsch. Chem. Ges.* **1898**, *31*, 2950.
2. Froissard, J.; Greiner, J.; Pastor, R.; Cambon, A. *J. Fluorine Chem.* **1984**, *26*, 47.
3. Aoyama, T.; Inoue, S.; Shioiri, T. *Tetrahedron Lett.* **1984**, *25*, 433.
4. Asaki, T.; Aoyama, T.; Shioiri, T. *Heterocycles* **1988**, *27*, 343.
5. Farina, F.; Fernandez, P.; Teresa Fraile, M.; Martin, M. V.; Martin, M. R. *Heterocycles* **1989**, *29*, 967.
6. Sauer, D. R.; Schneller, S. W. *J. Org. Chem.* **1990**, *55*, 5535.
7. Yuraev, A. D.; Makhsumov, A. G.; Yuldasheva, K.; Tolipova, M. A. *J. Organomet. Chem.* **1992**, *431*, 129.
8. Sibous, L.; Tipping, A. E. *J. Fluorine Chem.* **1993**, *62*, 39.

Perkin reaction (cinnamic acid synthesis)

Cinnamic acid synthesis from aryl aldehyde and acetic anhydride.

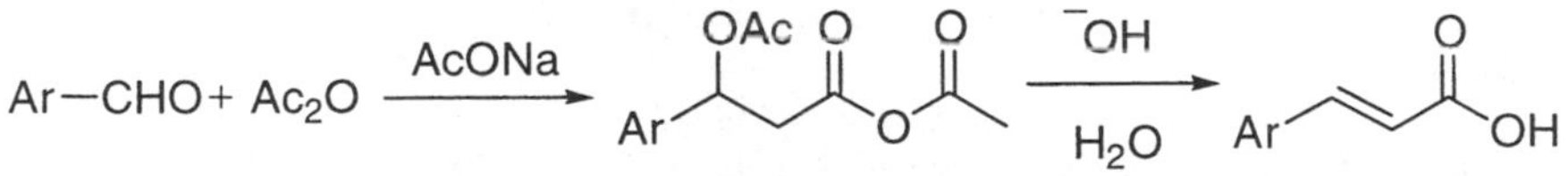
Ar—CHO+ Ac2O
AcONa
OAc
Ar
⁻OH
H2O
OH

cinnamic acid

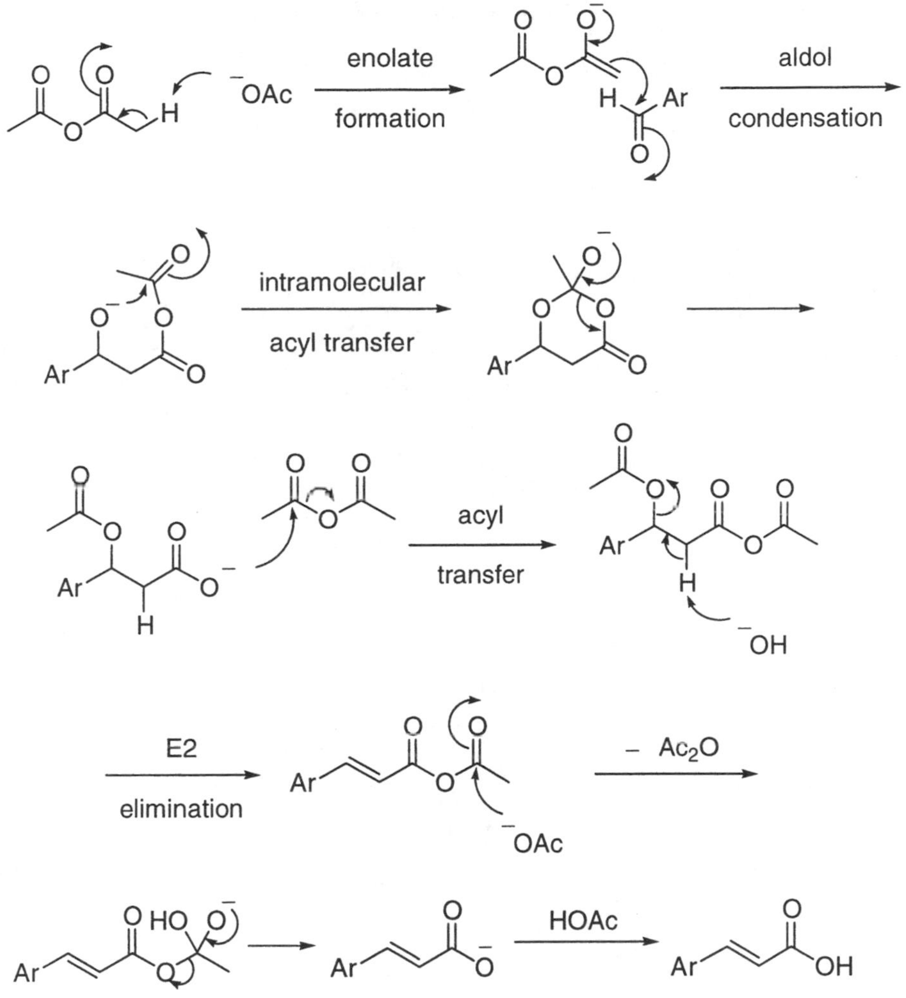
⁻OAc
enolate
formation
aldol
condensation
intramolecular
acyl transfer
acyl
transfer
⁻OH
E2
elimination
– Ac2O
⁻OAc
HOAc
Ar
OH

306

References

1. Perkin, W. H. *J. Chem. Soc.* **1868**, *21*, 53.
2. Pohjala, E. *Heterocycles* **1975**, *3*, 615.
3. Poonia, N. S.; Sen, S.; Porwal, P. K.; Jayakumar, A. *Bull. Chem. Soc. Jpn.* **1980**, *53*, 3338.
4. Gaset, A.; Gorrichon, J. P. *Synth. Commun.* **1982**, *12*, 71.
5. Kinastowski, S.; Nowacki, A. *Tetrahedron Lett.* **1980**, *23*, 3723.
6. Koepp, E.; Voegtle, F. *Synthesis* **1987**, 177.
7. Brady, W. T.; Gu, Y.-Q. *J. Heterocycl. Chem.* **1988**, *25*, 969.
8. Palinko, I.; Kukovecz, A.; Torok, B.; Kortvelyesi, T. *Monatsh. Chem.* **2001**, *131*, 1097.

Perkow vinyl phosphate synthesis

Enol phosphate synthesis from α-halocarbonyls and trialkylphosphites.

General scheme:

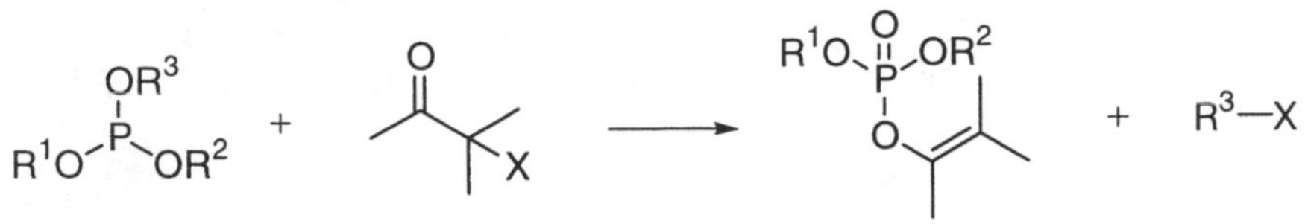

X = Cl, Br, I, secondary or tertiary halides are required to retard the Michaelis–Arbuzov reaction (page 233).

e.g.

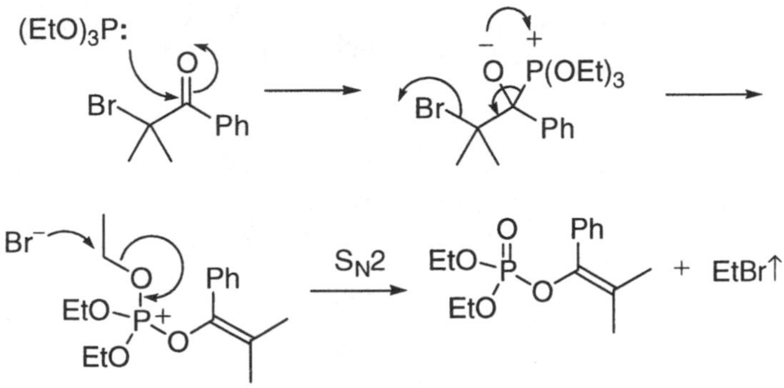

References

1. Perkow, W.; Ullrich, K.; Meyer, F. *Nasturwiss.* **1952**, *39*, 353.
2. Perkow, W. *Ber. Dtsch. Chem. Ges.* **1954**, *87*, 755.
3. Borowitz, G. B.; Borowitz, I. J. *Handb. Organophosphorus Chem.* **1992**, 115.
4. Hudson, H. R.; Matthews, R. W.; McPartlin, M.; Pryce, M. A.; Shode, O. O. *J. Chem. Soc., Perkin Trans. 2* **1993**, 1433.
5. Janecki, T.; Bodalski, R. *Heteroat. Chem.* **2000**, *11*, 115.

Peterson olefination

Alkenes from α-silyl carbanion and carbonyl compounds. Also known as sila-Wittig reaction.

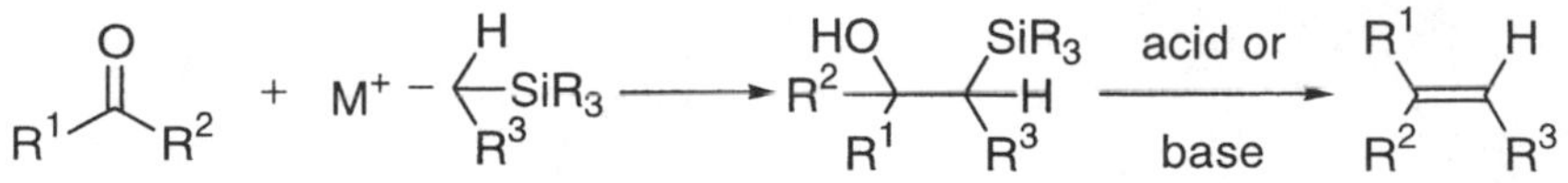

Basic conditions:

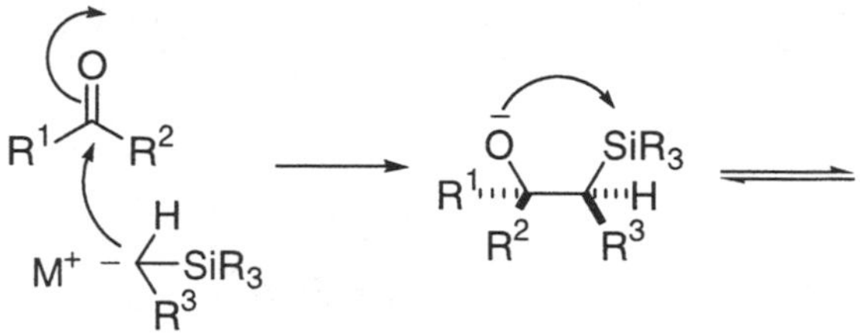

β-silylalkoxide intermediate

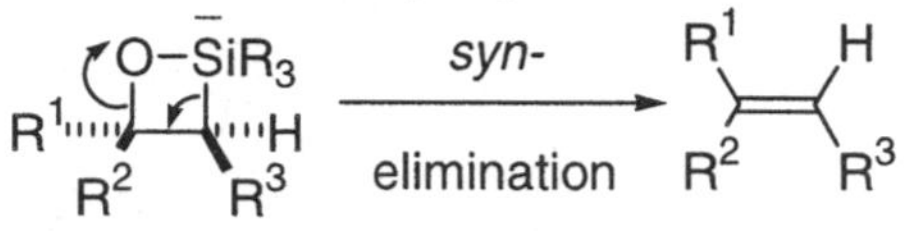

Acidic conditions:

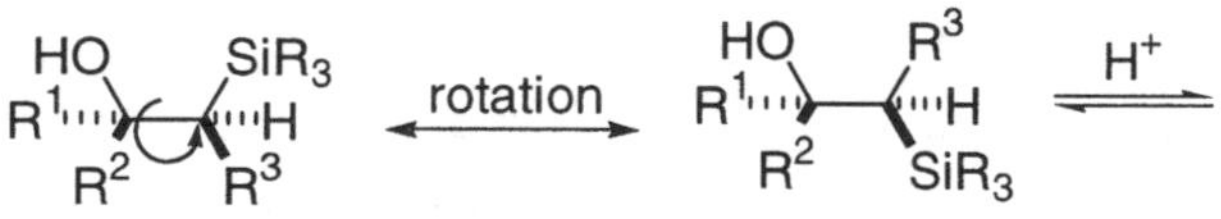

β-hydroxysilane

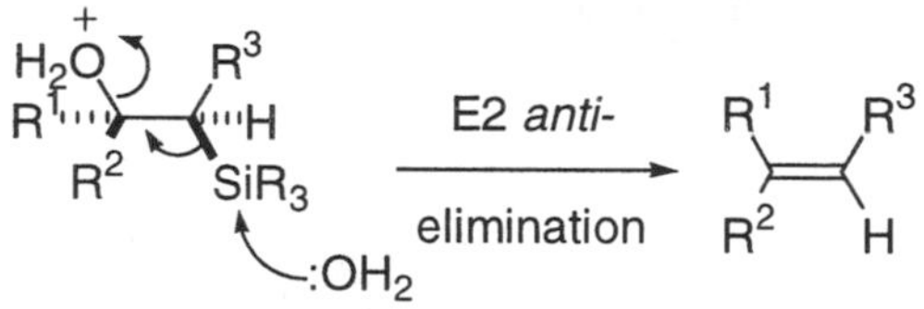

References

1. Peterson, D. J. *J. Org. Chem.* **1968**, *33*, 780.
2. Ager, D. J. *Synthesis* **1984**, 384–98. (Review).
3. Ager, D. J. *Org. React.* **1990**, *38*, 1. (Review).
4. Barrett, A. G. M.; Hill, J. M.; Wallace, E. M.; Flygare, J. A. *Synlett* **1991**, 764–770. (Review).
5. Waschbusch, R.; Carran, J.; Savignac, P. *Tetrahedron* **1996**, *52*, 14199.

6.   Barrett, A. G. M.; Hill, J. M.; Wallace, E. M.; Flygare, J. A. *Synlett* **1991**, 764.
7.   Fassler, J.; Linden, A.; Bienz, S. *Tetrahedron* **1999**, *55*, 1717.
8.   Chiang, C.-C.; Chen, Y.-H.; Hsieh, Y.-T.; Luh, T.-Y. *J. Org. Chem.* **2000**, *65*, 4694.
9.   Galano, J.-M.; Audran, G.; Monti, H. *Tetrahedron Lett.* **2001**, *42*6125.
10.   van Staden, L. F.; Gravestock, D.; Ager, D. J. *Chem. Soc. Rev.* **2002**, *31*, 195–200. (Review).
11.   Ager, D. J. *Science of Synthesis* **2002**, *4*, 789–809. (Review).
12.   Adam, W.; Ortega-Schulte, C. M. *Synlett* **2003**, 414.

Pfau–Plattner azulene synthesis

Azulene formation from indanes and diazoacetates. *Cf.* Buchner ring expansion.

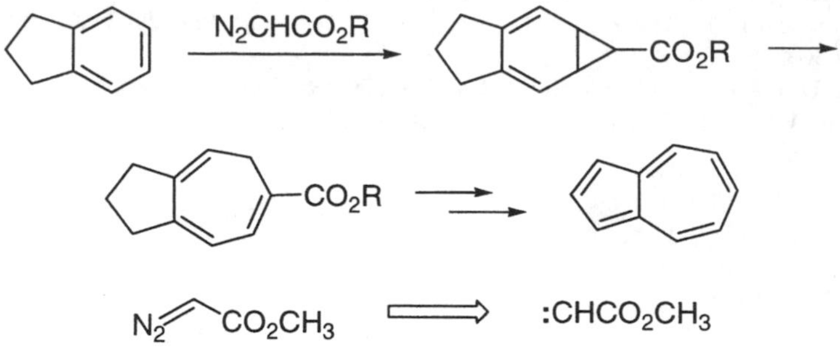

The diazo compound is depicted as a carbene equivalent in the mechanism

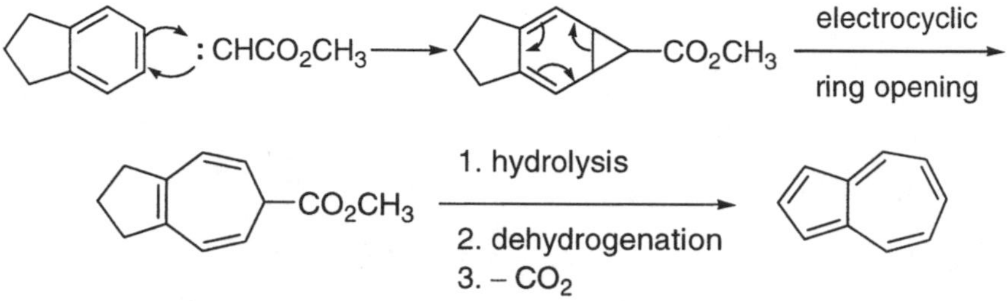

References

1. St. Pfau, A.; Plattner, P. A. *Helv. Chim. Acta* **1939**, *22*, 202.
2. Huzita, Y. J. *Chem. Soc. Jpn.* **1940**, *61*, 729.
3. Gunthard, H.; Plattner, Pl. A.; Brandenberger, E. *Experientia* **1948**, *4*, 425.
4. Sorenson, N. A.; Hougen, F. *Acta Chem. Scand.* **1948**, *2*, 447.
5. Hansen, H. J. *Chimia* **1996**, *50*, 489.
6. Hansen, H. J. *Chimia* **1997**, *51*, 147.

Pfitzinger quinoline synthesis

Quinoline-4-carboxylic acids from the condensation of isatic acids and α-methylene carbonyl compounds using base.

References

1. Buu-Hoi, N. P.; Royer, R.; Nuong, N. D.; Jacquhnos, P. *J. Org. Chem.* **1953**, *18*, 1209.
2. Cragoe, E. J., Jr.; Robb, C. M. *Org. Synth.* **1973**, *Coll. Vol. 5*, 635.
3. Lutz, R. E.; Sanders, J. M. *J. Med. Chem.* **1976**, *19*, 407.
4. Cragoe, E. J., Jr.; Robb, C. M.; Bealor, M. D. *J. Am. Chem. Soc.* **1982**, *53*, 552.
5. Gainer, J. A.; Weinreb, S. M. *J. Org. Chem.* **1982**, *47*, 2833.
6. Baldwin, M. A.; Langley, G. J. *J. Labeled Compd. Radiopharm.* **1985**, *22*, 1233.
7. Lasikova, A.; Vegh, D. *Chem. Pap.* **1997**, *51*, 408.
8. Pardasani, R. T.; Pardasani, P.; Sherry, D.; Chaturvedi, V. *Indian J. Chem., Sect. B* **2001**, *40B*, 1275.
9. Wang, J.-J.; Wang, Z.-Y.; Sun, G.-Q.; Zhao, Y.; Jiang, G.-J. *Yingyong Huaxue* **2002**, *19*, 1174.

Pictet–Gams isoquinoline synthesis

Isoquinolines from acylated aminomethyl phenylcarbinols using phosphorus pentoxide.

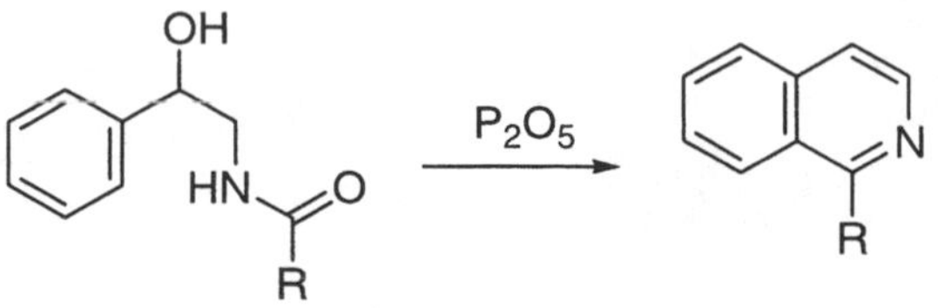

P_2O_5 actually exists as P_4O_{10}, an adamantane-like structure.

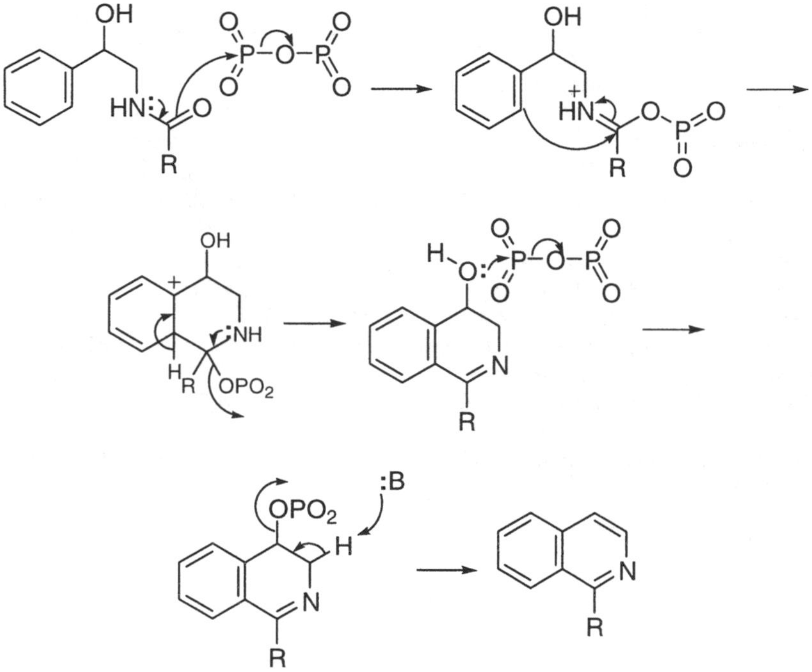

Reference

1. Pictet, A.; Gams, A. *Ber. Dtsch. Chem. Ges.* **1910**, *43*, 2384.
2. Kulkarni, S. N.; Nargund, K. S. *Indian J. Chem., Sect. B* **1967**, *5*, 294.
3. Ardabilchi, N.; Fitton, A. O.; Frost, J. R.; Oppong-Boachie, F. *Tetrahedron Lett.* **1977**, 4107.
4. Ardabilchi, N.; Fitton, A. O.; Frost, J. R.; Oppong-Boachie, F. K.; Hadi, A. Hamid, A.; Sharif, A. .M. *J. Chem. Soc., Perkin Trans. 1* **1979**, 539.
5. Ardabilchi, N.; Fitton, A. O. *J. Chem. Soc. (S)* **1979**, 310.

6. Cerri, A.; Mauri, P.; Mauro, M.; Melloni, P. *J. Heterocycl. Chem.* **1993**, *30*, 1581.
7. Dyker, G.; Gabler, M.; Nouroozian, M.; Schulz, P. *Tetrahedron Lett.* **1994**, *35,* 9697.
8. Poszavacz, L.; Simig, G. *J. Heterocycl. Chem.* **2000**, *37*, 343.
9. Poszavacz, L.; Simig, G. *Tetrahedron* **2001**, *57*, 8573.

Pictet–Spengler tetrahydroisoquinoline synthesis

Tetrahydroisoquinolines from condensation of β-arylethylamines and carbonyl compounds followed by cyclization.

References

1. Pictet, A.; Spengler, T. *Ber. Dtsch. Chem. Ges.* **1911**, *44*, 2030.
2. Hudlicky, T.; Kutchan, T. M.; Shen, G.; Sutliff, V. E.; Coscia, C. J. *J. Org. Chem.* **1981**, *46*, 1738.
3. Miller, R. B.; Tsang, T. *Tetrahedron Lett.* **1988**, *29*, 6715.
4. Rozwadowska, M. D. *Heterocycles* **1994**, *39*, 903.
5. Cox, E. D.; Cook, J. M. *Chem. Rev.* **1995**, *95*, 1797. (Review).
6. Yokoyama, A.; Ohwada, T.; Shudo, K. *J. Org. Chem.* **1999**, *64*, 611.
7. Singh, K.; Deb, P. K.; Venugopalan, P. *Tetrahedron* **2001**, *57*, 7939.
8. Kang, I.-J.; Wang, H.-M.; Su, C.-H.; Chen, L.-C. *Heterocycles* **2002**, *57*, 1.
9. Yu, J.; Wearing, X. Z.; Cook, J. M. *Tetrahedron Lett.* **2003**, *44*, 543.
10. Tsuji, R.; Nakagawa, M.; Nishida, A. *Tetrahedron: Asymmetry* **2003**, *14*, 177.

Pinacol rearrangement

Acid-catalyzed rearrangement of vicinyl diols (pinacols) to carbonyl compounds.

References

1. Fittig, R. *Justus Liebigs Ann. Chem.* **1860**, *114*, 54.
2. Toda, F.; Shigemasa, T. *J. Chem. Soc., Perkin Trans. 1* **1989**, 209.
3. Nakamura, K.; Osamura, Y. *J. Am. Chem. Soc.* **1993**, *115*, 9112.
4. Paquette, L. A.; Lord, M. D.; Negri, J. T. *Tetrahedron Lett.* **1993**, *34*, 5693.
5. Jabur, F. A.; Penchev, V. J.; Bezoukhanova, C. P. *J. Chem. Soc., Chem. Commun.* **1994**, 1591.
6. Patra, D.; Ghosh, S. *J. Org. Chem.* **1995**, *60*, 2526.
7. Magnus, P.; Diorazio, L.; Donohoe, T. J.; Giles, M.; Pye, P.; Tarrant, J.; Thom, S. *Tetrahedron* **1996**, *52*, 14147.
8. Bach, T.; Eilers, F. *J. Org. Chem.* **1999**, *64*, 8041.
9. Razavi, H.; Polt, R. *J. Org. Chem.* **2000**, *65*, 5693.
10. Rashidi-Ranjbar, P.; Kianmehr, E. *Molecules* **2001**, *6*, 442.
11. Marson, C. M.; Oare, C. A.; McGregor, J.; Walsgrove, T.; Grinter, T. J.; Adams, H. *Tetrahedron Lett.* **2003**, *44*, 141.

Pinner synthesis

Transformation of a nitrile into an imino ether, which can be converted to either an ester or an amidine.

common intermediate

References

1. Pinner, A.; Klein, F. *Ber. Dtsch. Chem. Ges.* **1877**, *10*, 1889.
2. Poupaert, J.; Bruylants, A.; Crooy, P. *Synthesis* **1972**, 622.
3. Wagner, G.; Horn, H. *Pharmazie* **1975**, *30*, 353.
4. Lee, Y. B.; Goo, Y. M.; Lee, Y. Y.; Lee, J. K. *Tetrahedron Lett.* **1990**, *31*, 1169.
5. Cheng, C. C. *Org. Prep. Proced. Int.* **1990**, *22*, 643.
6. Neugebauer, W.; Pinet, E.; Kim, M.; Carey, P. R. *Can. J. Chem.* **1996**, *74*, 341.
7. Spychala, J. *Synth. Commun.* **2000**, *30*, 1083.
8. Kigoshi, H.; Hayashi, N.; Uemura, D. *Tetrahedron Lett.* **2001**, *42*, 7469.
9. Siskos, A. P.; Hill, A. M. *Tetrahedron Lett.* **2003**, *44*, 789.

Polonovski reaction

Treatment of a tertiary *N*-oxide with an activating agent such as acetic anhydride, resulting in rearrangement where an *N,N*-disubstituted acetamide and an aldehyde are generated.

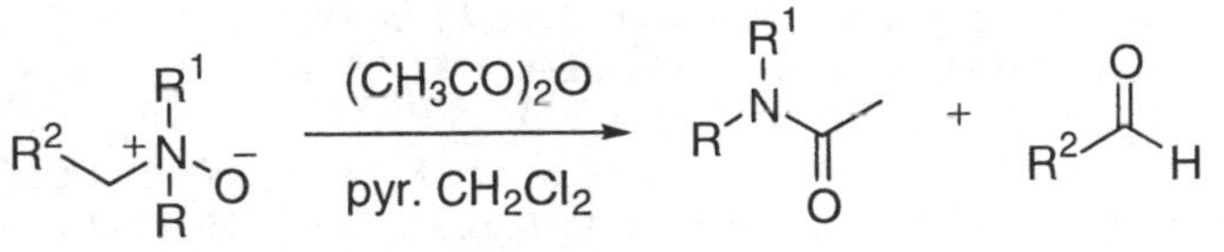

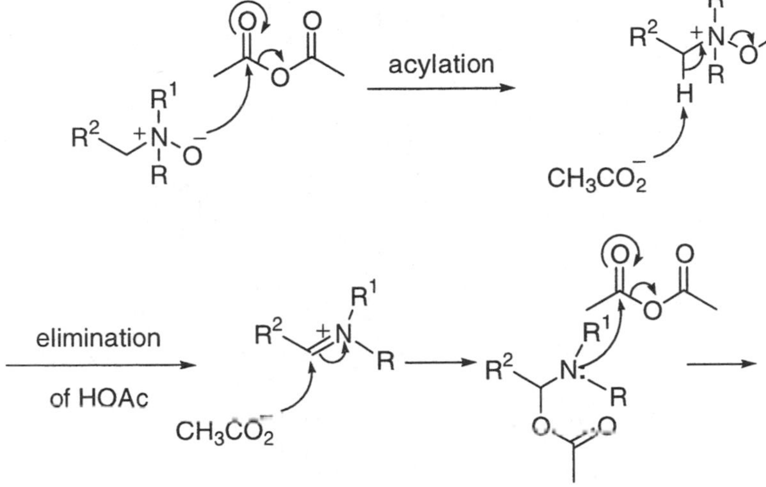

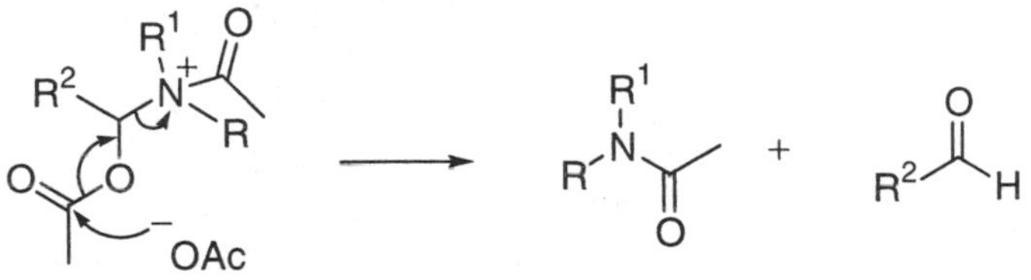

The intramolecular pathway is also possible:

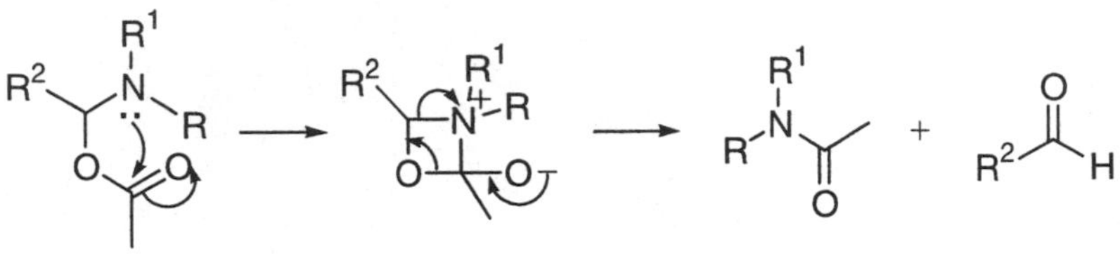

318

References

1. Polonovski, M.; Polonovski, M. *Bull. Soc. Chim. Fr.* **1927**, *41*, 1190.
2. Michelot, R. *Bull. Soc. Chim. Fr.* **1969**, 4377.
3. Volz, H.; Ruchti, L. *Ann.* **1972**, *763*, 184.
4. Hayashi, Y.; Nagano, Y.; Hongyo, S.; Teramura, K. *Tetrahedron Lett.* **1974**, 1299.
5. M'Pati, J.; Mangeney, P.; Langlois, Y. *Tetrahedron Lett.* **1981**, *22*, 4405.
6. Lounasmaa, M.; Koskinen, A. *Tetrahedron Lett.* **1982**, *23*, 349.
7. Grierson, D. *Org. React.* **1990**, *39*, 85. (Review).
8. Lounasmaa, M.; Jokela, R.; Halonen, M.; Miettinen, J. *Heterocycles* **1993**, *36*, 2523.
9. Thomas, O. P.; Zaparucha, A.; Husson, H.-P. *Tetrahedron Lett.* **2001**, *42*, 3291.
10. Morita, H.; Kobayashi, J. *J. Org. Chem.* **2002**, *67*, 5378.

Polonovski–Potier reaction

A modification of the Polonovski reaction where trifluoroacetic anhydride is used in place of acetic anhydride.

tertiary *N*-oxide

acylation

$CF_3CO_2^-$

iminium ion

enamine

References

1. Lewin, G.; Poisson, J.; Schaeffer, C.; Volland, J. P. *Tetrahedron* **1990**, *46*, 7775.
2. Kende, A. S.; Liu, K.; Jos Brands, K. M. *J. Am. Chem. Soc.* **1995**, *117*, 10597.
3. Sundberg, R. J.; Gadamasetti, K. G.; Hunt, P. J. *Tetrahedron* **1992**, *48*, 277.
4. Lewin, G.; Schaeffer, C.; Morgant, G.; Nguyen-Huy, D. *J. Org. Chem.* **1996**, *61*, 9614.
5. Renko, D.; Mary, A.; Guillou, C.; Potier, P.; Thal, C. *Tetrahedron Lett.* **1998**, *39*, 4251.
6. Suau, R.; Najera, F.; Rico, R. *Tetrahedron* **2000**, *56*, 9713.
7. Thomas, O. P.; Zaparucha, A.; Husson, H.-P. *Tetrahedron Lett.* **2001**, *42*, 3291.

Pomeranz–Fritsch reaction

Isoquinoline synthesis from benzaldehyde and aminoacetal.

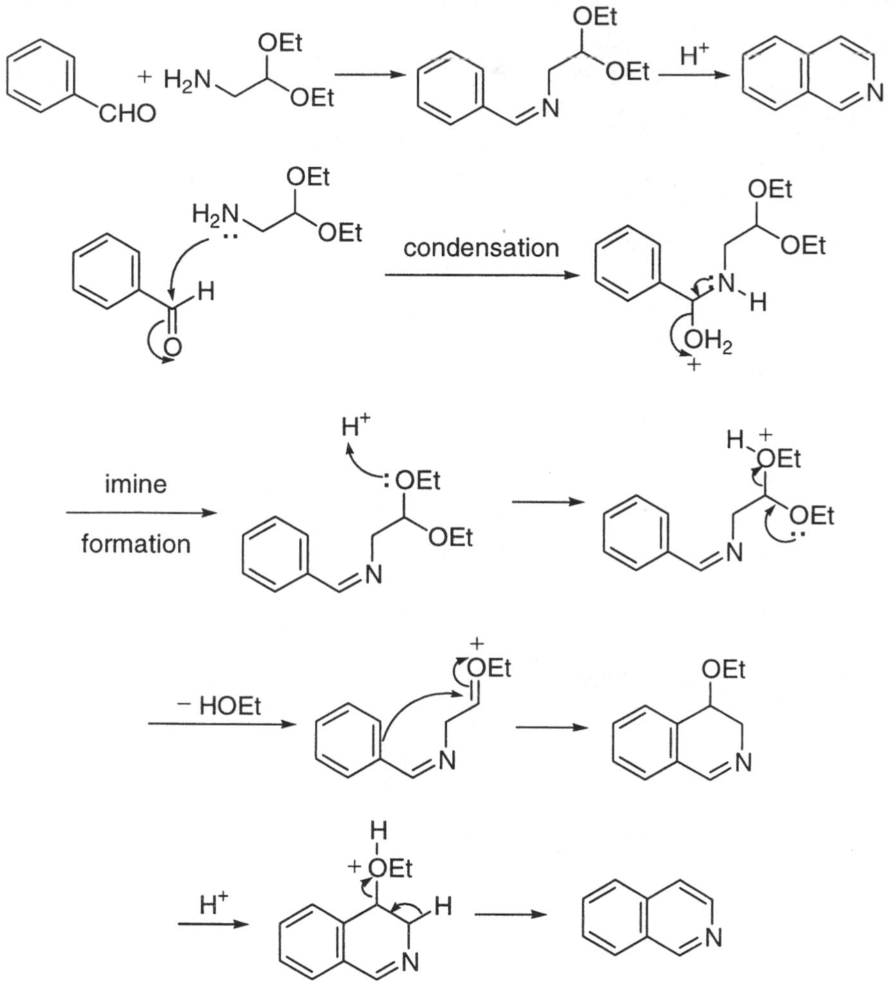

Schilittle–Müller modification

References

1. Bevis, M. J.; Forbes, Eric J.; Uff, B. C. *Tetrahedron* **1969**, *25*, 1585.
2. Bevis, M. J.; Forbes, E. J.; Naik, N. N.; Uff, B. C. *Tetrahedron* **1971**, *27*, 1253.
3. Birch, A. J.; Jackson, A. H.; Shannon, P. V. R. *J. Chem. Soc., Perkin Trans. 1* **1974**, 2185.
4. Birch, A. J.; Jackson, A. H.; Shannon, P. V. R. *J. Chem. Soc., Perkin Trans. 1* **1974**, 2190.
5. Brown, E. V. *J. Org. Chem.* **1977**, *42*, 3208.
6. Gill, E. W.; Bracher, A. W. *J. Heterocycl. Chem.* **1983**, *20*, 1107.
7. Ishii, H.; Ishida, T. *Chem. Pharm. Bull.* **1984**, *32*, 3248.
8. Katritzky, A. R.; Yang, Z.; Cundy, D. J. *Heteroat. Chem.* **1994**, *5*, 103.
9. Schlosser, M.; Simig, G.; Geneste, H. *Tetrahedron* **1998**, *54*, 9023.
10. Poli, G.; Baffoni, S. C.; Giambastiani, G.; Reginato, G. *Tetrahedron* **1998**, *54*, 10403.
11. Gluszynska, A.; Rozwadowska, M. D. *Tetrahedron: Asymmetry* **2000**, *11*, 2359.

322

Prévost *trans*-dihydroxylation

Cf. Woodward *cis*-dihydroxylation.

cyclic iodonium ion intermediate

neighboring group assistance

References

1. Prévost, C. *Compt. Rend.* **1933**, *196* 1129.
2. Campbell, M. M.; Sainsbury, M.; Yavarzadeh, R. *Tetrahedron* **1984**, *40*, 5063.
3. Campi, E. M.; Deacon, G. B.; Edwards, G. L.; Fitzroy, M. D.; Giunta, N.; Jackson, W. R.; Trainor, R. *J. Chem. Soc., Chem. Commun.* **1989**, 407.
4. Prasad, K. J. R.; Subramaniam, M. *Indian J. Chem., Sect. B* **1994**, *33B*, 696.
5. Ciganek, E.; Calabrese, J. C. *J. Org. Chem.* **1995**, *60*, 4439.
6. Deota, P. T.; Singh, V. *J. Chem. Res. (S)*, **1996**, 258.
7. Katoch, R.; Baig, M. H. A.; Trivedi, G. K. *J. Chem. Res. (S)*, **1998**, 2401.
8. Brimble, M. A.; Nairn, M. R. *J. Org. Chem.* **1996**, *61*, 4801.
9. Zajc, B. *J. Org. Chem.* **1999**, *64*, 1902.
10. Hamm, S.; Hennig, L.; Findeisen, M.; Muller, D.; Welzel, P. *Tetrahedron* **2000**, *56*, 1345.
11. Ray, J. K.; Gupta, S.; Kar, G. K.; Roy, Bidhan C.; Lin, J.-M.; Amin, S. *J. Org. Chem.* **2000**, *65*, 8134.
12. Sabat, M.; Johnson, C. R. *Tetrahedron Lett.* **2001**, *42*, 1209.

Prilezhaev reaction

Epoxidation of olefins using peracids.

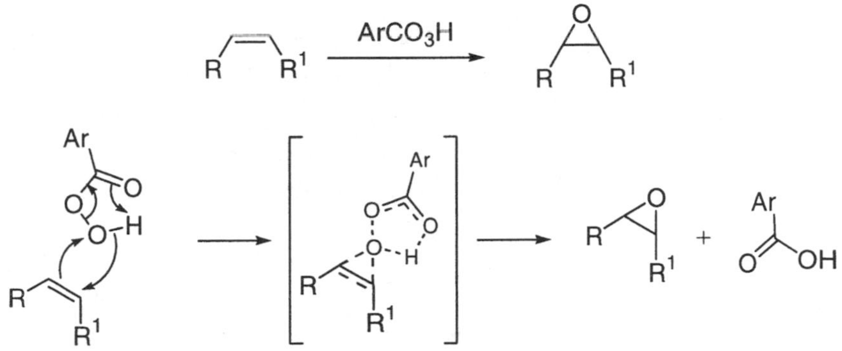

The "butterfly" transition state

References

1. Prilezhaev, N. *Ber. Dtsch. Chem. Ges.* **1909**, *64*, 8041.
2. Rebek, J., Jr.; Marshall, L.; McManis, J.; Wolak, R. *J. Org. Chem.* **1986**, *51*, 1649.
3. Kaneti, I. *Tetrahedron* **1986**, *42*, 4017.
4. De Cock, C. J. C.; De Keyser, J. L.; Poupaert, J. H.; Dumont, P. *Bull. Soc. Chim. Belg.* **1987**, *96*, 783.
5. Hilker, I.; Bothe, D.; Pruss, J.; Warnecke, H.-J. *Chem. Eng. Sci.* **2001**, *56*, 427.

Prins reaction

Addition of alkene to formaldehyde.

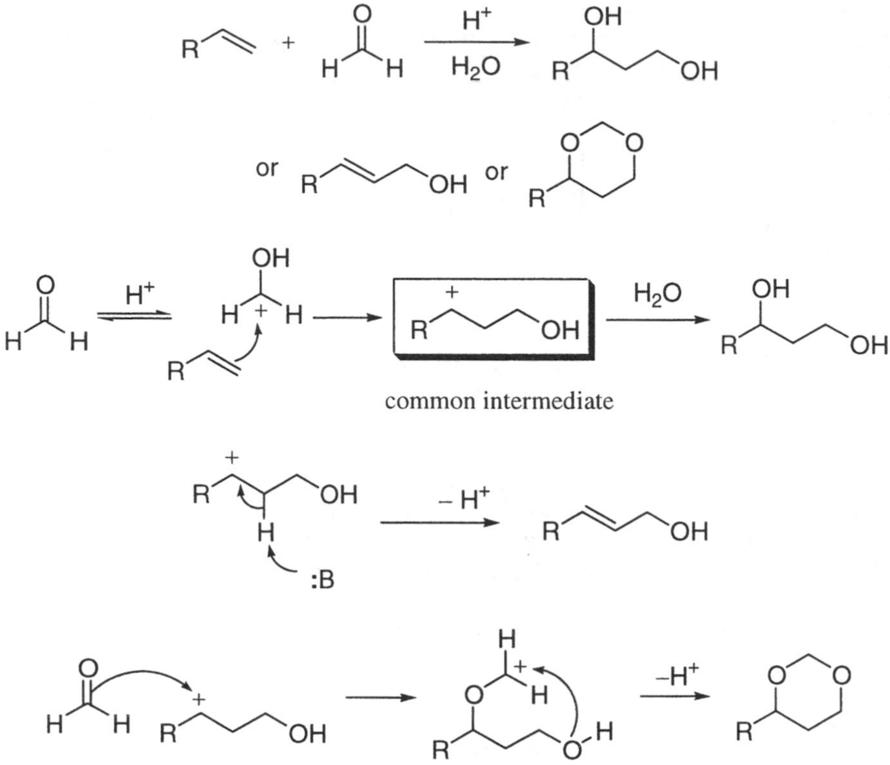

common intermediate

References

1. Prins, H. J. *Chem. Weekblad* **1919**, *16*, 64, 1072.
2. Adam, D. R.; Bhtnagar, S. P. *Synthesis* **1977**, 661.
3. El Gharbi, R. *Synthesis* **1981**, 361.
4. Hanaki, N.; Link, J. T.; MacMillan, D. W. C.; Overman, L. E.; Trankle, W. G.; Wurster, J. A. *Org. Lett.* **2000**, *2*, 223.
5. Yadav, J. S.; Reddy, B. V. S.; Kumar, G. M.; Murthy, C. V. S. R. *Tetrahedron Lett.* **2001**, *42*, 89.
6. Cho, Y. S.; Kim, H. Y.; Cha, J. H.; Pae, A. N.; Koh, H. Y.; Choi, J. H.; Chang, M. H. *Org. Lett.* **2002**, *4*, 2025.
7. Davis, C. E.; Coates, R. M. *Angew. Chem., Int. Ed.* **2002**, *41*, 472.
8. Braddock, D. C.; Badine, D. M.; Gottschalk, T.; Matsuno, A.; Rodrihuez-Lens, M. *Synlett* **2003**, 345.

Pschorr ring closure

The intramolecular version of the Gomberg–Bachmann reaction.

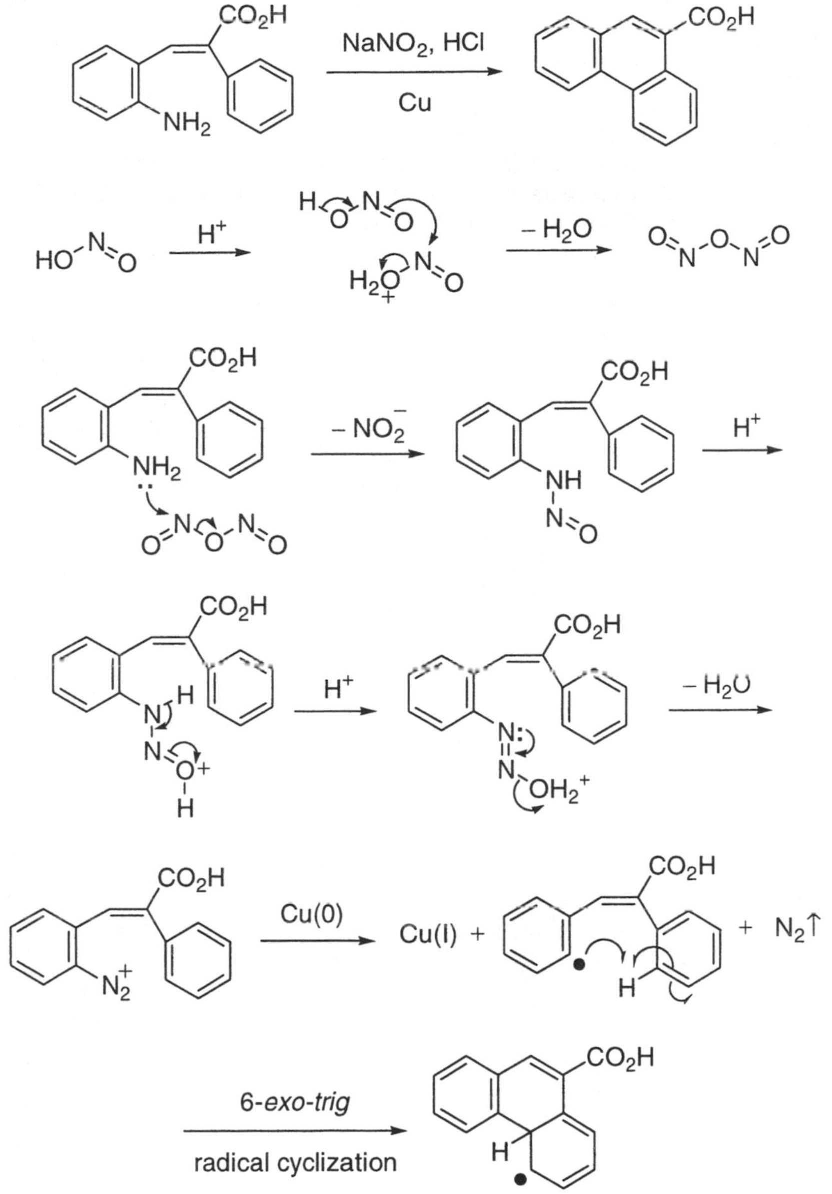

$$\xrightarrow[{-\ H^+}]{Cu(I)} \quad Cu(0) \ + \quad \text{(phenanthrene-CO}_2\text{H)}$$

References

1. Pschorr, R. *Ber. Dtsch. Chem. Ges.* **1896**, *29*, 496.
2. Kametani, T.; Fukumoto, K. *J. Heterocycl. Chem.* **1971**, *8*, 341.
3. Kupchan, S. M.; Kameswaran, V.; Findlay, J. W. A. *J. Org. Chem.* **1973**, *38*, 405.
4. Daidone, G. *J. Heterocycl. Chem.* **1980**, *17*, 1409.
5. Buck, K. T.; Edgren, D. L.; Blake, G. W.; Menachery, M. D. *Heterocycles* **1993**, *36*, 2489.
6. Wassmundt, F. W.; Kiesman, W. F. *J. Org. Chem.* **1995**, *60*, 196.
7. Qian, X.; Cui, J.; Zhang, R. *Chem. Commun.* **2001**, 2656.
8. Hassan, J.; Sevignon, M.; Gozzi, C.; Schulz, E.; Lemaire, M. *Chem. Rev.* **2002**, *102*, 1359–1469. (Review).
9. Karady, S.; Cummins, J. M.; Dannenberg, J. J.; del Rio, E.; Dormer, P. G.; Marcune, B. F.; Reamer, R. A.; Sordo, T. L. *Org. Lett.* **2003**, *5*, 1175.

Pummerer rearrangement

The transformation of sulfoxides into α-acyloxythioethers using acetic anhydride.

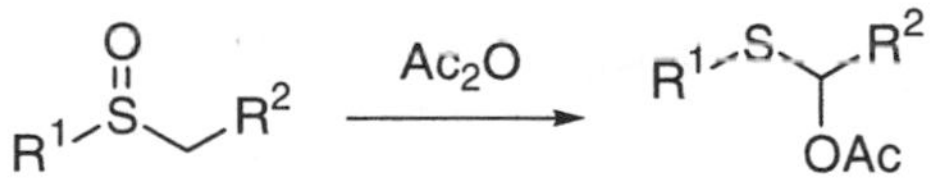

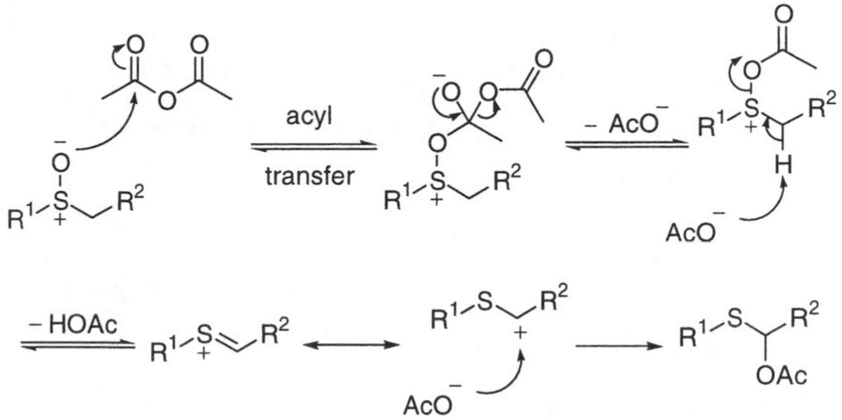

References

1. Pummerer, R. *Ber. Dtsch. Chem. Ges.* **1910**, *43*, 1401.
2. De Lucchi, O.; Miotti, U.; Modena, G. *Org. React.* **1991**, *40*, 157. (Review).
3. Kita, Y. *Phosphorus, Sulfur Silicon Relat. Elem.* **1991**, *120 & 121*, 145.
4. Padwa, A.; Gunn, D. E., Jr.; Osterhout, M. H. *Synthesis* **1997**, 1353.
5. Padwa, A.; Waterson, A. G. *Curr. Org. Chem.* **2000**, *4*, 175.
6. Marchand, P.; Gulea, M.; Masson, S.; Averbuch-Pouchot, M.-T. *Synthesis* **2001**, 1623.
7. Padwa, A.; Bur, S. K.; Danca, M. D.; Ginn, J. D.; Lynch, S. M. *Synlett* **2002**, 851–862. (Review).
8. Padwa, A.; Danca, M. D.; Hardcastle, K. I.; McClure, M. S. *J. Org. Chem.* **2003**, *68*, 929.

Ramberg–Bäcklund olefin synthesis

Olefin synthesis *via* α-halosulfone extrusion.

episulfone intermediate

References

1. Ramberg, L.; Bäcklund, B. *Arkiv. Kemi, Mineral Geol.* **1940**, *13A*, 50.
2. Paquette, L. A. *Acc. Chem. Res.* **1968**, *1*, 209. (Review).
3. Paquette, L. A. *Org. React.* **1977**, *25*, 1–71. (Review).
4. Braveman, S.; Zafrani, Y. *Tetrahedron* **1998**, *54*, 1901.
5. Taylor, R. J. K. *Chem. Commun.* **1999**, 217.
6. McGee, D. I.; Beck, E. J. *Can. J. Chem.* **2000**, *78*, 1060.
7. McAllister, G. D.; Taylor, R. J. K. *Tetrahedron Lett.* **2001**, *42*, 1197.
8. Murphy, P. V.; McDonnell, C.; Hamig, L.; Paterson, D. E.; Taylor, R. J. K. *Tetrahedron: Asymmetry* **2003**, *14*, 79.
9. W., X.-L.; Cao, X.-P.; Zhou, Z.-L. *Youji Huaxue* **2003**, *23*, 120.
10. Wei, C.; Mo, K.-F.; Chan, T.-L. *J. Org. Chem.* **2003**, *68*, 2948.

Reformatsky reaction

Nucleophilic addition of organozinc reagents generated from α-haloesters to carbonyls.

References

1. Reformatsky, S. *Ber. Dtsch. Chem. Ges.* **1887**, *20*, 1210.
2. Gaudemar, M. *Organometal. Chem. Rev., Sect. A* **1972**, *8*, 183. (Review).
3. Fürstner, A. *Synthesis* **1989**, 571. (Review).
4. Fürstner, A. In *Organozinc Reagents* Knochel, P.; Jones, P. eds.; Oxford University Press: New York, **1999**, pp 287–305. (Review).
5. Hirashita, T.; Kinoshita, K.; Yamamura, H.; Kawai, M.; Araki, S. *J. Chem. Soc., Perkin Trans. 1* **2000**, 825.
6. Kurosawa, T.; Fujiwara, M.; Nakano, H.; Sato, M.; Yoshimura, T.; Murai, T. *Steroids* **2001**, *66*, 499.
7. Ocampo, R.; Dolbier, W. R.; Abboud, K. A.; Zuluga, F. *J. Org. Chem.* **2002**, *67*, 72.
8. Obringer, M.; Colobert, F.; Neugnot, B.; Solladié, G. *Org. Lett.* **2003**, *5*, 629.

Regitz diazo synthesis

Synthesis of 2-diazo-1,3-dicarbonyl or 2-diazo-3-ketoesters using tosyl azide or mesyl azide.

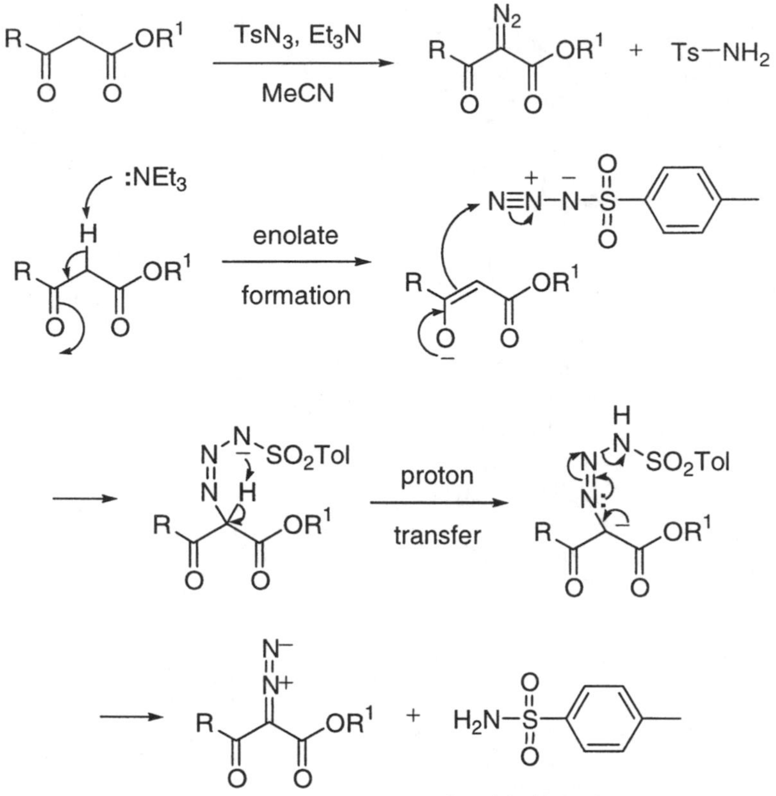

tosyl amide is the by-product

When only one carbonyl is present, ethylformate can be used as an activating auxiliary [6–9]:

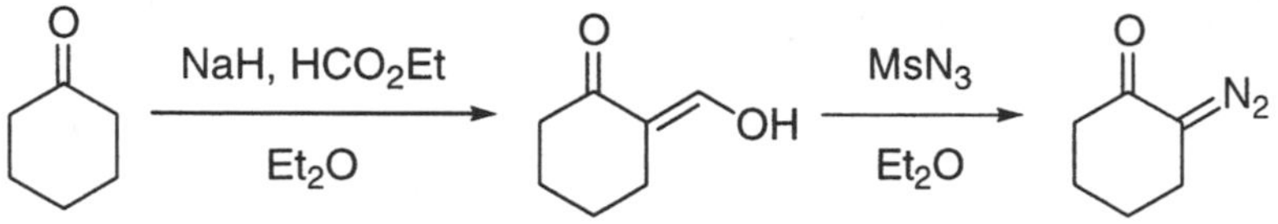

Alternatively, the triazole intermediate may be assembled via a 1,3-dipolar cycloaddition of the enol and mesyl azide:

References

1. Regitz, M. *Angew. Chem., Int. Ed.* **1967**, *6*, 733.
2. Regitz, M.; Anschuetz, W.; Bartz, W.; Liedhegener, A. *Tetrahedron Lett.* **1968**, 3171.
3. Regitz, M. *Synthesis* **1972**, 351. (Review).
4. Taber, D. F.; Ruckle, R. E., Jr.; Hennessy, M. J. *J. Org. Chem.* **1986**, *51*, 4077.
5. Taber, D. F.; Schuchardt, J. L. *Tetrahedron* **1987**, *43*, 5677.
6. Pudleiner, H.; Laatsch, H. *Justus Liebigs Ann. Chem.* **1990**, 423.
7. Evans, D. A.; Britton, T. C.; Ellman, J. A.; Dorow, R. L. *J. Am. Chem. Soc.* **1990**, *112*, 4011.
8. Ihara, M.; Suzuki, T.; Katogi, M.; Taniguchi, N.; Fukumoto, K. *J. Chem. Soc., Perkin Trans. 1* **1992**, 865.
9. Charette, A. B.; Wurz, R. P.; Ollevier, T. *J. Org. Chem.* **2000**, *65*, 9252.

Reimer–Tiemann reaction

Synthesis of *o*-formylphenol from phenols and chloroform in alkaline medium.

$$\text{C}_6\text{H}_5\text{OH} + \text{CHCl}_3 + 3\,\text{KOH} \longrightarrow \text{(2-HOC}_6\text{H}_4)\text{CHO} + 3\,\text{KCl} + 2\,\text{H}_2\text{O}$$

a. Carbene generation:

$$\text{Cl}_3\text{C–H} \quad \xrightleftharpoons{\text{fast}} \quad \text{H}_2\text{O} + {}^-\text{CCl}_2\,(\text{Cl}) \quad \xrightarrow[\alpha\text{-elimination}]{-\,\text{Cl}^-,\ \text{slow}} \quad :\text{CCl}_2$$

b. Addition of dichlorocarbene and hydrolysis:

References

1. Reimer, K.; Tiemann, F. *Ber. Dtsch. Chem. Ges.* **1876**, *9*, 824.
2. Wynberg, H.; Meijer, E. W. *Org. React.* **1982**, *28*, 1. (Review).
3. Smith, K. M.; Bobe, F. W.; Minnetian, O. M.; Hope, H.; Yanuck, M. D. *J. Org. Chem.* **1985**, *50*, 790.
4. Bird, C. W.; Brown, A. L.; Chan, C. C. *Tetrahedron* **1985**, *41*, 4685.
5. Cochran, J. C.; Melville, M. G. *Synth. Commun.* **1990**, *20*, 609.
6. Langlois, B. R. *Tetrahedron Lett.* **1991**, *32*, 3691.

7. Jimenez, M. Co.; Miranda, M. A.; Tormos, R. *Tetrahedron* **1995**, *51*, 5825.
8. Jung, M. E.; Lazarova, T. I. *J. Org. Chem.* **1995**, *62*, 1553.
9. Pan, J.; Huang, Y. *225th ACS National Meeting*, New Orleans, LA, USA, March 23–27, (**2003**), ORGN-399.

Reissert reaction (aldehyde synthesis)

Aldehyde synthesis from the corresponding acid chloride, quinoline, and KCN.

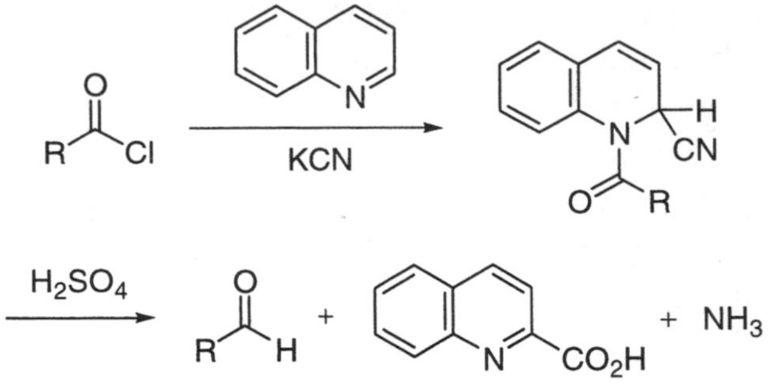

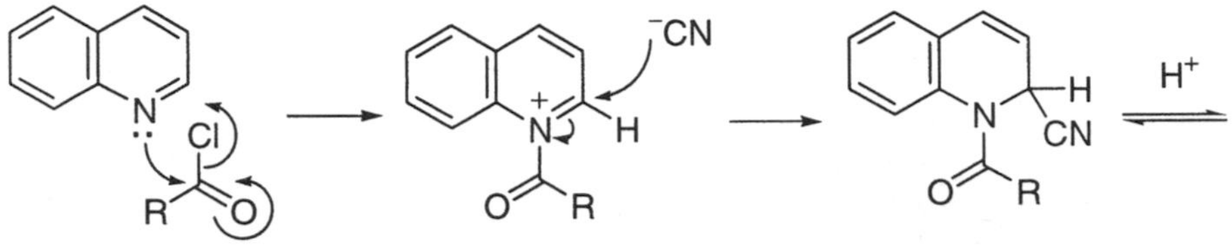

Reissert compound

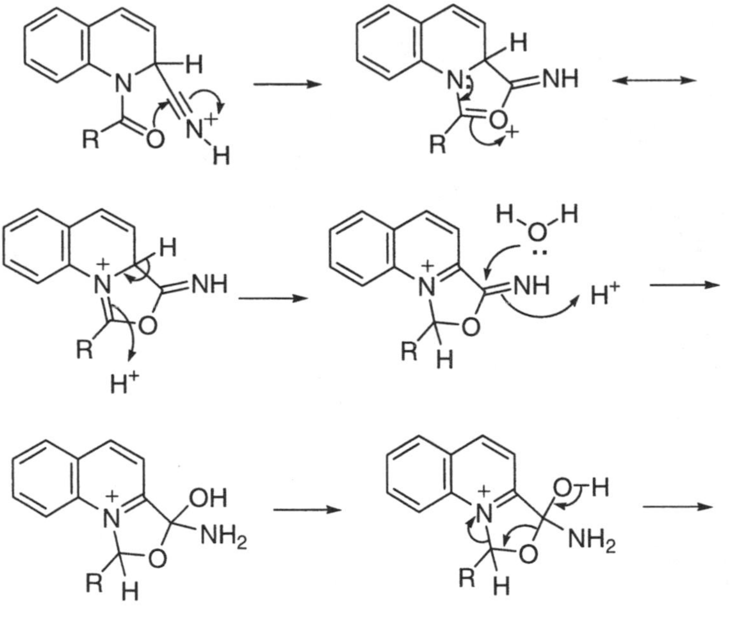

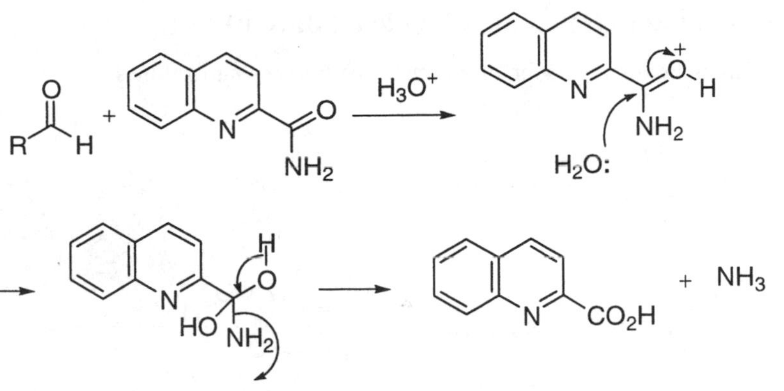

References

1. Reissert, A. *Ber. Dtsch. Chem. Ges.* **1905**, *38*, 1603, 3415.
2. Popp, F. D. *Adv. Heterocyclic Chem.* **1979**, *24*, 187.
3. Fife, W. K.; Scriven, E. F. V. *Heterocycles* **1984**, *22*, 2375.
4. Popp, F. D.; Uff, B. C. *Heterocycles* **1985**, *23*, 731.
5. Lorsbach, B. A.; Bagdanoff, J. T.; Miller, R. B.; Kurth, M. J. *J. Org. Chem.* **1998**, *63*, 2244.
6. Perrin, S.; Monnier, K.; Laude, B.; Kubicki, M.; Blacque, O. *Eur. J. Org. Chem.* **1999**, 297.
7. Takamura, M.; Funabashi, K.; Kanai, M.; Shibasaki, M. *J. Am. Chem. Soc.* **2001**, *123*, 6801.
8. Sieck, O.; Schaller, S.; Grimme, S.; Liebscher, J. *Synlett* **2003**, 337.

Riley oxidation (selenium dioxide oxidation)

A selenium dioxide oxidation of activated methylenes into ketones.

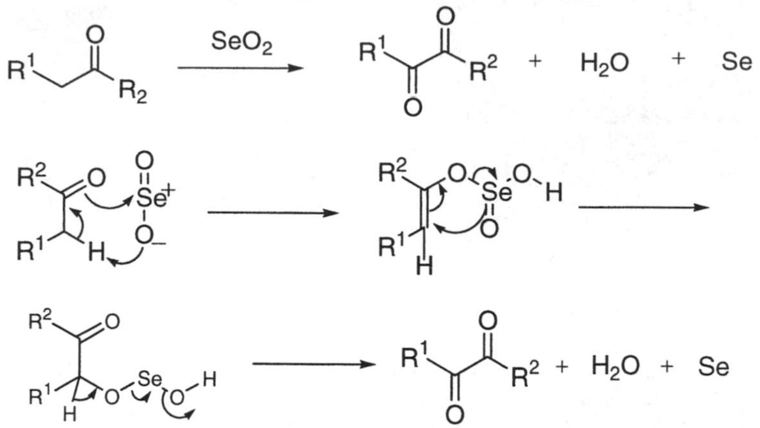

References

1. Riley, H. L.; Morley, J. F.; Friend, N. A. C. *J. Chem. Soc.* **1932**, 1875.
2. Rabjohn, N. *Org. React.* **1976**, *24*, 261. (Review).
3. Goudgaon, N. M.; Nayak, U. R. *Indian J. Chem., Sect. B* **1985**, *24B*, 589.
4. Dalavoy, V. S.; Deodhar, V. B.; Nayak, U. R. *Indian J. Chem., Sect. B* **1987**, *26B*, 1.

Ring-closing metathesis (RCM)
using Grubbs and Schrock catalysts

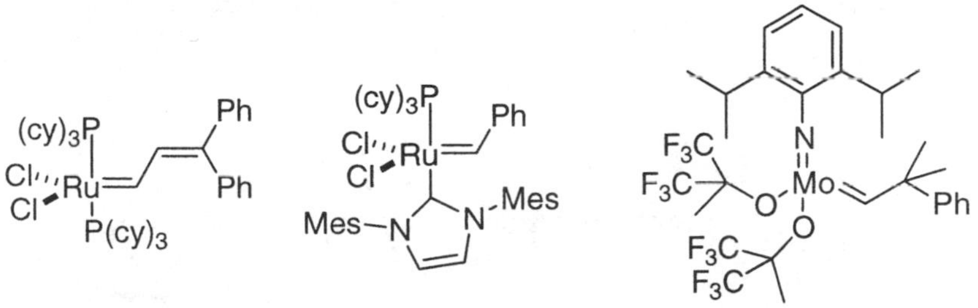

Grubbs' reagents
Mes = mesityl

Schrock's reagent

All three catalysts are illustrated as "L$_n$M=CHR" in the mechanism below.

Generation of the catalyst from the precatalysts:

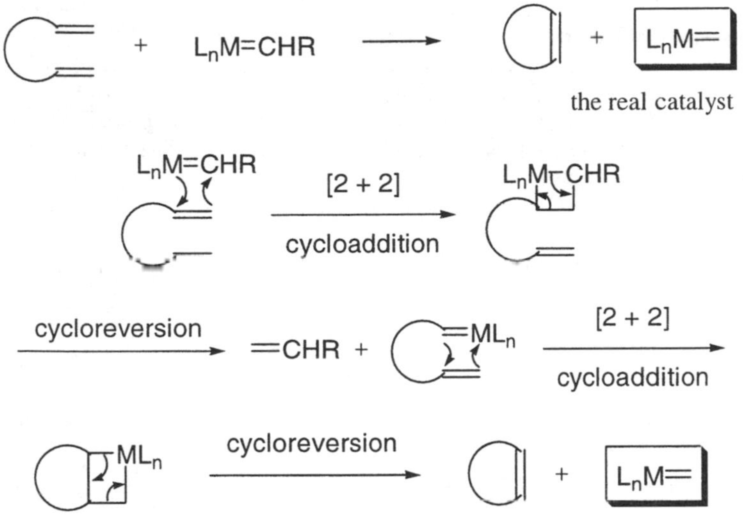

the real catalyst

Catalytic cycle:

338

References

1. Schrock R. R.; Murdzek, J. S.; Bazan, G. C.; Robbins, J.; DiMare, M.; O'Reagan, M. *J. Am. Chem. Soc.* **1990**, *112*, 3875.
2. Grubbs, R. H.; Miller, S. J.; Fu, G. C. *Acc. Chem. Res.* **1995**, *28,* 446. (Review).
3. Armstrong, S. K. *J. Chem. Soc., Perkin Trans. 1* **1998**, 371.
4. Morgan, J. P.; Grubbs, R. H. *Org. Lett.* **2000**, *2*, 3153.
5. Renaud, J.; Graf, C.-D.; Oberer, L. *Angew. Chem., Int. Ed.* **2000**, *39*, 3101.
6. Lane, C.; Snieckus, V. *Synlett* **2000**, 1294.
7. Fellows, I. M.; Kaelin, D. E., Jr.; Martin, S. F. *J. Am. Chem. Soc.* **2000**, *122*, 10781.
8. Timmer, M. S. M.; Ovaa, H.; Filippov, D. V.; Van der Marel, G. A.; Van Boom, J. H. *Tetrahedron Lett.* **2000**, *41*, 8635.
9. Lee, C. W.; Grubbs, R. H. *J. Org. Chem.* **2001**, *66*, 7155.
10. Morgan, J. P.; Morrill, C.; Grubbs, R. H. *Org. Lett.* **2002**, *4*, 67.
11. van Otterlo, W. A. L.; Ngidi, E. L.; Coyanis, E. M.; de Koning, C. B. *Tetrahedron Lett.* **2003**, *44*, 311.

Ritter reaction

Amides from nitriles and alcohols in strong acids.
General scheme:

$$R^1\text{-OH} + R^2\text{-CN} \xrightarrow{H^+} R^1\underset{H}{\overset{}{N}}\overset{O}{\underset{}{C}}R^2$$

e.g.:

$$\text{(CH}_3)_3\text{C-OH} + H_3\text{C-CN} \xrightarrow[H_2O]{H_2SO_4}$$

Similarly:

$$\text{(CH}_3)_2\text{C=CH}_2 + H_3\text{C-CN} \xrightarrow[H_2O]{H_2SO_4}$$

References

1. Ritter, J. J.; Minieri, P. P. *J. Am. Chem. Soc.* **1948**, *70*, 4045.
2. Ritter, J. J.; Kalish, J. *J. Am. Chem. Soc.* **1948**, *70*, 4048.
3. Krimen, L. I.; Cota, D. J. *Org. React.* **1969**, *17*, 2123. (Review).
4. Djaidi, D.; Leung, I. S. H.; Bishop, R.; Craig, D. C.; Scudder, M. L. *Perkin 1* **2000**, 2037.
5. Jirgensons, A.; Kauss, V.; Kalvinsh, I.; Gold, M. R. *Synthesis* **2001**, 1709.
6. Le Goanic, D.; Lallemand, M.-C.; Tillequin, F.; Martens, T. *Tetrahedron Lett.* **2001**, *42*, 5175.
7. Nair, V.; Rajan, R.; Rath, N. P. *Org. Lett.* **2002**, *4*, 1575.
8. Reddy, K. L. *Tetrahedron Lett.* **2003**, *44*, 1453.

340

Robinson annulation

Michael addition of cyclohexanones to methyl vinyl ketone followed by intramolecular aldol condensation to afford six-membered α,β-unsaturated ketones.

methyl vinyl ketone (MVK)

enolate
formation

Michael
addition

isomerization

aldol
addition

− H₂O

dehydration

References

1. Rapson, W. S.; Robinson, R. *J. Chem. Soc.* **1935**, 1285.
2. Gawley, R. E. *Synthesis* **1996**, 777. (Review).
3. Bui, T.; Barbas, C. F., III *Tetrahedron Lett.* **2000**, *41*, 6951.
4. Jansen, B. J. M.; Hendrix, C. C. J.; Masalov, N.; Stork, G. A.; Meulemans, T. M.; Macaev, F. Z.; De Groot, A. *Tetrahedron* **2000**, *56*, 2075.
5. Guarna, A.; Lombardi, E.; Machetti, F.; Occhiato, E. G.; Scarpi, D. *J. Org. Chem.* **2000**, *65*, 8093.
6. Tai, C.-L.; Ly, T. W.; Wu, J.-D.; Shia, K.-S,; Liu, H.-J. *Synlett* **2001**, 214.
7. Jung, M. E.; Piizzi, G. *Org. Lett.* **2003**, *5*, 137.
8. Liu, H.-J.; Ly, T. W.; tai, C.-L.; Wu, J.-D. *et al. Tetrahedron* **2003**, *59*, 1209.

Robinson–Schöpf reaction

Tropinone synthesis.

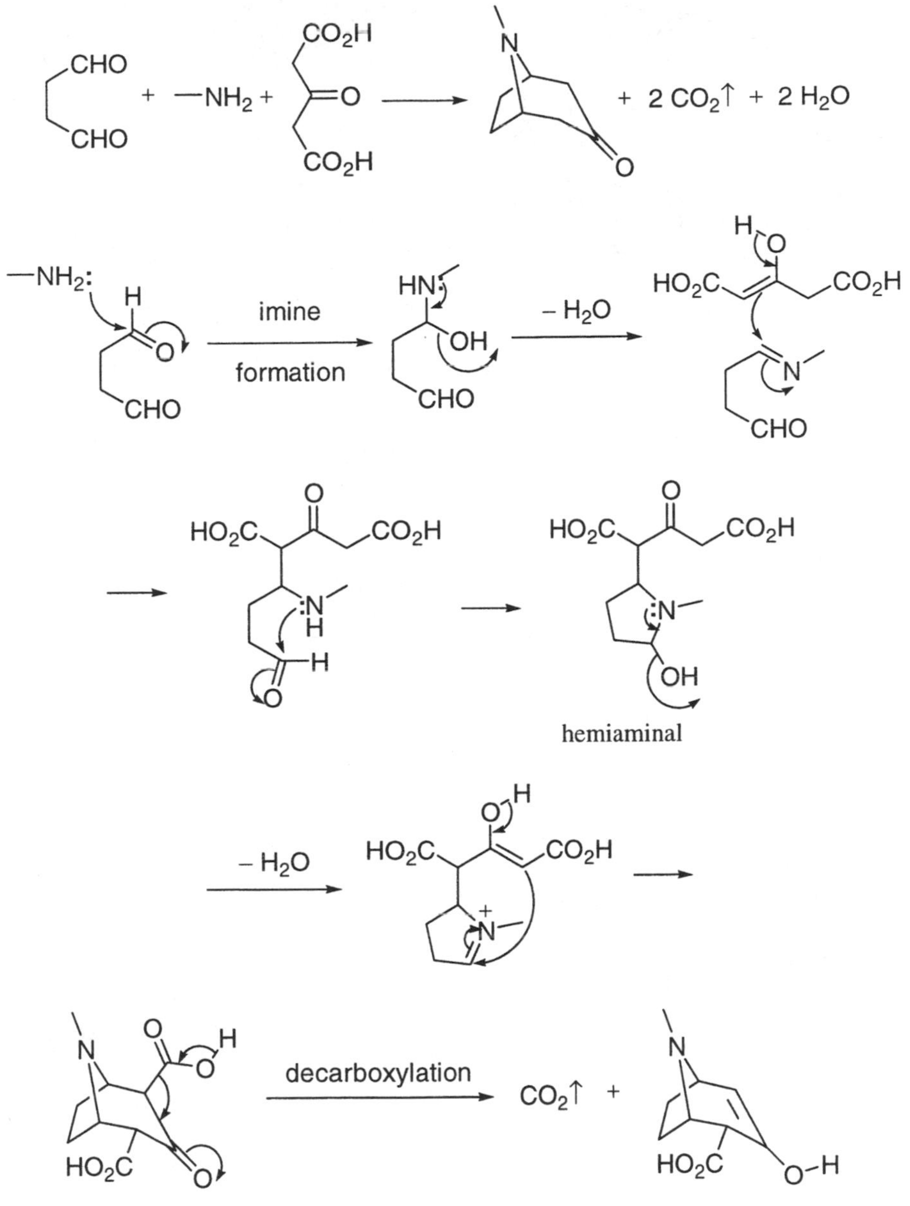

hemiaminal

decarboxylation

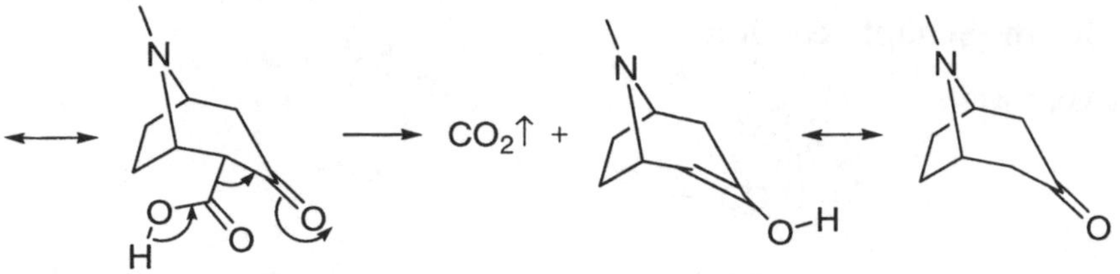

References

1. Robinson, R. *J. Chem. Soc.* **1917**, *111*, 762.
2. Büchi, G.; Fliri, H.; Shapiro, R. *J. Org. Chem.* **1978**, *43*, 4765.
3. Guerrier, L.; Royer, J.; Grierson, D. S.; Husson, H. P. *J. Am. Chem. Soc.* **1983**, *105*, 7754.
4. Royer, J.; Husson, H. P. *Tetrahedron Lett.* **1987**, *28*, 6175.
5. Langlois, M.; Yang, D.; Soulier, J. L.; Florac, C. *Synth. Commun.* **1992**, *22*, 3115.
6. Jarevang, T.; Anke, H.; Anke, T.; Erkel, G.; Sterner, O. *Acta Chem. Scand.* **1998**, *52*, 1350.

Rosenmund reduction

Hydrogenation reduction of acid chloride to aldehyde using $BaSO_4$-poisoned palladium catalyst. Without poison, the resulting aldehyde may be further reduced to alcohol.

References

1. Rosenmund, K. W. *Ber. Dtsch. Chem. Ges.* **1918**, *51*, 585.
2. Mosettig, E.; Mozingo, R. *Org. React.* **1948**, *4*, 362. (Review).
3. Burgstahler, A. W.; Weigel, L. O.; Schäfer, C. G. *Synthesis* **1976**, 767.
4. McEwen, A. B.; Guttieri, M. J.; Maier, W. L.; Laine, R. M.; Shvo, Y. *J. Org. Chem.* **1983**, *48*, 4436.
5. Bold, G.; Steiner, H.; Moesch, L.; Walliser, B. St. Pfau, A.; Plattner, P. A. *Helv. Chim. Acta* **1990**, *73*, 405.
6. Yadav, V. G.; Chandalia, S. B. *Org. Proc. Res. Dev.* **1997**, *1*, 226.
7. Chandnani, K. H.; Chandalia, S. B. *Org. Proc. Res. Dev.* **1999**, *3*, 416.
8. Chimichi, S.; Boccalini, M.; Cosimelli, B. *Tetrahedron* **2002**, *58*, 4851.

Roush allylboronate reagent

Tartrate allyl boronate, asymmetric allylation agent

References

1. Roush, W. R.; Walts, A. E.; Hoong, L. K. *J. Am. Chem. Soc.* **1985**, *107*, 8186.
2. Roush, W. R.; Adam, M. A.; Walts, A. E.; Harris, D, J. *J. Am. Chem. Soc.* **1986**, *108*, 3422.
3. Roush, W. R.; Ando, K.; Powers, D. B.; Halterman, R. L.; Palkowitz, A. D. *Tetrahedron Lett.* **1988**, *29*, 5579.
4. Brown, H. C.; Racherla, U. S.; Pellechia, P. J. *J. Org. Chem.* **1990**, *55*, 1868.
5. Kadota, I.; Yamamoto, Y. *Chemtracts: Org. Chem.* **1992**, *5*, 242. (Review).
6. White, J. D.; Tiller, T.; Ohba, Y.; Porter, W. J.; Jackson, R. W.; Wang, S.; Hanselmann, R. *Chem. Commun.* **1998**, 79.
7. Yamamoto, Y.; Takahashi, M.; Miyaura, N. *Synlett* **2002**, 128.
8. Mandal, A. K. *Org. Lett.* **2002**, *4*, 2043.
9. Kozlowski, M. C.; Panda, M. *J. Org. Chem.* **2003**, *68*, 2061.

Rubottom oxidation

α-Hydroxylation of enolsilanes.

The "butterfly" transition state

References

1. Rubottom, G. M.; Vazquez, M. A.; Pelegrina, D. R. *Tetrahedron Lett.* **1974**, 4319.
2. Brook, A. G.; Macrae, D. M. *J. Organomet. Chem.* **1974**, *77*, C19.
3. Hassner, A.; Reuss, R. H.; Pinnick, H. W. *J. Org. Chem.* **1975**, *40*, 3427.
4. Rubottom, G. M.; Gruber, J. M.; Boeckman, R. K., Jr.; Ramaiah, M.; Medwid, J. B. *Tetrahedron Lett.* **1978**, 4603.
5. Paquette, L. A.; Lin, H.-S.; Coghlan, M. J. *Tetrahedron Lett.* **1987**, *28*, 5017.
6. Hirota, H.; Yokoyama, A.; Miyaji, K.; Nakamura, T.; Takahashi, T. *Tetrahedron Lett.* **1987**, *28*, 435.
7. Gleiter, R.; Kraemer, R.; Irngartinger, H.; Bissinger, C. *J. Org. Chem.* **1992**, *57*, 252.
8. Johnson, C. R.; Golebiowski, A.; Steensma, D. H. *J. Am. Chem. Soc.* **1992**, *114*, 9414.
9. Jauch, J. *Tetrahedron* **1994**, *50*, 1203.
10. Gleiter, R.; Staib, M.; Ackermann, U. *Liebigs Ann.* **1995**, 1655.
11. Xu, Y.; Johnson, C. R. *Tetrahedron Lett.* **1997**, *38*, 1117.

Rupe rearrangement

The acid-catalyzed rearrangement of tertiary α-acetylenic (terminal) alcohols, leading to the formation of α,β-unsaturated ketones rather than the corresponding α,β-unsaturated aldehydes. *Cf.* Meyer–Schuster rearrangement.

References

1. Schmidt, C.; Thazhuthaveetil, J. *Tetrahedron Lett.* **1970**, 2653.
2. Swaminathan, S.; Narayanan, K. V. *Chem. Rev.* **1971**, *71*, 429. (Review).
3. Hasbrouck, R. W.; Anderson, A. D. *J. Org. Chem.* **1973**, *38*, 2103.
4. Barre, V.; Massias, F.; Uguen, D. *Tetrahedron Lett.* **1989**, *30*, 7389.
5. An, J.; Bagnell, L.; Cablewski, T.; Strauss, C. R.; Trainor, R. W. *J. Org. Chem.* **1997**, *62*, 2505.
6. Strauss, C. R. *Aust. J. Chem.* **1999**, *52*, 83.
7. Weinmann, H.; Harre, M.; Neh, H.; Nickisch, K.; Skötsch, C.; Tilstam, U. *Org. Proc. Res. Dev.* **2002**, *6*, 216.

Rychnovsky polyol synthesis

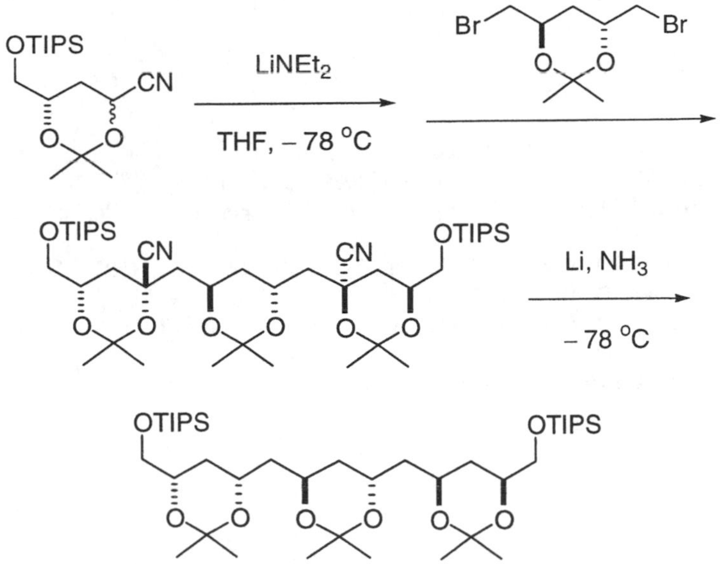

The stereochemical outcome of the reductive decyanation:

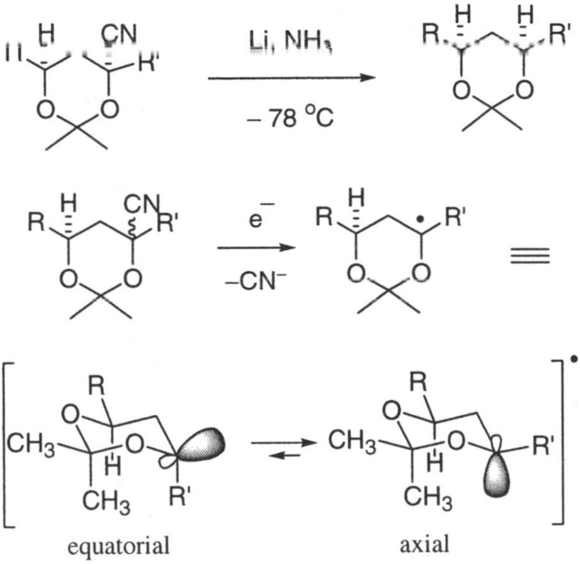

equatorial   axial

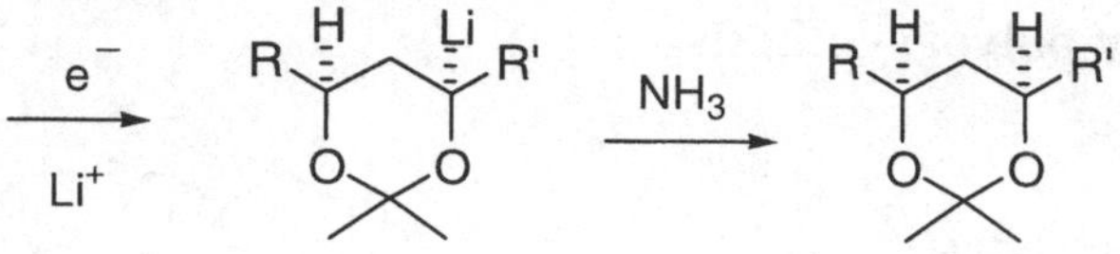

References

1. Cohen, T.; Lin, M. T. *J. Am. Chem. Soc.* **1984**, *106*, 1130.
2. Cohen, T.; Bhupathy, M. *Acc. Chem. Res.* **1989**, *22*, 152. (Review).
3. Rychnovsky, S. D.; Zeller, S.; Skalitzky, D. J.; Griesgraber, G. *J. Org. Chem.* **1990**, *55*, 5550.
4. Rychnovsky, S. D.; Powers, J. P.; Lepage, T. J. *J. Am. Chem. Soc.* **1992**, *114*, 8375.
5. Rychnovsky, S. D.; Hoye, R. C. *J. Am. Chem. Soc.* **1994**, *116*, 1753.
6. Rychnovsky, S. D.; Griesgraber, G.; Kim, J. *J. Am. Chem. Soc.* **1994**, *116*, 2621.
7. Rychnovsky, S. D. *Chem. Rev.* **1995**, *95*, 2021. (Review).
8. Richardson, T. I.; Rychnovsky, S. D. *J. Am. Chem. Soc.* **1997**, *119*, 12360.

Sakurai allylation reaction (Hosomi–Sakurai reaction)

Lewis acid-mediated addition of allylsilanes to carbon nucleophiles.

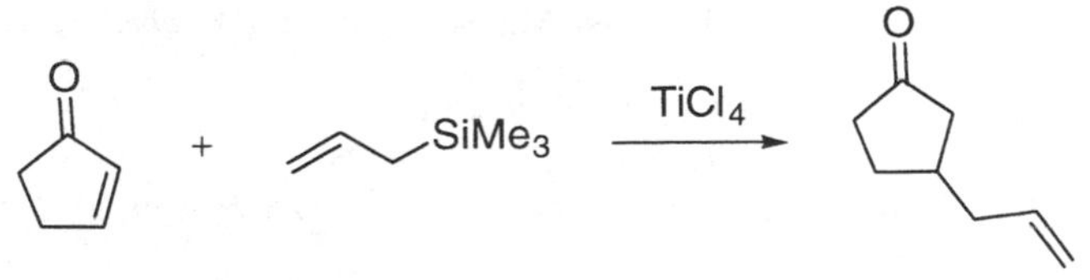

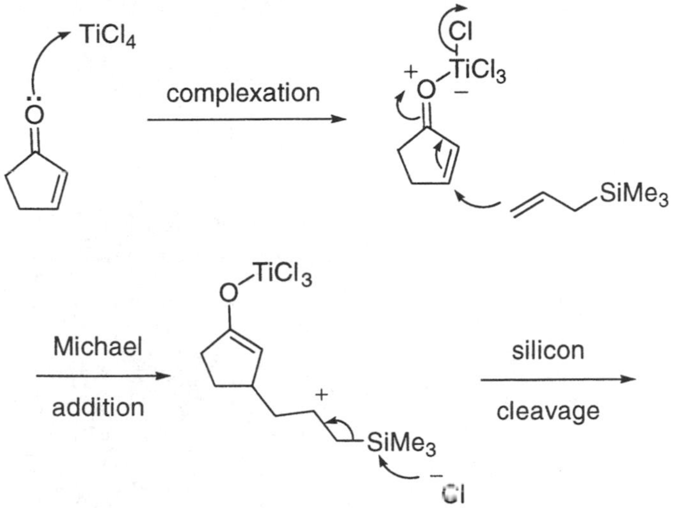

The β-carbocation is stabilized by the silicon group

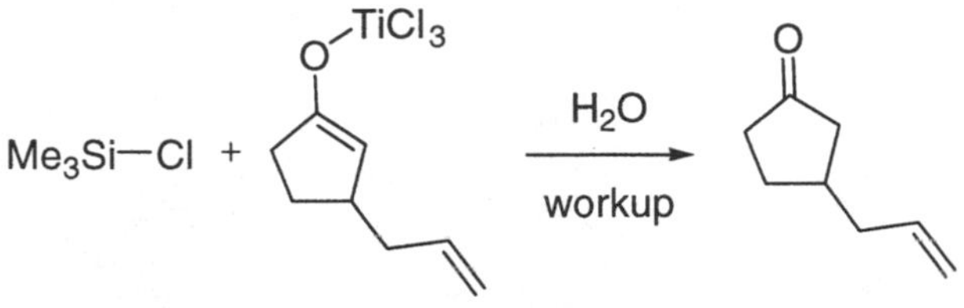

References

1. Hosomi, A.; Sakurai, H. *Tetrahedron Lett.* **1976**, 1295.
2. Marko, I. E.; Mekhalfia, A.; Murphy, F.; Bayston, D. J.; Bailey, M.; Janoouusek, Z.; Dolan, S. *Pure Appl. Chem.* **1997**, *69*, 565.
3. Bonini, B. F.; Comes-Franchini, M.; Fochi, M.; Mazzanti, G.; Ricci, A.; Varchi, G. *Tetrahedron: Asymmetry* **1998**, *9*, 2979.
4. Wang, D.-K.; Zhou, Y.-G.; Tang, Y.; Hou, X.-L.; Dai, L.-X. *J. Org. Chem.* **1999**, *64*, 4233.

5. Sugita, Y.; Kimura, Y.; Yokoe, I. *Tetrahedron Lett.* **1999**, *40*, 5877.
6. Wang, M. W.; Chen, Y. J.; Wang, D. *Synlett* **2000**, 385.
7. Organ, M. G.; Dragan, V.; Miller, M.; Froese, R. D. J.; Goddard, J. D. *J. Org. Chem.* **2000**, *65*, 3666.
8. Tori, M.; Makino, C.; Hisazumi, K.; Sono, M.; Nakashima, K. *Tetrahedron: Asymmetry* **2001**, *12*, 301.
9. Leroy, B.; Marko, I. E. *J. Org. Chem.* **2002**, *67*, 8744.
10. Itsuno, S.; Kumagai, T. *Helvet. Chim. Acta* **2002**, *85*, 3185.
11. Nosse, B.; Chhor, R. B.; Jeong, W. B.; Boehm, C.; Reiser, O. *Org. Lett.* **2003**, *5*, 941.

Sandmeyer reaction

Haloarenes from the reaction of a diazonium salt with CuX.

$$ArN_2^+ \ Y^- \ \xrightarrow{\text{CuX}} \ Ar-X \qquad X = Cl, \ Br, \ CN$$

e.g.:

$$ArN_2^+ \ Cl^- \ \xrightarrow{\text{CuCl}} \ Ar-Cl$$

$$ArN_2^+ \ Cl^- \ \xrightarrow{\text{CuCl}} \ N_2\uparrow \ + \ Ar\bullet \ + \ CuCl_2 \ \longrightarrow \ Ar-Cl \ + \ CuCl$$

References

1. Sandmeyer, T. *Ber. Dtsch. Chem. Ges.* **1884**, *17*, 1633.
2. Galli, C. *J. Chem. Soc., Perkin Trans. 2* **1984**, 897.
3. Suzuki, N.; Azuma, T.; Kaneko, Y.; Izawa, Y.; Tomioka, H.; Nomoto, T. *J. Chem. Soc., Perkin Trans. 1* **1987**, 645.
4. Merkushev, E. B. *Synthesis* **1988**, 923.
5. Obushak, M. D.; Lyakhovych, M. B.; Ganushchak, M. I. *Tetrahedron Lett.* **1998**, *39*, 9567.
6. Hanson, P.; Lovenich, P. W.; Rowell, S. C.; Walton, P. H.; Timms, A. W. *J. Chem. Soc., Perkin Trans. 2* **1999**, 49.
7. Chandler, St. A.; Hanson, P.; Taylor, A. B.; Walton, P. H.; Timms, A. W. *J. Chem. Soc., Perkin Trans. 2* **2001**, 214.
8. Hanson, P.; Rowell, S. C.; Taylor, A. B.; Walton, P. H.; Timms, A. W. *J. Chem. Soc., Perkin Trans. 2* **2002**, 1126.
9. Hanson, P.; Jones, J. R.; Taylor, A. B.; Walton, P. H.; Timms, A. W. *J. Chem. Soc., Perkin Trans. 2* **2002**, 1135.
10. Daab, J. C.; Bracher, F. *Monatsh. Chem.* **2003**, *134*, 573.

Sarett oxidation

Oxidation of alcohols to the corresponding carbonyl compounds using chromium trioxide-pyridine complex.

The intramolecular mechanism is also operative:

The Collins oxidation, Jones oxidation, and Corey's PCC (pyridinium chlorochromate) and PDC (pyridinium dichromate) oxidations follow a similar pathway.

References

1. Poos, G. I.; Arth, G. E.; Beyler, R. E.; Sarett, L. H. *J. Am. Chem. Soc.* **1953**, *75*, 422.
2. Jones, R. E.; Kocher, F. W. *J. Am. Chem. Soc.* **1953**, *76*, 3682.
3. Ratcliffe, R. W. *Org. Syn.* **1973**, *53*, 1852.
4. Andrieux, J.; Bodo, B.; Cunha, H.; Deschamps-Vallet, C.; Meyer-Dayan, M.; Molho, D. *Bull. Soc. Chim. Fr.* **1976**, 1975.
5. Gomez-Garibay, F.; Quijano, L.; Pardo, J. S. C.; Aguirre, G.; Rios, T. *Chem. Ind.* **1986**, 827.
6. Glinski, J. A.; Joshi, B. S.; Jiang, Q. P.; Pelletier, S. W. *Heterocycles* **1988**, *27*, 185.

7. Turjak-Zebic, V.; Makarevic, J.; Skaric, V. *J. Chem. Soc. (S)* **1991**, 132.
8. Luzzio, F. A. *Org. React.* **1998**, *53*, 1–222. (Review).
9. Caamano, O.; Fernandez, F.; Garcia-Mera, X.; Rodriguez-Borges, J. E. *Tetrahedron Lett.* **2000**, *41*, 4123.

Schiemann reaction (Balz–Schiemann reaction)

Fluoroarene formation from arylamines.

$$Ar-NH_2 + HNO_2 + HBF_4 \longrightarrow ArN_2^+ \ BF_4^- \xrightarrow{\Delta} Ar-F + N_2\uparrow + BF_3$$

$$H_2O + N{\equiv}O^+ \xrightarrow{\quad HO^{N}{\scriptstyle\gtrless}O \quad} O{\scriptstyle\lessgtr}N{\scriptstyle\diagdown}O{\scriptstyle\diagup}N{\scriptstyle\gtrless}O$$

$$ArN_2^+ \ BF_4^- \xrightarrow{\Delta} N_2\uparrow + Ar^+ + \overset{-}{F}-BF_3 \longrightarrow Ar-F + BF_3$$

References

1. Balz, G.; Schiemann. G. *Ber. Dtsch. Chem. Ges.* **1927**, *60*, 1186.
2. Sharts, C. M. *J. Chem. Educ.* **1968**, *45*, 185.
3. Matsumoto, J.; Miyamoto, T.; Minamida, A.; Nishimura, Y.; Egawa, H.; Nishimura, H. *J. Heterocycl. Chem.* **1984**, *21*, 673.
4. Corral, C.; Lasso, A.; Lissavetzky, J.; Sanchez Alvarez-Insua, A.; Valdeolmillos, A. M. *Heterocycles* **1985**, *23*, 1431.
5. Tsuge, A.; Moriguchi, T.; Mataka, S.; Tashiro, M. *J. Chem. Res., (S)* **1995**, 460.
6. Saeki, K.-i.; Tomomitsu, M.; Kawazoe, Y.; Momota, K.; Kimoto, H. *Chem. Pharm. Bull.* **1996**, *44*, 2254.
7. Laali, K. K.; Gettwert, V. J. *J. Fluorine Chem.* **2001**, *107*, 31.
8. Gronheid, R.; Lodder, G.; Okuyama, T. *J. Org. Chem.* **2002**, *67*, 693-702.

Schlosser modification of the Wittig reaction

The normal Wittig reaction of nonstabilized ylides with aldehydes gives Z-olefins. The Schlosser modification of the Wittig reaction of nonstabilized ylides furnishes E-olefins instead.

phosphorus ylide

LiBr complex of β-oxide ylide

LiBr complex of *threo*-betaine

References

1. Schlosser, M.; Christmann, K. F. *Angew. Chem., Int. Ed. Engl.* **1966**, *5*, 126.
2. Schlosser, M.; Christmann, K. F. *Justus Liebigs Ann. Chem.* **1967**, *708*, 35.
3. Schlosser, M.; Christmann, K. F.; Piskala, A.; Coffinet, D. *Synthesis* **1971**, 29.
4. Deagostino, A.; Prandi, C.; Tonachini, G.; Venturello, P. *Trends Org. Chem.* **1995**, *5*, 103. (Review).
5. Celatka, C. A.; Liu, P.; Panek, J. S. *Tetrahedron Lett.* **1997**, *38*, 5449.
6. Duffield, J. J.; Pettit, G. R. *J. Nat. Products* **2001**, *64*, 472.

356

Schmidt reaction

Conversion of ketones to amides using HN_3.

nitrilium ion intermediate (*Cf.* Ritter intermediate)

References

1. Schmidt, R. F. *Ber. Dtsch. Chem. Ges.* **1924**, *57*, 704.
2. Richard, J. P.; Amyes, T. L.; Lee, Y.-G.; Jagannadham, V. *J. Am. Chem. Soc.* **1994**, *116*, 10833.
3. Kaye, P. T.; Mphahlele, M. J. *Synth. Commun.* **1995**, *25*, 1495.
4. Krow, G. R.; Szczepanski, S W.; Kim, J. Y.; Liu, N.; Sheikh, A.; Xiao, Y.; Yuan, J. *J. Org. Chem.* **1999**, *64*, 1254.
5. Mphahlele, M. J. *Phosphorus, Sulfur Silicon Relat. Elem.* **1999**, *144-146*, 351.
6. Mphahlele, M. J. *J. Chem. Soc., Perkin Trans. 1* **1999**, 3477.
7. Iyengar, R.; Schildknegt, K.; Aubé, J. *Org. Lett.* **2000**, *2*, 1625.
8. Pearson, W. H.; Hutta, D. A.; Fang, W.-k. *J. Org. Chem.* **2000**, *65*, 8326.
9. Pearson, W. H.; Walavalkar, R. *Tetrahedron* **2001**, *57*, 5081.
10. Golden, J. E.; Aube, J. *Angew. Chem., Int. Ed.* **2002**, *41*, 4316.
11. Cristau, H.-J.; Marat, X.; Vors, J.-P.; Pirat, J.-L. *Tetrahedron Lett.* **2003**, *44*, 3179.

Schmidt's trichloroacetimidate glycosidation reaction

Lewis acid-promoted glycosidation of trichloroacetimidates with alcohols or phenols.

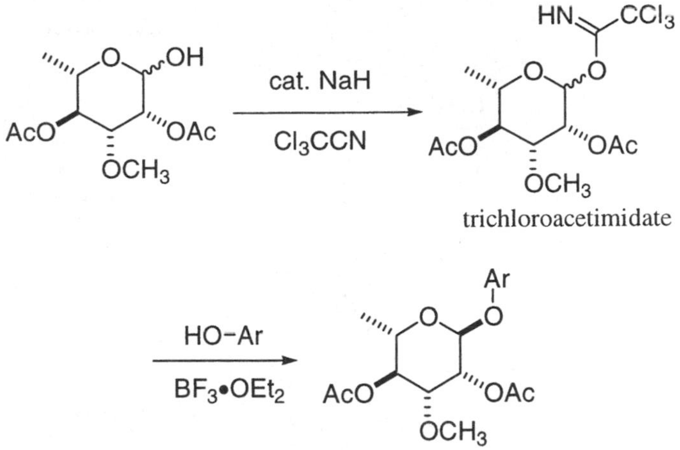

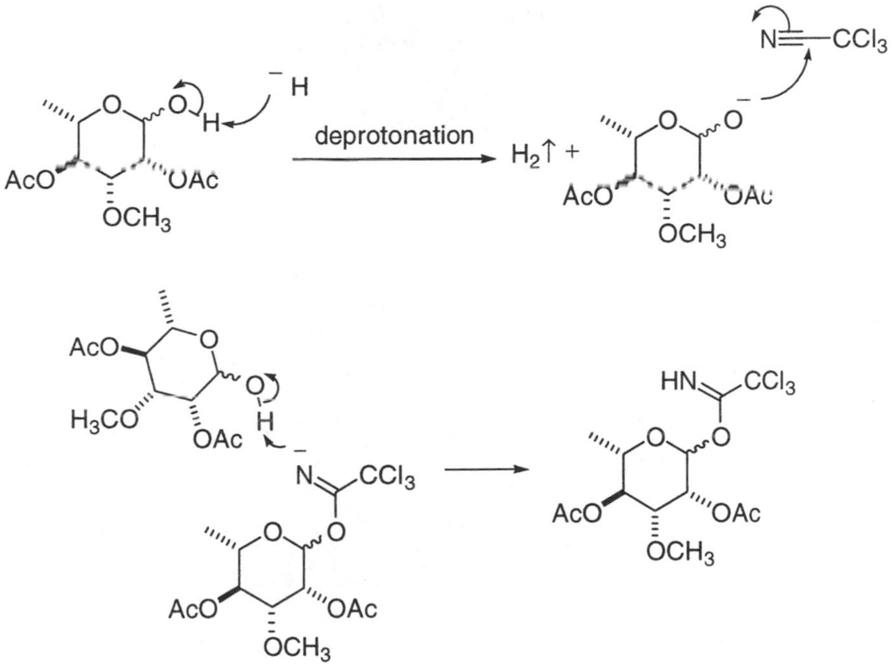

trichloroacetimidate

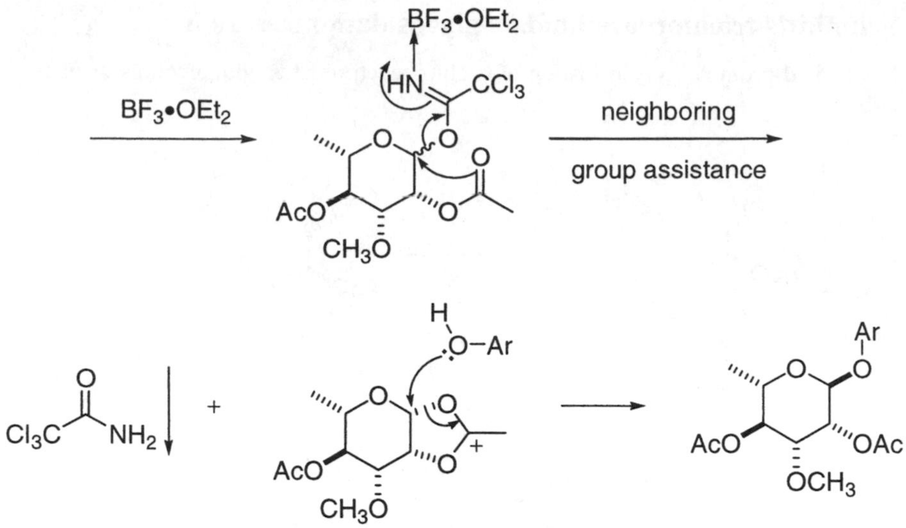

References

1. Grundler, G.; Schmidt, R. R. *Carbohydr. Res.* **1985**, *135*, 203.
2. Schmidt, R. R. *Angew. Chem., Int. Ed. Engl.* **1986**, *25*, 212.
3. Toshima, K.; Tatsuta, K. *Chem. Rev.* **1993**, *93*, 1503. (Review).
4. Nicolaou, K. C. *Angew. Chem., Int. Ed. Engl.* **1993**, *32*, 1377.
5. Weingart, R.; Schmidt, R. R. *Tetrahedron Lett.* **2000**, *41*, 8753.
6. Yan, L. Z.; Mayer, J. P. *Org. Lett.* **2003**, *5*, 1161.

Scholl reaction

The elimination of two aryl-bound hydrogens accompanied by the formation of an aryl-aryl bond under the influence of Friedel–Crafts catalysts. *Cf.* Friedel–Crafts reaction.

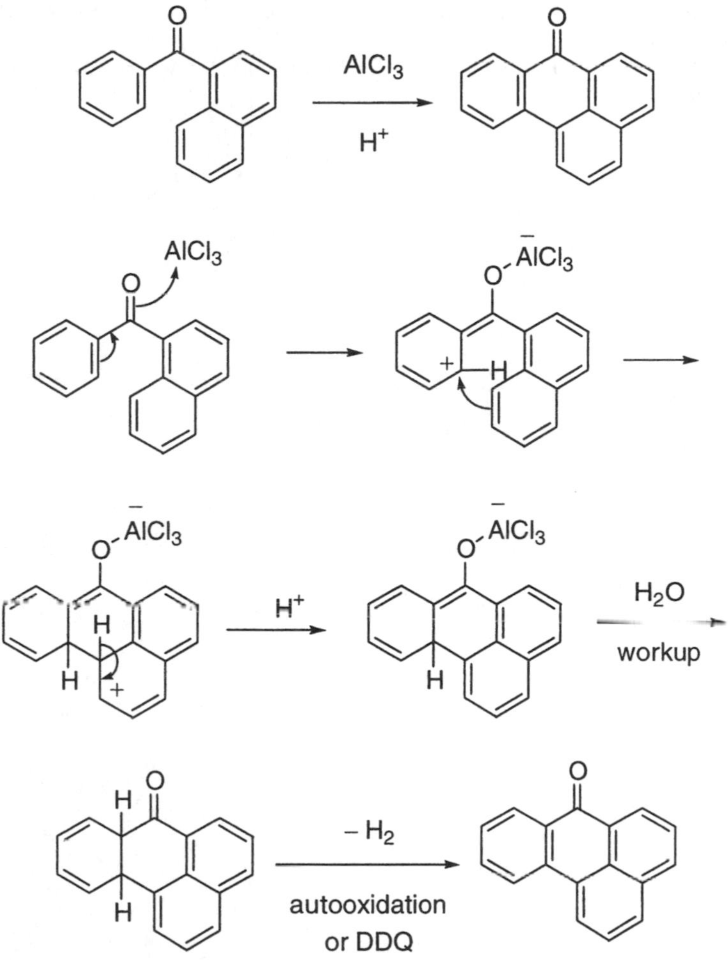

References

1. Scholl, R.; Seer, C. *Ann,* **1912**, *394*, 111.
2. Clowes, G. A. *J. Chem. Soc., C* **1968**, 2519.
3. Olah, G. A.; Schilling, P.; Gross, I. M. *J. Am. Chem. Soc.* **1974**, *96*, 876.
4. Dopper, J. H.; Oudman, D.; Wynberg, H. *J. Org. Chem.* **1975**, *40*, 3398.

5. Poutsma, M. L.; Dworkin, A. S.; Brynestad, J.; Brown, L. L.; Benjamin, B. M.; Smith, G. P. *Tetrahedron Lett.* **1978**, 873.

6. Youssef, A. K.; Vingiello, F. A.; Ogliaruso, M. A. *Org. Prep. Proced. Int.* **1979**, *11*, 17.

7. Pritchard, R. G.; Steele, M.; Watkinson, M.; Whiting, A. *Tetrahedron Lett.* **2000**, *41*, 6915.

8. Ma, C.; Liu, X.; Li, X.; Flippen-Anderson, J.; Yu, S.; Cook, J. M. *J. Org. Chem.* **2001**, *66*, 4525.

9. Rozas, M. F.; Piro, O. E.; Castellano, E. E.; Mirifico, M. V.; Vasini, E. J. *Synthesis* **2002**, 2399.

Schöpf reaction

Keto-piperidine from cyclic imine and β-ketoacetate.

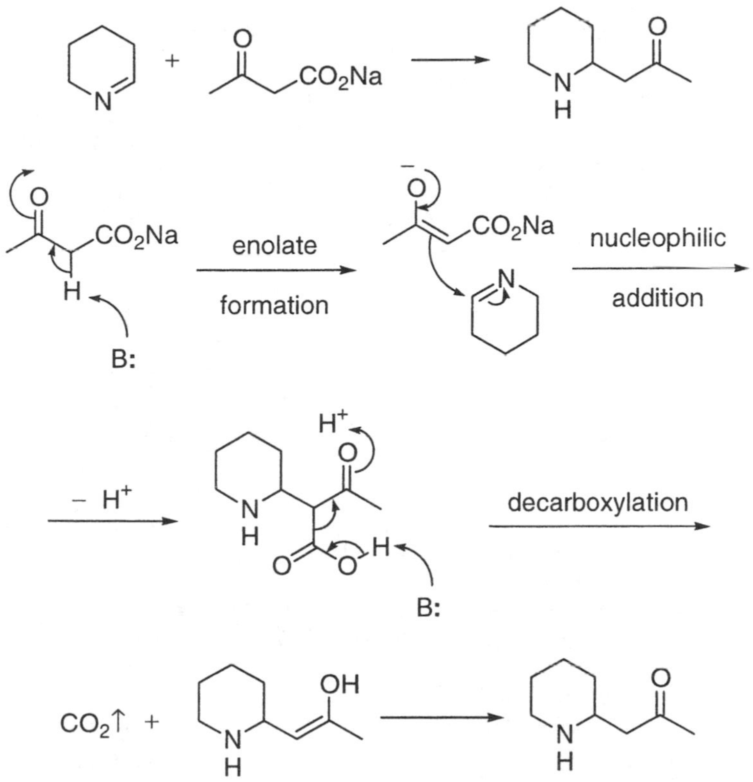

References

1. Schöpf, C.; Braun, F.; Burkhardt, K.; Dummer, G.; Müller, H. *Ann,* **1959**, *626*, 123.
2. Bender, D. R.; Bjelfdanes, L. F.; Knapp, D. R.; Rapopport, H. *J. Org. Chem.* **1975**, *40*, 1264.
3. Guerrier, L.; Royer, J.; Grierson, D. S.; Husson, H. P. *J. Am. Chem. Soc.* **1983**, *105*, 7754.
4. Bermudez, J.; Gregory, J. A.; King, F. D.; Starr, S.; Summersell, R. J. *Bioorg. Med. Chem. Lett.* **1992**, *2*, 519.
5. Jarevang, T.; Anke, H.; Anke, T.; Erkel, G.; Sterner, O. *Acta Chem. Scand.* **1998**, *52*, 1350.

Schotten–Baumann reaction

Esterification or amidation of acid chloride with alcohol or amine under basic conditions.

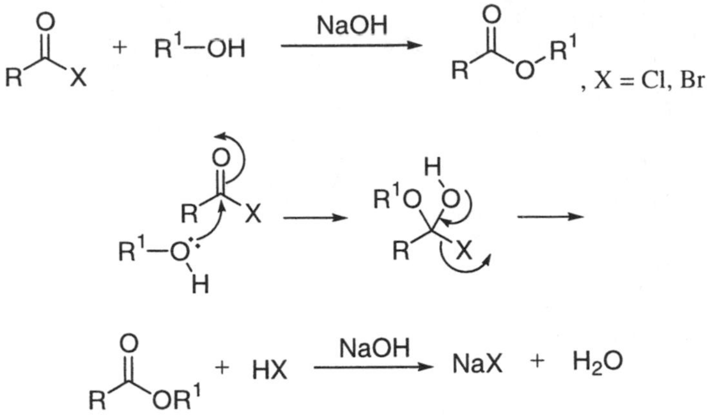

, X = Cl, Br

References

1. Schotten, C. *Ber. Dtsch. Chem. Ges.* **1884**, *17*, 2544.
2. Altman, J.; Ben-Ishai, D. *J. Heterocycl. Chem.* **1968**, *5*, 679.
3. Babad, E.; Ben-Ishai, D. *J. Heterocycl. Chem.* **1969**, *6*, 235.
4. Tsuchiya, M.; Yoshida, H.; Ogata, T.; Inokawa, S. *Bull. Chem. Soc. Jpn.* **1969**, *42*, 1756.
5. Gutteridge, N. J. A.; Dales, J. R. M. *J. Chem. Soc., C* **1971**, 122.
6. Low, C. M. R.; Broughton, H. B.; Kalindjian, S. B.; McDonald, I. M. *Bioorg. Med. Chem. Lett.* **1992**, *2*, 325.
7. Sano, T.; Sugaya, T.; Inoue, K.; Mizutaki, S.-i.; Ono, Y.; Kasai, M. *Org. Process Res. Dev.* **2000**, *4*, 147.

Shapiro reaction

The Shapiro reaction is a variant of the Bamford–Stevens reaction. The former uses bases such as alkyl lithiums and Grignard reagents whereas the latter employs bases such as Na, NaOMe, LiH, NaH, NaNH$_2$, *etc.* Consequently, the Shapiro reaction generally affords the less-substituted olefins as the kinetic products, while the Bamford–Stevens reaction delivers the more-substituted olefins as the thermodynamic products.

References

1. Bamford, W. R.; Stevens, T. S. M. *J. Chem. Soc.* **1952**, 4735.
2. Casanova, J.; Waegell, B. *Bull. Soc. Chim. Fr.* **1975**, 922.
3. Shapiro, R. H. *Org. React.* **1976**, *23*, 405. (Review).
4. Adlington, R. M.; Barrett, A. G. M. *Acc. Chem. Res.* **1983**, *16*, 55. (Review).
5. Corey, E. J.; Lee, J.; Roberts, B. E. *Tetrahedron Lett.* **1997**, *38*, 8915.
6. Corey, E. J.; Roberts, B. E. *Tetrahedron Lett.* **1997**, *38*, 8919.
7. Kurek-Tyrlik, A.; Marczak, S.; Michalak, K.; Wicha, J. *Synlett* **2000**, 547.
8. Kurek-Tyrlik, A.; Marczak, S.; Michalak, K.; Wicha, J.; Zarecki, A. *J. Org. Chem.* **2001**, *65*, 6994.
9. Tormakangas, O. P.; Toivola, R. J.; Karvinen, E. K.; Koskinen, A. M. P. *Tetrahedron* **2002**, *58*, 2175.

Sharpless asymmetric amino hydroxylation

Osmium-mediated *cis*-addition of nitrogen and oxygen to olefins. Nitrogen sources (X–NClNa) include:

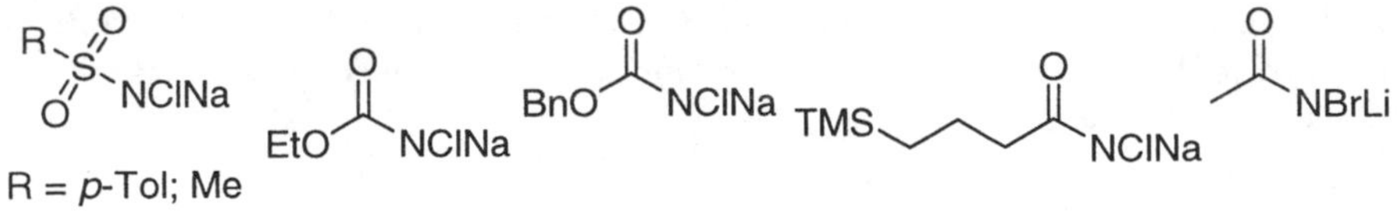
R = p-Tol; Me
NClNa
EtO
BnO
TMS
NBrLi

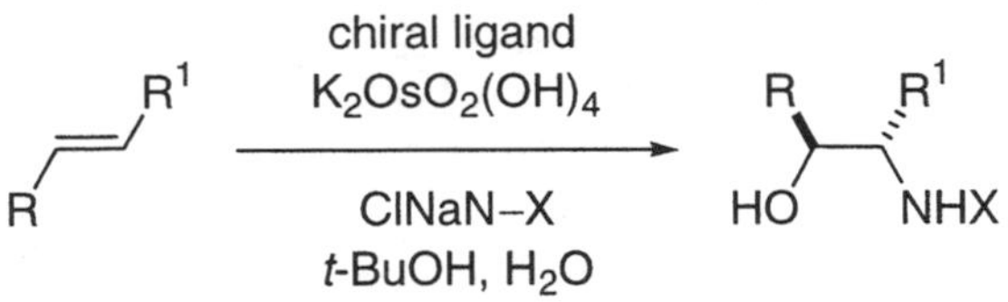
chiral ligand
$K_2OsO_2(OH)_4$
ClNaN–X
t-BuOH, H_2O
HO
NHX

The catalytic cycle:

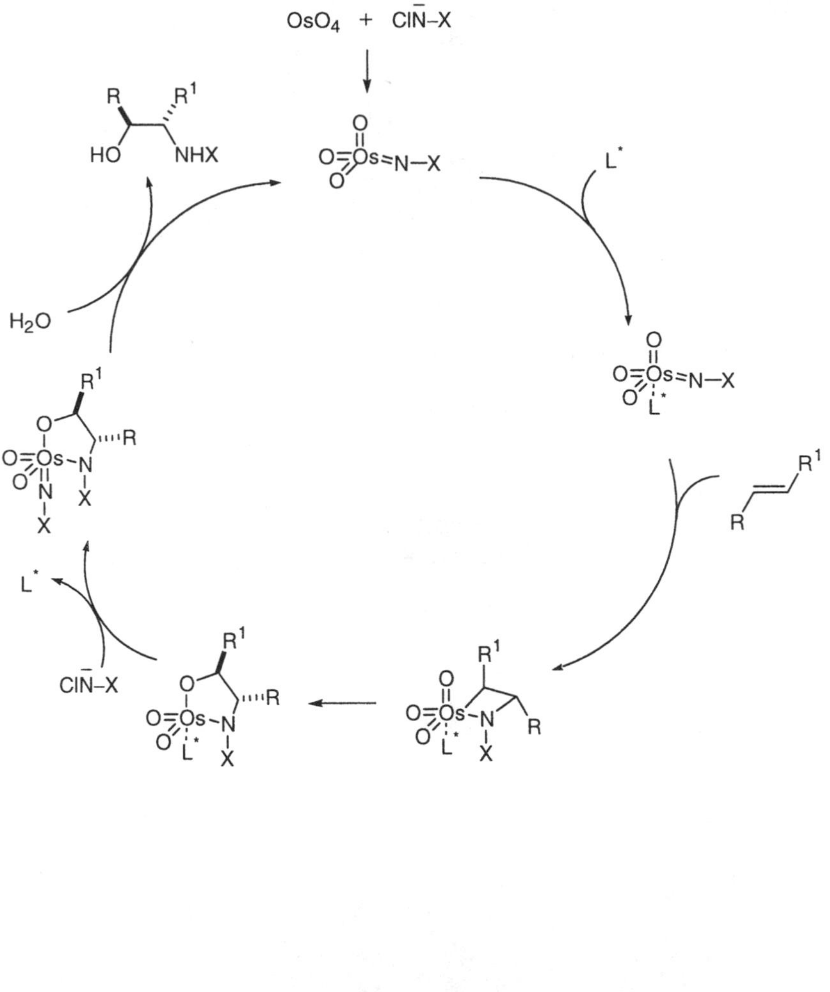
OsO_4 + ClN–X
H_2O
L*
NHX

References

1. Herranz, E.; Sharpless, K. B. *J. Org. Chem.* **1978**, *43*, 2544.
2. Mangatal, L.; Adeline, M. T.; Guenard, D.; Gueritte-Voegelein, F.; Potier, P. *Tetrahedron* **1989**, *45*, 4177.
3. Engelhardt, L. M.; Skelton, B. W.; Stick, R. V.; Tilbrook, D. M. G.; White, A. H. *Aust. J. Chem.* **1990**, *43*, 1657.
4. Rubin, A. E.; Sharpless, K. B. *Angew. Chem., Int. Ed. Engl.* **1997**, *36*, 2637.
5. Kolb, H. C.; Sharpless, K. B. *Transition Met. Org. Synth.* **1998**, *2*, 243. (Review).
6. Thomas, A.; Sharpless, K. B. *J. Org. Chem.* **1999**, *64*, 8279.
7. Gontcharov, A. V.; Liu, H.; Sharpless, K. B. *Org. Lett.* **1999**, *1*, 783.
8. Demko, Z. P.; Bartsch, M.; Sharpless, K. B. *Org. Lett.* **2000**, *2*, 2221.
9. Bolm, C.; Hildebrand, J. P.; Muñiz, K. In *Catalytic Asymmetric Synthesis;* 2nd ed., Ojima, I., ed.; Wiley–VCH: New York, **2000**, 399. (Review).
10. Bodkin, J. A.; McLeod, M. D. *Perkin 1* **2002**, 2733–2746. (Review).

Sharpless asymmetric epoxidation

Enantioselective epoxidation of allylic alcohols using *t*-butyl peroxide, titanium tetra-*iso*-propoxide, and optically pure diethyl tartrate.

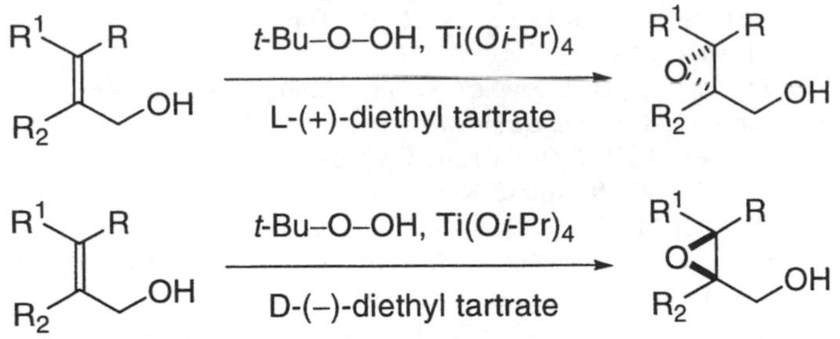

The putative active catalyst [2]:

E = CO$_2$Et

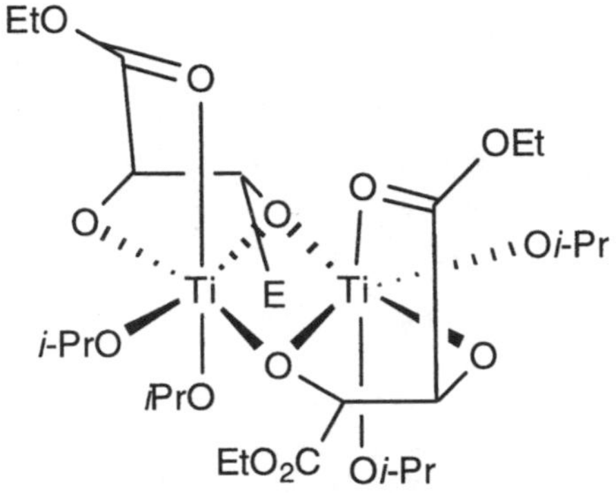

The transition state:

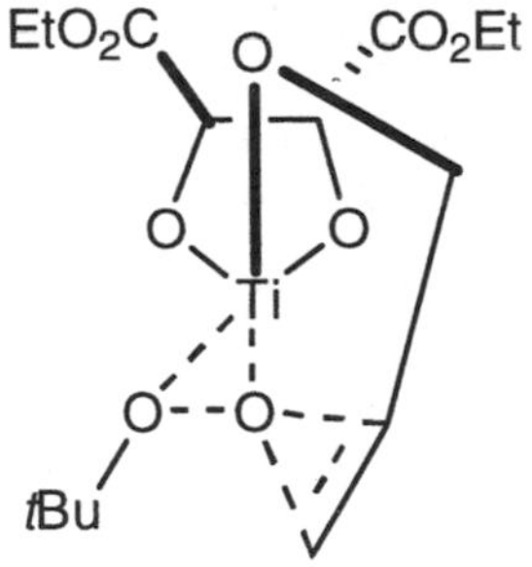

The catalytic cycle:

OiPr
* LnTi
OiPr

OH   tBu–O–OH

O
OH

O
* LnTi
O
O–tBu

* LnTi–O
O *
tBu

O
LnTi–O *
O *
tBu

O
LnTi–O *
O
tBu

O   O
* LnTi
O
tBu

References

1. Katsuki, T.; Sharpless, K. B. *J. Am. Chem. Soc.* **1980**, *102*, 5974.
2. Williams, I. D.; Pedersen, S. F.; Sharpless, K. B.; Lippard, S. J. *J. Am. Chem. Soc.* **1984**, *106*, 6430.
3. Rossiter, B. E. *Chem. Ind.* **1985**, *22(Catal. Org. React.)*, 295. (Review).
4. Pfenninger, A. *Synthesis* **1986**, 89. (Review).
5. Corey, E. J. *J. Org. Chem.* **1990**, *55*, 1693.
6. Woodard, S. S.; Finn, M. G.; Sharpless, K. B. *J. Am. Chem. Soc.* **1991**, *113*, 106.
7. Yamamoto, K.; Kawanami, Y.; Miyazawa, M. *J. Chem. Soc., Chem. Commun.* **1993**, 436.
8. Schinzer, D. *Org. Synth. Highlights II* **1995**, 3. (Review).
9. Katsuki, T.; Martin, V. S. *Org. React.* **1996**, *48*, 1–299. (Review).

10. Honda, T.; Mizutani, H.; Kanai, K. *J. Chem. Soc., Perkin Trans. 1* **1996**, 1729.
11. Honda, T.; Ohta, M.; Mizutani, H. *J. Chem. Soc., Perkin Trans. 1* **1999**, 23.
12. Johnson, R. A.; Sharpless, K. B. In *Catalytic Asymmetric Synthesis;* 2nd ed., Ojima, I., ed.; Wiley-VCH: New York, **2000**, 231.
13. Black, P. J.; Jenkins, K.; Williams, J. M. J. *Tetrahedron: Asymmetry* **2002**, *13*, 317.
14. Ghosh, A. K.; Lei, H. *Tetrahedron: Asymmetry* **2003**, *14*, 629.

Sharpless dihydroxylation

Enantioselective *cis*-dihydroxylation of olefins using osmium catalyst in the presence of cinchona alkaloid ligands.

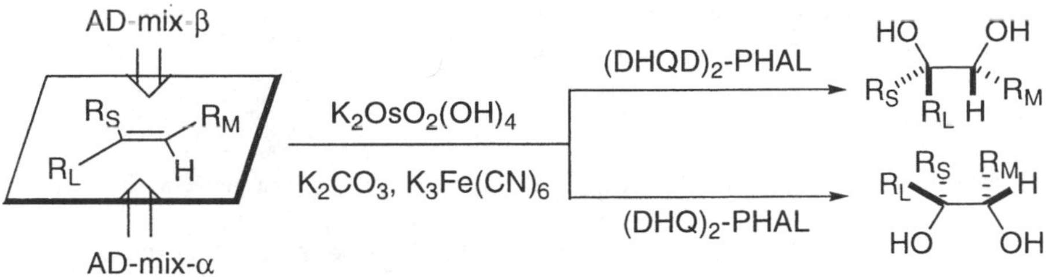

(DHQD)$_2$-PHAL = 1,4-bis(9-*O*-dihydroquinidine)phthalazine:

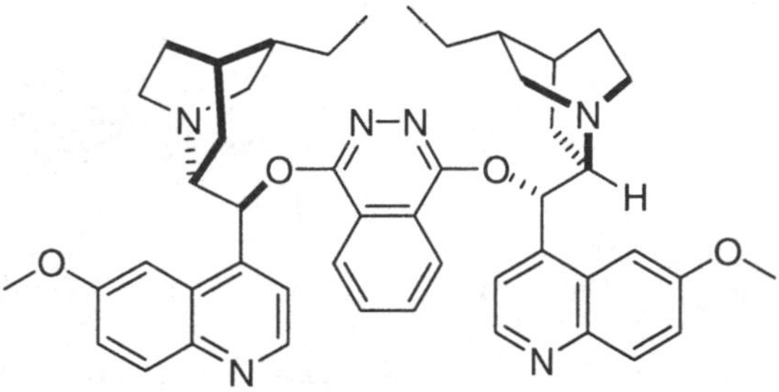

(DHQ)$_2$-PHAL = 1,4-bis(9-*O*-dihydroquinine)phthalazine:

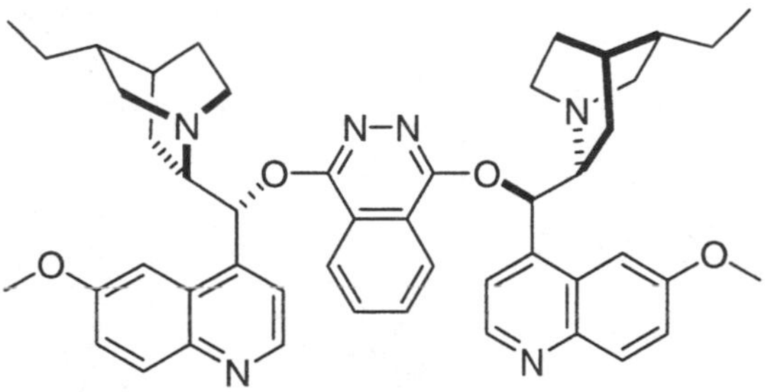

A stepwise mechanism involving osmaoxetane seems to be more consistent with the experimental data than the corresponding concerted [3 + 2] mechanism:

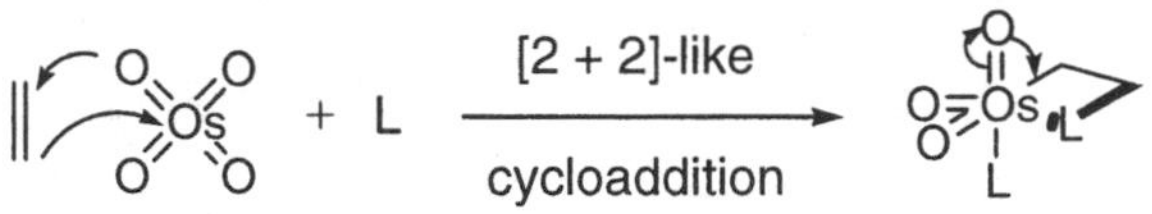

$$\xrightarrow{\text{rearrangement}} \quad \xrightarrow{\text{hydrolysis}}$$

The catalytic cycle is shown on the next page (the secondary cycle is shut off by maintaining a low concentration of olefin): page 371.

References

1. Jacobsen, E. N.; Markó, I.; Mungall, W. S.; Schröder, G.; Sharpless, K. B. *J. Am. Chem. Soc.* **1988**, *110*, 1968.
2. Wai, J. S. M.; Markó, I.; Svenden, J. S.; Finn, M. G.; Jacobsen, E. N.; Sharpless, K. B. *J. Am. Chem. Soc.* **1989**, *111*, 1123.
3. Kolb, H. C.; VanNiewenhze, M. S.; Sharpless, K. B. *Chem. Rev.* **1994**, *94*, 2483. (Review).
4. Bolm, C.; Gerlach, A. *Eur. J. Org. Chem.* **1998**, 21.
5. Balachari, D.; O'Doherty, G. A. *Org. Lett.* **2000**, *2*, 863.
6. Liang, J.; Moher, E. D.; Moore, R. E.; Hoard, D. W. *J. Org. Chem.* **2000**, *65*, 3143.
7. Mehltretter, G. M.; Dobler, C.; Sundermeier, U.; Beller, M. *Tetrahedron Lett.* **2000**, *41*, 8083.
8. Sharpless, K. B. *Angew. Chem., Int. Ed.* **2002**, *41*, 2024. (Review, Nobel Prize Address).
9. Moitessier, N.; Henry, C.; Len, C.; Postel, D.; Chapleur, Y. *J. Carbohydrate Chem.* **2003**, *22*, 25.
10. Choudary, B. M.; Chowdari, N. S.; Madhi, S.; Kantam, M. L. *J. Org. Chem.* **2003**, *68*, 1736.

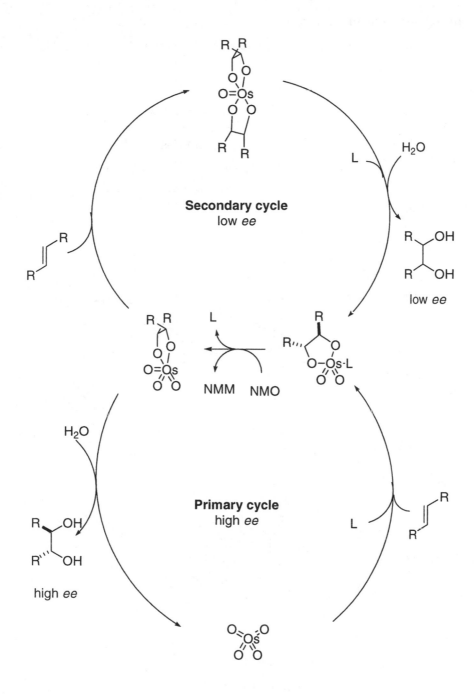

Shi asymmetric epoxidation

An asymmetric epoxidation using fructose-derived chiral ketone.

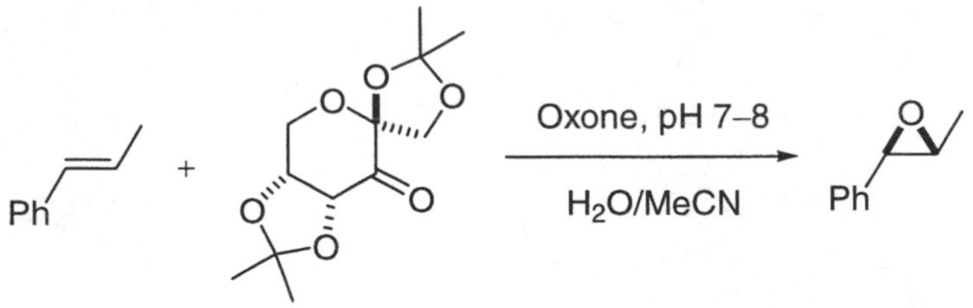

The catalytic cycle:

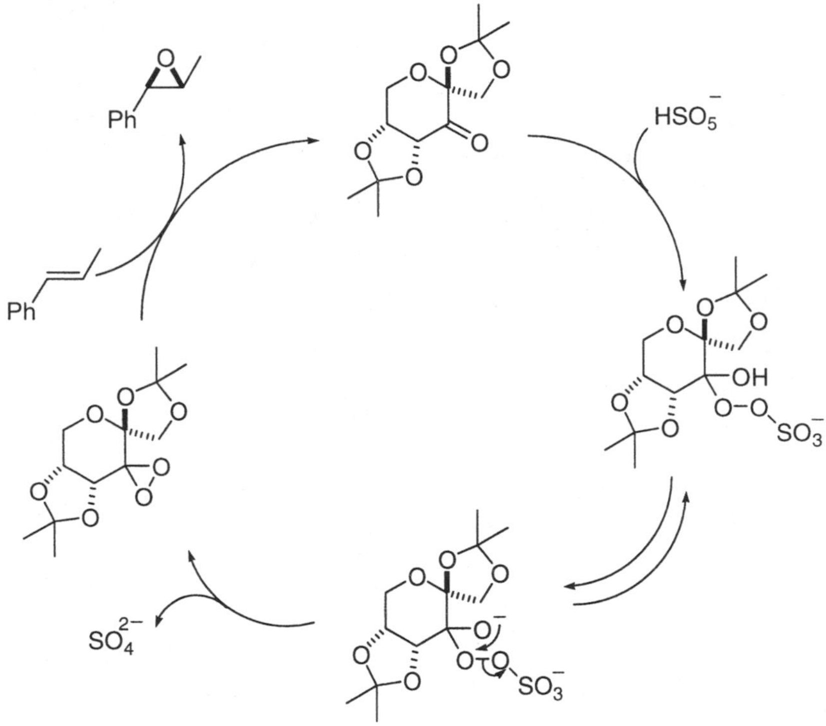

References

1. Wang, Z.-X.; Tu, Y.; Frohn, M.; Zhang, J.-R.; Shi, Y. *J. Am. Chem. Soc.* **1997**, *119*, 11224.
2. Wang, Z.-X.; Shi, Y. *J. Org. Chem.* **1997**, *62*, 8622.

3. Tu, Y.; Wang, Z.-X.; Frohn, M.; He, M.; Yu, H.; Tang, Y.; Shi, Y. *J. Org. Chem.* **1998**, *63*, 8475.
4. Tian, H.; She, X.; Shu, L.; Yu, H.; Shi, Y. *J. Am. Chem. Soc.* **2000**, *1229*, 11551.
5. Katsuki, T. In *Catalytic Asymmetric Synthesis;* 2nd ed., Ojima, I., ed.; Wiley–VCH: New York, **2000**, 287.
6. Tian, H.; She, X.; Yu, H.; Shu, L.; Shi, Y. *J. Org. Chem.* **2002**, *67*, 2435.
7. Hoard, D. W.; Moher, E. D.; Martinelli, M. J.; Norman, B. H. *Org. Lett.* **2002**, *4*, 1813.
8. Shu, L.; Wang, P.; Gan, Y.; Shi, Y. *Org. Lett.* **2003**, *5*, 293.

374

Simmons–Smith reaction

Cyclopropanation of olefins using CH_2I_2 and $Zn(Cu)$.

$$CH_2I_2 \ + \ Zn(Cu) \ \longrightarrow \ ICH_2ZnI \ \xrightarrow{\rlap{$\diagup\!\!\!\!\diagdown$}} \ \text{(cyclopropane)}$$

$$I{-}CH_2{-}I \ \xrightarrow[\text{addition}]{\text{Zn, Oxidative}} \ ICH_2ZnI$$

Simmons–Smith reagent

$$2 \ ICH_2ZnI \ \rightleftharpoons \ (ICH_2)_2Zn \ + \ ZnI_2$$

$$\underset{H_2}{\overset{I}{\underset{}{\diagdown}}}\overset{ZnI}{C} \quad \longrightarrow \quad \left[\ \overset{I{-}{-}Zn I}{\underset{\times}{\triangle}} \ \right] \quad \longrightarrow \quad \triangle \ + \ ZnI_2$$

References

1. Simmons, H. E.; Smith, R. D. *J. Am. Chem. Soc.* **1958**, *80*, 5323.
2. Kaltenberg, O. P. *Wiad. Chem.* **1972**, *26*, 285.
3. Takai, K.; Kakiuchi, T.; Utimoto, K. *J. Org. Chem.* **1994**, *59*, 2671.
4. Takahashi, H.; Yoshioka, M.; Shibasaki, M.; Ohno, M.; Imai, N.; Kobayashi, S. *Tetrahedron* **1995**, *51*, 12013.
5. Nakamura, E.; Hirai, A.; Nakamura, M. *J. Am. Chem. Soc.* **1998**, *120*, 5844.
6. Kaye, P. T.; Molema, W. E. *Chem. Commun.* **1998**, 2479.
7. Kaye, P. T.; Molema, W. E. *Synth. Commun.* **1999**, *29*, 1889.
8. Baba, Y.; Saha, G.; Nakao, S.; Iwata, C.; Tanaka, T.; Ibuka, T.; Ohishi, H.; Takemoto, Y. *J. Org. Chem.* **2001**, *66*, 81.
9. Charette, A. B.; Beauchemin, A. *Org. React.* **2001**, *58*, 1–415. (Review).
10. Nakamura, M.; Hirai, A.; Nakamura, E. *J. Am. Chem. Soc.* **2003**, *125*, 2341.
11. Mahata, P. K.; Syam Kumar, U. K.; Sriram, V.; Ila, H.; Junjappa, H. *Tetrahedron* **2003**, *59*, 2631.

Simonini reaction

Ester formation when silver carboxylate is treated with iodine. Alternatively, when silver carboxylate is treated with bromine, the product is alkyl bromide, R–Br (**Hunsdiecker reaction**, page 202).

References

1. Simonini, A. *Monatsh. Chem.* **1892**, *13*, 320.
2. Oldham, J. W. H. *J. Chem. Soc.* **1950**, 100.
3. Darzens, G.; Meyer, M. *Compt. Rend.* **1953**, *237*, 1334.
4. Wasserman, H. H.; Precopio, F. M. *J. Am. Chem. Soc.* **1954**, *76*, 1242.
5. Wasserman, H. H.; Precopio, F. M. *J. Am. Chem. Soc.* **1954**, *76*, 1242.
6. Wiberg, K. B.; Motell, E. L. *Tetrahedron* **1963**, *19*, 2009.
7. Chalmers, D. J.; Thomson, R. H. *J. Chem. Soc. (C)* **1968**, 848.
8. Bunce, N. J.; Murray, N. G. *Tetrahedron* **1971**, *27*, 5323.

Simonis chromone cyclization

Chromones from phenols and β-ketoesters using phosphorus pentoxide.

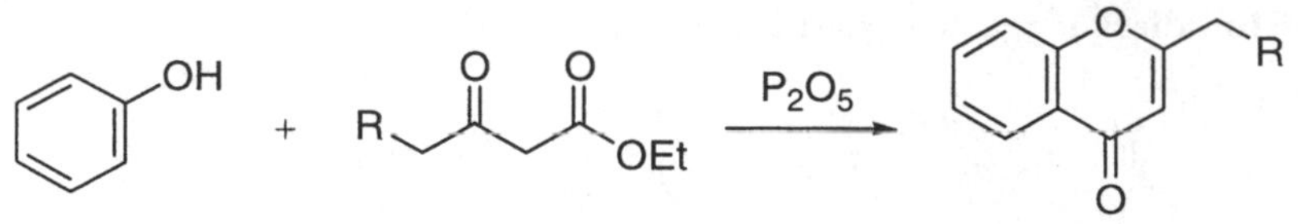

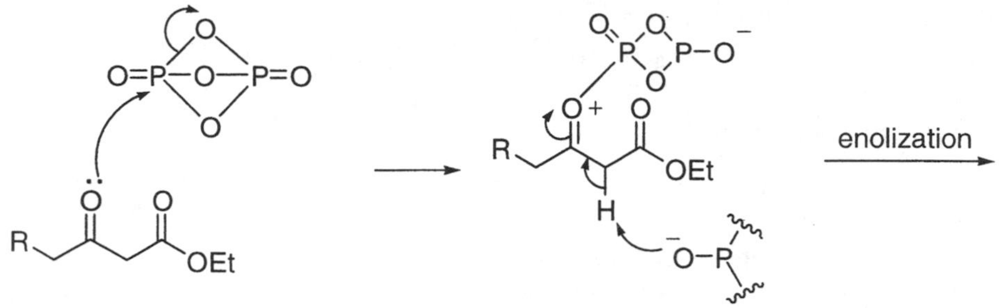

P$_2$O$_5$ actually exists as P$_4$O$_{10}$, an adamantane-like structure.

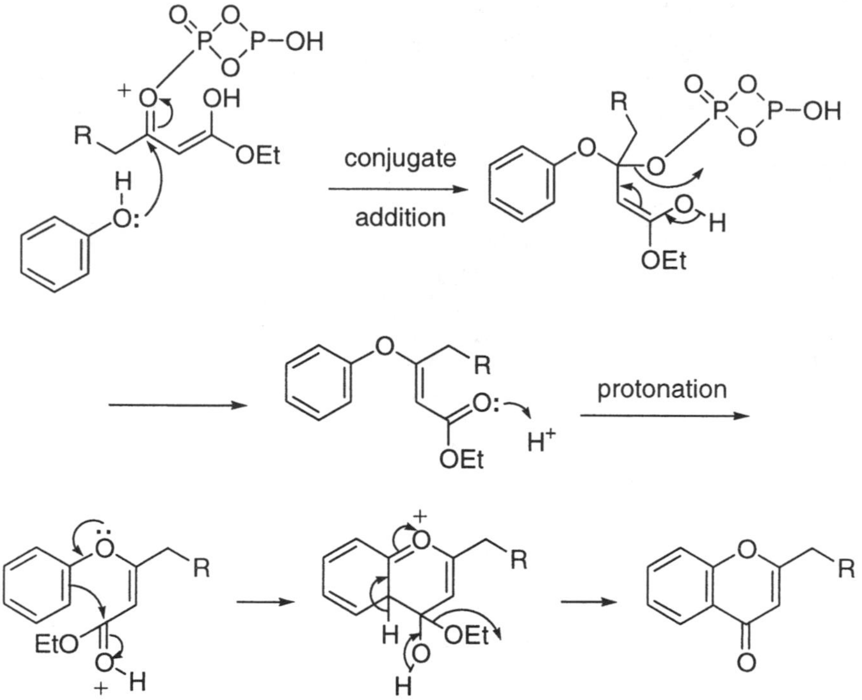

References

1. Petschek, E.; Simonis, H. *Ber. Dtsch. Chem. Ges.* **1913**, *46*, 2014.
2. Chakravarti, D. *J. Indian Chem. Soc.* **1931**, *8*, 129.
3. Chakravarti, D.; Banerjee, B. C. *J. Indian Chem. Soc.* **1936**, *13*, 619.
4. Chakravarti, D.; Banerjee, B. *Science Culture* **1936**, *1*, 783
5. Sethna, S. M.; Shah, N. M. *Chem. Rev.* **1945**, *36*, 14. (Review).
6. Dallemagne, M. J.; Martinet, J. *Bull. Soc. Chim. Fr.* **1950**, 1132.
7. Sethna, S. M.; Phadke, P. *Org. React.* **1953**, *7*, 15.
8. Ruwet, A.; Janne, D.; Renson, M. *Bull. Soc. Chim. Belg.* **1970**, *79*, 81.
9. Tan, S. F. *Aus. J. Chem.* **1972**, *25*, 1367.
10. Oyman, U.; Gunaydin, K. *Bull. Soc. Chim. Belg.* **1994**, *103*, 763.

Skraup quinoline synthesis

Quinoline from aniline, glycerol, sulfuric acid and oxidizing agent (e.g. $PhNO_2$).

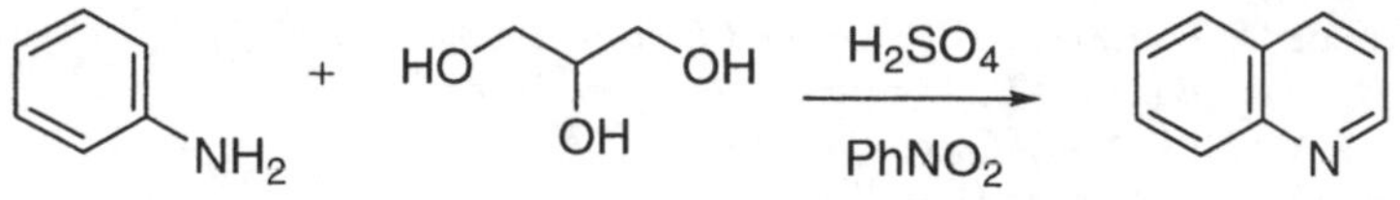
NH2
HO
OH
OH
H2SO4
PhNO2
N

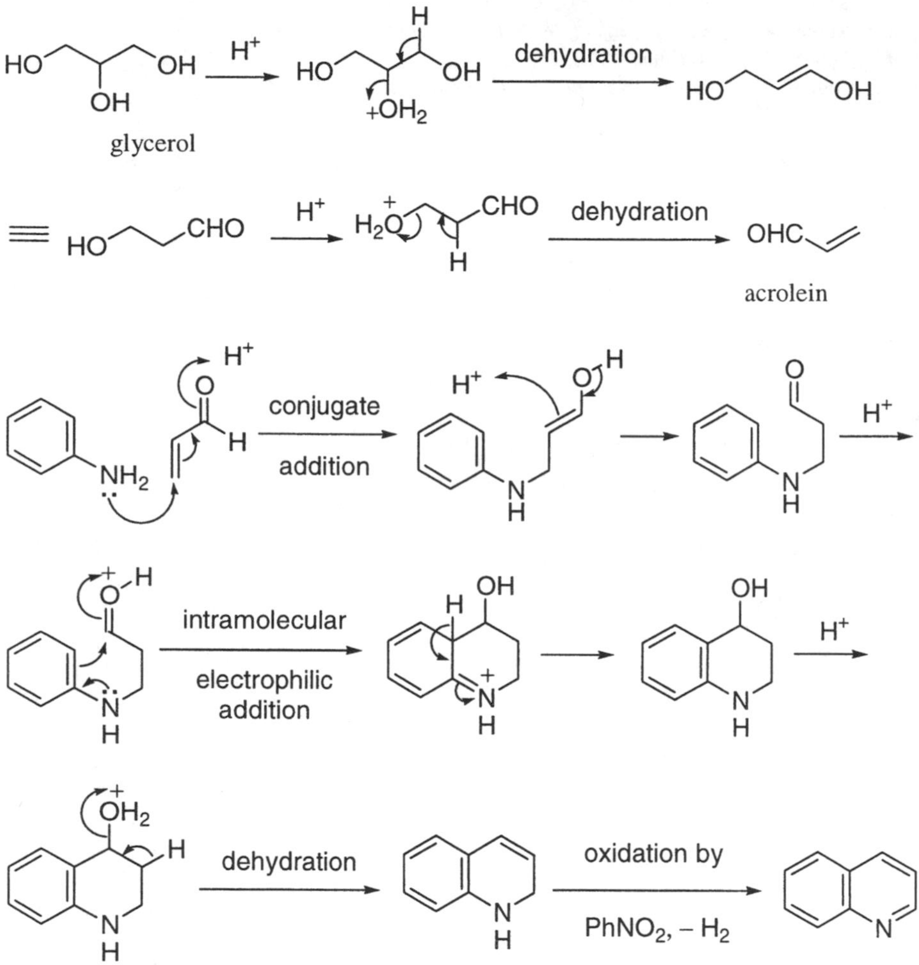
glycerol
dehydration
acrolein
conjugate
addition
intramolecular
electrophilic
addition
dehydration
oxidation by
PhNO2, – H2

For an alternative mechanism, see that of the Doebner–von Miller reaction (page 117).

References

1. Skraup, Z. H. *Ber. Dtsch. Chem. Ges.* **1880**, *13*, 2086.
2. Eisch, J. J.; Dluzniewski, T. *J. Org. Chem.* **1989**, *54,* 1269.
3. Takeuchi, I.; Hamada, Y.; Hirota, M. *Chem. Pharm. Bull.* **1993**, *41*, 747.
4. Fujiwara, H.; Okabayashi, I. *Chem. Pharm. Bull.* **1994**, *42*, 1322.
5. Fujiwara, H. *Heterocycles* **1997**, *45*, 119.
6. Fujiwara, H.; Kitagawa, K. *Heterocycles* **2000**, *53*, 409.
7. Theoclitou, M.-E.; Robinson, L. A. *Tetrahedron Lett.* **2002**, *43,* 3907.

Smiles rearrangement

General scheme:

X = S, SO, SO$_2$, O, CO$_2$
YH = OH, NHR, SH, CH$_2$R, CONHR
Z = NO$_2$, SO$_2$R

e.g.

spirocyclic anion intermediate
(Meisenheimer complex)

References

1. Evans, W. J.; Smiles, S. *J. Chem. Soc.* **1935**, 181.
2. Truce, W. E.; Kreider, E. M.; Brand, W. W. *Org. React.* **1970**, *18*, 99. (Review).
3. Gerasimova, T. N.; Kolchina, E. F. *J. Fluorine Chem.* **1994**, *66*, 69.
4. Boschi, D.; Sorba, G.; Bertinaria, M.; Fruttero, R.; Calvino, R.; Gasco, A. *J. Chem. Soc., Perkin Trans. 1* **2001**, 1751.
5. Hirota, T.; Tomita, K.-I.; Sasaki, K.; Okuda, K.; Yoshida, M.; Kashino, S. *Heterocycles* **2001**, *55*, 741.
6. Selvakumar, N.; Srinivas, D.; Azhagan, A. M. *Synthesis* **2002**, 2421.
7. Kumar, G.; Gupta, V.; Gautam, D. C.; Gupta, R. R. *Heterocycl. Commun.* **2002**, *8*, 447.

Sommelet reaction

Transformation of benzyl halides to the corresponding benzaldehydes with the aide of hexamethylenetetramine.

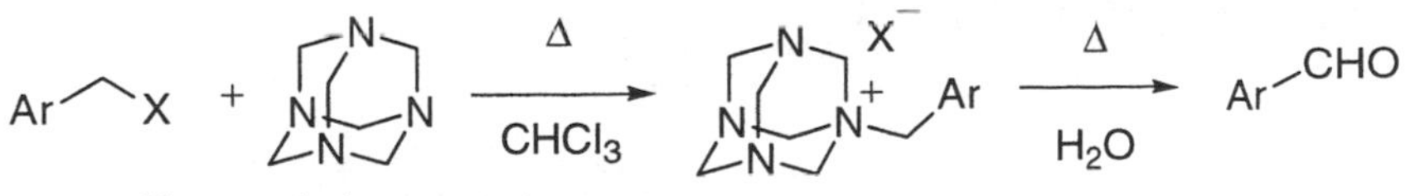

Hexamethylenetetramine

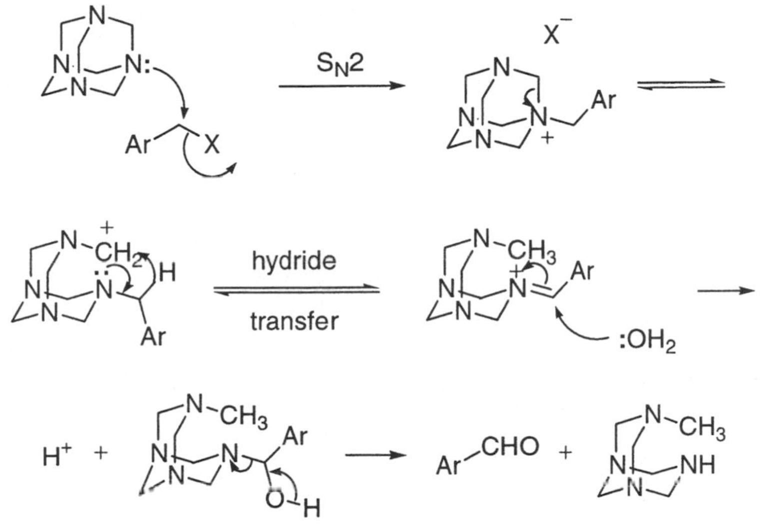

hemiaminal

The hydride transfer and the ring-opening of hexamethylenetetramine may occur in a synchronized fashion:

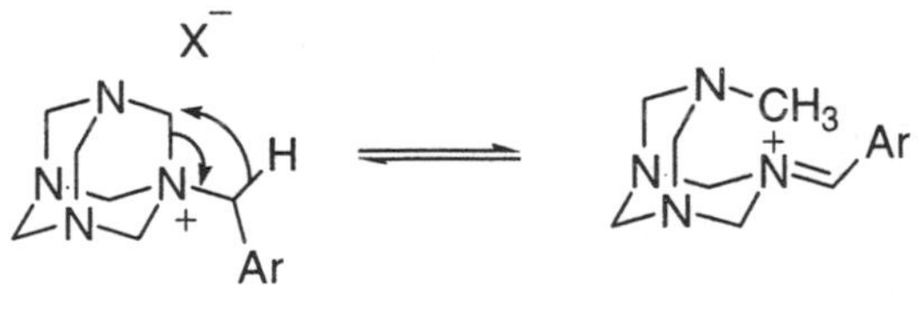

References

1. Sommelet, M. *Compt. Rend.* **1913**, *157*, 852.
2. Le Henaff, P. *Annals Chim. Phys.* **1962**, 367.
3. Zaluski, M. C.; Robba, M.; Bonhomme, M. *Bull. Soc. Chim. Fr.* **1970**, 1445.
4. Smith, W. E. *J. Org. Chem.* **1972**, *37*, 3972.

382

5. Simiti, I.; Chindris, E. *Arch. Pharm.* **1975**, *308*, 688.
6. Stokker, G. E.; Schultz, E. M. *Synth. Commun.* **1982**, *12*, 847.
7. Armesto, D.; Horspool, W. M.; Martin, J. A. F.; Perez-Ossorio, R. *Tetrahedron Lett.* **1985**, *26*, 5217.
8. Simiti, I.; Oniga, O. *Monatsh. Chem.* **1996**, *127*, 733.
9. Liu, X.; He, W. *Huaxue Shiji* **2001**, *23*, 237.

Sommelet–Hauser (ammonium ylide) rearrangement

Rearrangement of benzylic quaternary ammonium salts upon treatment with alkali metal amides.

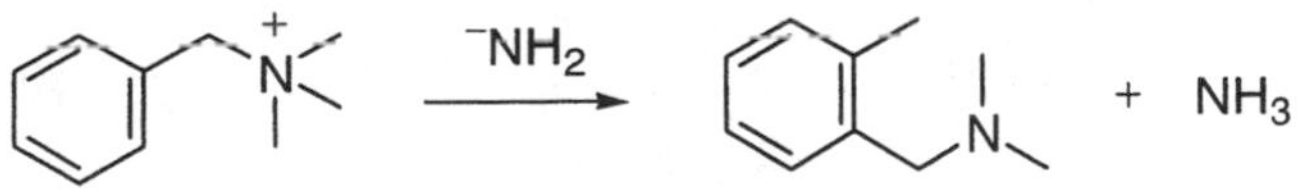

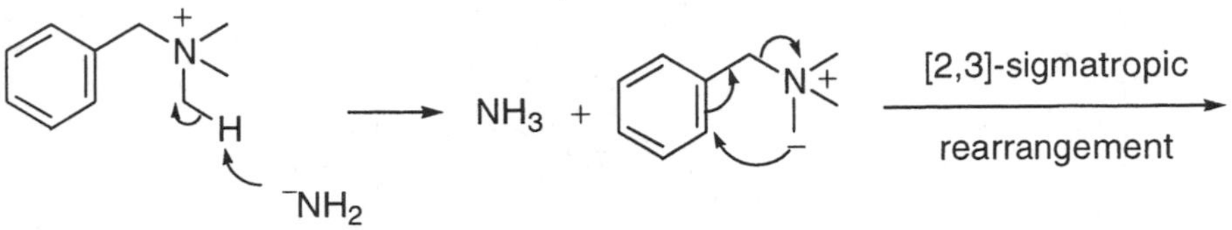

ammonium ylide

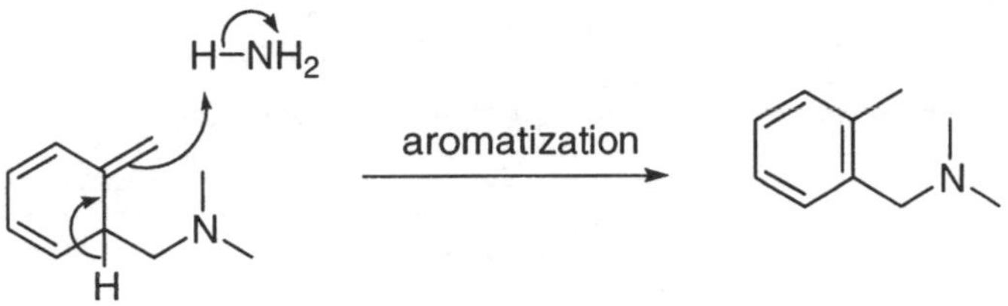

References

1. Sommelet, M. *Compt. Rend.* **1937**, *205*, 56.
2. Pine, S. H. *Tetrahedron Lett.* **1967**, 3393.
3. Wittig, G. *Bull. Soc. Chim. Fr.* **1971**, 1921.
4. Robert, A.; Lucas-Thomas, M. T. *J. Chem. Soc., Chem. Commun.* **1980**, 629.
5. Shirai, N.; Sumiya, F.; Sato, Y.; Hori, M. *J. Chem. Soc., Chem. Commun.* **1988**, 370.
6. Tanaka, T.; Shirai, N.; Sugimori, J.; Sato, Y. *J. Org. Chem.* **1992**, *57*, 5034.
7. Klunder, J. M. *J. Heterocycl. Chem.* **1995**, *32*, 1687.
8. Maeda, Y.; Sato, Y. *J. Org. Chem.* **1996**, *61*, 5188.
9. Endo, Y.; Uchida, T.; Shudo, K. *Tetrahedron Lett.* **1997**, *38*, 2113.

Sonogashira reaction

Pd/Cu-catalyzed cross-coupling of organohalides with terminal alkynes.
Cf. Castro–Stephens reaction.

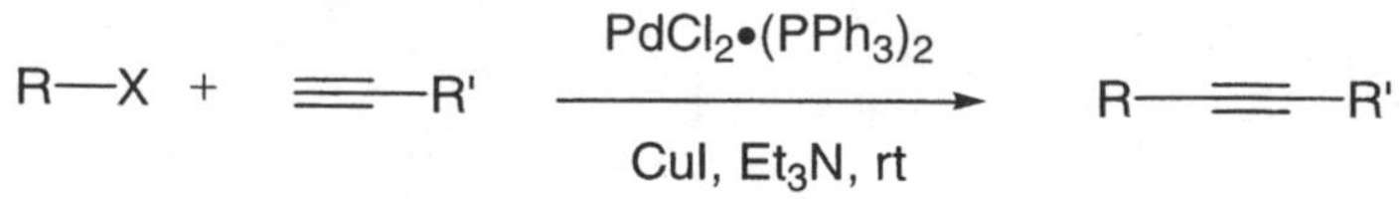

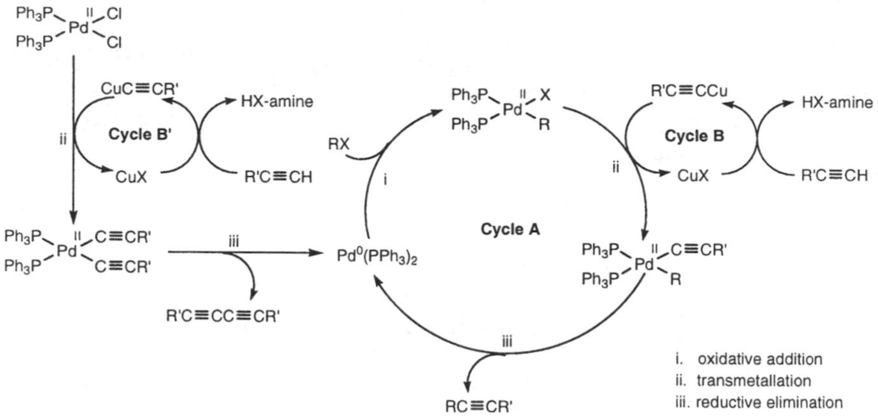

Note that Et_3N may reduce Pd(II) to Pd(0) as well, where Et_3N is oxidized to iminium ion at the same time:

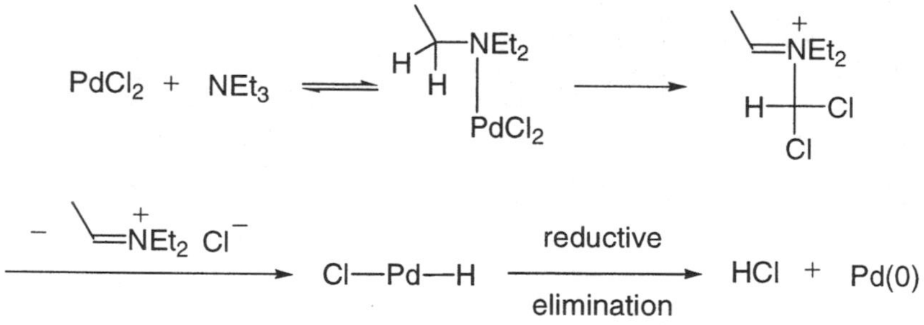

References

1. Sonogashira K.; Tohda, Y.; Hagihara, N. *Tetrahedron Lett.* **1975**, 4467.
2. McCrindle, R.; Ferguson, G.; Arsenaut, G. J.; McAlees, A. J.; Stephenson, D. K. *J. Chem. Res. (S)* **1984**, 360.
3. Rossi, R. Carpita, A.; Belina, F. *Org. Prep. Proc. Int.* **1995**, 27, 129.

4. Campbell, I. B. In *Organocopper Reagents*; Taylor, R. J. K. Ed.; IRL Press: Oxford, UK, **1994**, 217. (Review).
5. Hundermark, T.; Littke, A.; Buchwald, S. L.; Fu, G. C. *Org. Lett.* **2000**, *2*, 1729.
6. Dai, W.-M.; Wu, A. *Tetrahedron Lett.* **2001**, *42*, 81.
7. Alami, M.; Crousse, B.; Ferri, F. *J. Organomet. Chem.* **2001**, *624*, 114.
8. Bates, R. W.; Boonsombat, J. *J. Chem. Soc., Perkin Trans. 1* **2001**, 654.
9. Batey, R. A.; Shen, M.; Lough, A. J. *Org. Lett.* **2002**, *4*, 1411.
10. Balova, I. A.; Morozkina, S. N.; Knight, D. W.; Vasilevsky, S. F. *Tetrahedron Lett.* **2003**, *44*, 107.

386

Staudinger reaction

Phosphazo compounds (e.g. iminophosphoranes) from the reaction of tertiary phosphine (e.g. Ph_3P) with organic azides.

$$X{-}N_3 \xrightarrow{PR_3} X{-}N{=}N{-}N{=}PR_3 \xrightarrow{-N_2} X{-}N{=}PR_3$$
$$\text{phosphazide}$$

$$X{-}\overset{-}{N}{-}\overset{+}{N}{\equiv}N \quad :PR_3 \longrightarrow X{-}\overset{-}{N}{-}N{\equiv}\overset{+}{N}{-}PR_3 \equiv X{-}N{=}N{-}N{=}PR_3$$
$$\text{phosphazide}$$

$$\equiv \begin{matrix} N{=}PR_3 \\ | \\ N{\equiv}N_{\diagdown X} \end{matrix} \equiv \begin{matrix} N{-}\overset{+}{P}R_3 \\ \| \\ N{-}N_{\diagdown X} \end{matrix} \longrightarrow \begin{bmatrix} N{\cdot\cdot}PR_3 \\ \| \quad | \\ N{\cdot\cdot}N_{\diagdown X} \end{bmatrix}^{\ddagger}$$

4-membered ring transition state

$$\longrightarrow X{-}N{=}PR_3 + N_2{\uparrow}$$

References

1. Staudinger, H.; Meyer, J. *Helv. Chim. Acta* **1919**, *2*, 635.
2. Leffler, J. E.; Temple, R. D. *J. Am. Chem. Soc.* **1967**, *89*, 5235.
3. Gololobov, Y. G.; Zhmurova, I. N.; Kasukhin, L. F. *Tetrahedron* **1981**, *37*, 437.
4. Gololobov, Y. G.; Kasukhin, L. F. *Tetrahedron* **1992**, *48*, 1353.
5. Kovács, J. Pinter, I.; Kajtar-Peredy, M.; Sowsák, L. *Tetrahedron* **1997**, *53*, 15041.
6. Velasco, M. D.; Molina, P.; Fresneda, P. M.; Sanz, M. A. *Tetrahedron* **2000**, *56*, 4079.
7. Bongini, A.; Panunzio, M.; Piersanti, G.; Bandini, E.; Martelli, G.; Spunta, G.; Venturini, A. *Eur. J. Org. Chem.* **2000**, 2379.
8. Balakrishna, M. S.; Abhyankar, R. M.; Walawalker, M. G. *Tetrahedron Lett.* **2001**, *42*, 2733.
9. Conroy, K. D.; Thompson, A. *Chemtracts* **2002**, *15*, 514.
10. Venturini, A.; Gonzalez, Jr. *J. Org. Chem.* **2002**, *67*, 9089.
11. Chen, J.; Forsyth, C. J. *Org. Lett.* **2003**, *5*, 1281.

Stetter reaction (Michael–Stetter reaction)

1,4-Dicarbonyl derivatives from aldehydes and α,β-unsaturated ketones. The thiazolium catalyst serves as a safe surrogate for $^-$CN. *Cf.* Benzoin condensation.

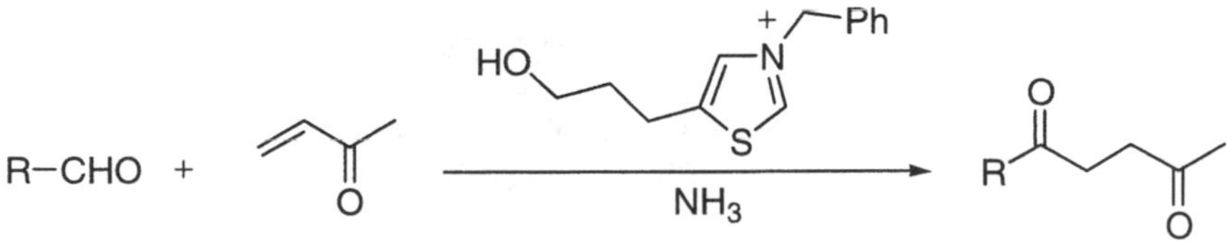

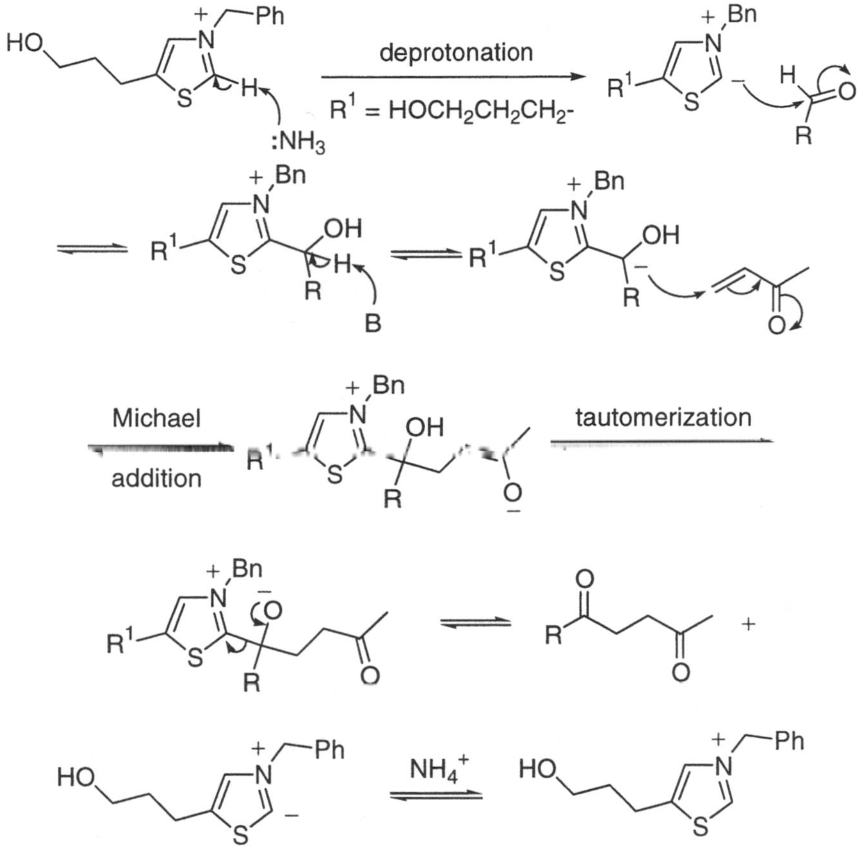

References

1. Stetter, H. *Angew. Chem.* **1973**, *85*, 89.
2. Stetter, H. *Angew. Chem., Int. Ed.* **1976**, *15*, 639.

388

3. Castells, J.; Dunach, E.; Geijo, F.; Lopez-Calahorra, F.; Prats, M.; Sanahuja, O.; Villanova, L. *Tetrahedron Lett.* **1980**, *21*, 2291.
4. Ho, T. L.; Liu, S. H. *Synth. Commun.* **1983**, *13*, 1125.
5. Phillips, R. B.; Herbert, S. A.; Robichaud, A. J. *Synth. Commun.* **1986**, *16*, 411.
6. Stetter, H.; Kuhlmann, H.; Haese, W. *Org. Synth.* **1987**, *65*, 26.
7. Ciganek, E. *Synthesis* **1995**, 1311.
8. Enders, D.; Breuer, K.; Runsink, J.; Teles, J. H. *Helv. Chim. Acta* **1996**, *79*, 1899.
9. Harrington, P. E.; Tius, M. A. *Org. Lett.* **1999**, *1*, 649.
10. Kobayashi, N.; Kaku, Y.; Higurashi, K.; Yamauchi, T.; Ishibashi, A.; Okamoto, Y. *Bioorg. Med. Chem. Lett.* **2002**, *12*, 1747.

Stevens rearrangement

A quaternary ammonium salt containing an electron-withdrawing group Z on one of the carbons attached to the nitrogen is treated with a strong base to give a rearranged tertiary amine.

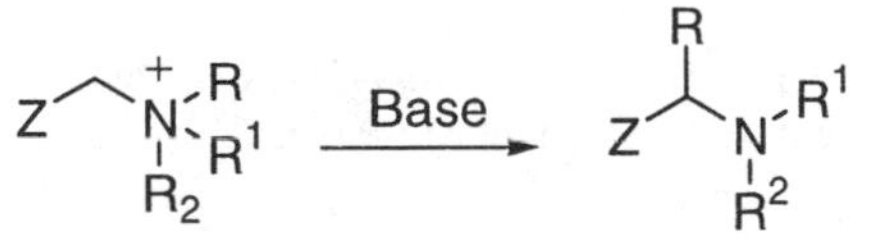

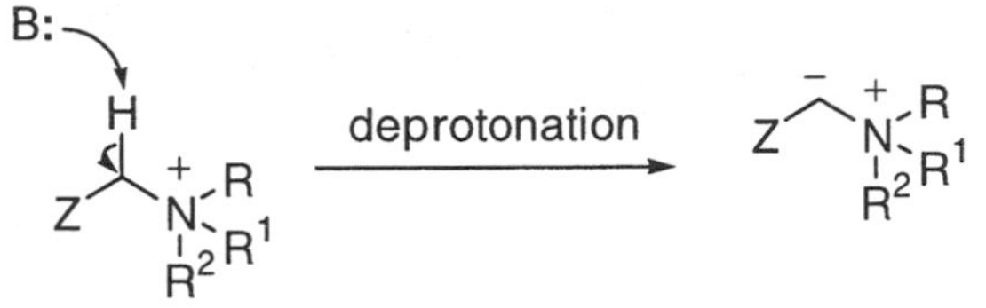

The contemporary radical mechanism:

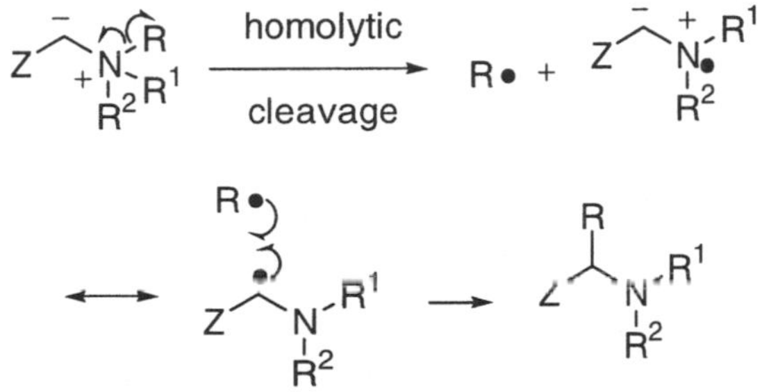

The original ionic mechanism:

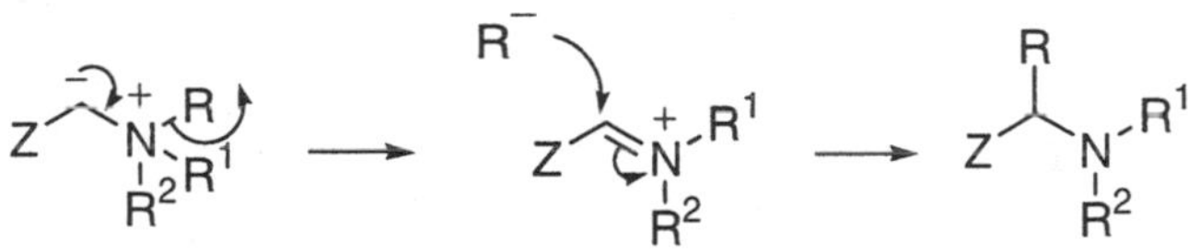

References

1. Stevens, T. S.; Creighton, E. M.; Gordon, A. B.; MacNicol, M. *J. Chem. Soc.* **1928**, 3193.
2. Schöllkopf, U.; Ludwig, U.; Ostermann, G.; Paysch, M. *Tetrahedron Lett.* **1969**, 3415.
3. Pine, S. H.; Catto, B. A.; Yamagishi, F. G. *J. Org. Chem.* **1970**, *35*, 3663.
4. Lepey, A. R.; Giumanini, A. G. *Mech. Mol. Migr.* **1971**, *3*, 297.

5. Doyle, M. P.; Ene, D. G.; Forbes, D. C.; Tedrow, J. S. *Tetrahedron Lett.* **1997**, *38*, 4367.
6. Makita, K.; Koketsu, J.; Ando, F.; Ninomiya, Y.; Koga, N. *J. Am. Chem. Soc.* **1998**, *120*, 5764.
7. Feldman, K. S.; Wrobleski, M. L. *J. Org. Chem.* **2000**, *65*, 8659.
8. Kitagaki, S.; Yanamoto, Y.; Tsutsui, H.; Anada, M.; Nakajima, M.; Hashimoto, S. *Tetrahedron Lett.* **2001**, *42*, 6361.
9. Knapp, S.; Morriello, G. J.; Doss, G. A. *Tetrahedron Lett.* **2002**, *43*, 5797.
10. Hanessian, S.; Parthasarathy, S.; Mauduit, M.; Payza, K. *J. Med. Chem.* **2003**, *46*, 34.

Stieglitz rearrangement

Rearrangement of trityl *N*-haloamines.

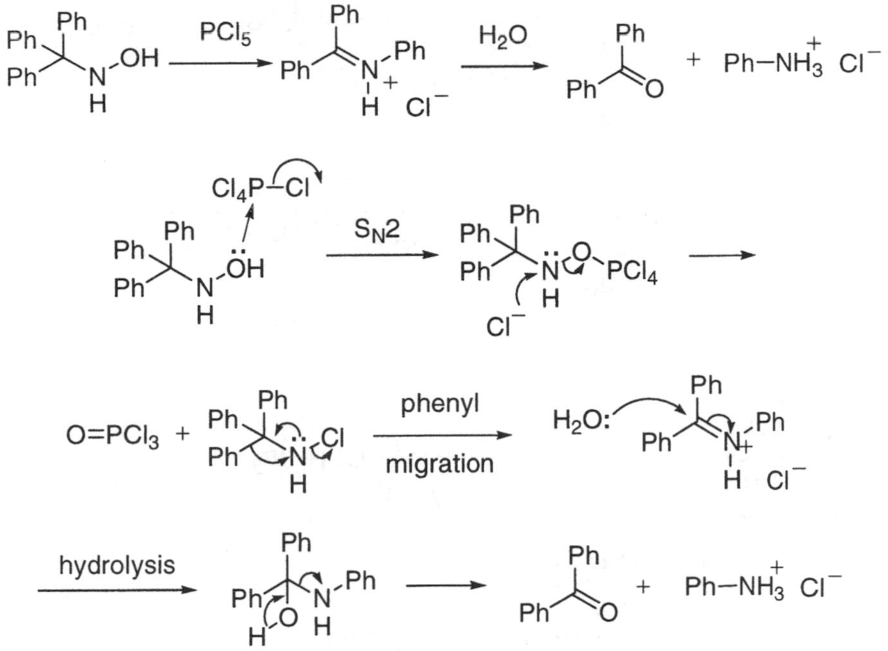

References

1. Stieglitz, J.; Leech, P. N. *Ber. Dtsch. Chem. Ges.* **1913**, *46*, 2147.
2. Pinck, L. A.; Hilbert, G. E. *J. Am. Chem. Soc.* **1937**, *59*, 8.
3. Berg, S. S.; Petrow, V. *J. Chem. Soc.* **1952**, 3713.
4. Newman, M. S.; Hay, P M. *J. Am. Chem. Soc.* **1953**, *75*, 2322.
5. Koga, N.; Anselme, J. P. *Tetrahedron Lett.* **1969**, 4773.
6. Sisti, A. J.; Milstein, S. R. *J. Org. Chem.* **1974**, *39*, 3932.
7. Hoffman, R. V.; Poelker, D. J. *J. Org. Chem.* **1979**, *44*, 2364.
8. Renslo, A. R.; Danheiser, R. L. *J. Org. Chem.* **1998**, *63*, 7840.

Still–Gennari phosphonate reaction

Horner–Emmons reaction using bis(trifluoroethyl)phosphonate to give Z-olefins.

erythro isomer, kinetic adduct

References

1. Still, W. C.; Gennari, C. *Tetrahedron Lett.* **1983**, *24*, 4405.
2. Ralph, J.; Zhang, Y. *Tetrahedron* **1998**, *54*, 1349.
3. Mulzer, J.; Mantoulidis, A.; Ohler, E. *Tetrahedron Lett.* **1998**, *39*, 8633.
4. Jung, M. E.; Marquez, R. *Org. Lett.* **2000**, *2*, 1669.
5. White, J. D.; Blakemore, P. R.; Browder, C. C.; *et al. J. Am. Chem. Soc.* **2001**, *123*, 8593.
6. Paterson, I.; Florence, G. J.; Gerlach, K.; Scott, J. P.; Sereinig, N. *J. Am. Chem. Soc.* **2001**, *123*, 9535.
7. Mulzer, J.; Ohler, E. *Angew. Chem., Int. Ed. Engl.* **2001**, *40*, 3842.
8. Beaudry, C. M.; Trauner, D. *Org. Lett.* **2002**, *4*, 2221.
9. Sano, S.; Yokoyama, K.; Shiro, M.; Nagao, Y. *Chem. Pharm. Bull.* **2002**, *50*, 706.
10. Dakin, L. A.; Langille, N. F.; Panek, J. S. *J. Org. Chem.* **2002**, *67*, 6812.

Stille coupling

Palladium-catalyzed cross-coupling reaction of organostannanes with organic halides, triflates, *etc.* For the catalytic cycle, see Kumada coupling on page 234.

$$R-X \ + \ R^1-Sn(R^2)_3 \ \xrightarrow{\text{Pd(0)}} \ R-R^1 \ + \ X-Sn(R^2)_3$$

$$R-X + L_2Pd(0) \ \xrightarrow[\text{addition}]{\text{oxidative}} \ \underset{L}{\overset{R}{\underset{\diagup}{Pd}}}{\overset{L}{\diagdown}}_X \ \xrightarrow[\substack{\text{transmetallation} \\ \text{isomerization}}]{R^1-Sn(R^2)_3}$$

$$X-Sn(R^2)_3 \ + \ \underset{R}{\overset{L}{\underset{\diagup}{Pd}}}\overset{L}{\diagdown}_{R^1} \ \xrightarrow[\text{elimination}]{\text{reductive}} \ R-R^1 \ + \ L_2Pd(0)$$

References

1. Milstein, D.; Stille, J. K. *J. Am. Chem. Soc.* **1978**, *100*, 3636.
2. Milstein, D.; Stille, J. K. *J. Am. Chem. Soc.* **1979**, *101*, 4992.
3. Stille, J. K. *Angew. Chem., Int. Ed. Engl.* **1986**, *25*, 508.
4. Farina, V.; Krishnamurphy, V.; Scott, W. J. *Org. React.* **1997**, *50*, 1–652. (Review).
5. For an excellent review on the intramolecular Stille reaction, see, Duncton, M. A. J.; Pattenden, G. *J. Chem. Soc., Perkin Trans. 1* **1999**, 1235.
6. Nakamura, H.; Bao, M.; Yamamoto, Y. *Angew. Chem., Int. Ed.* **2001**, *40*, 3208.
7. Heller, M.; Schubert, U. S. *J. Org. Chem.* **2002**, *67*, 8269.
8. Lin, S.-Y.; Chen, C.-L.; Lee, Y.-J. *J. Org. Chem.* **2003**, *68*, 2968.
9. Samuelsson, L.; Langstrom, B. *J. Labeled Compounds Radiopharm.* **2003**, *46*, 263.

Stille–Kelly reaction

Palladium-catalyzed intramolecular cross-coupling reaction of bis-aryl halides using ditin reagents.

References

1. Kelly, T. R.; Li, Q.; Bhushan, V. *Tetrahedron Lett.* **1990**, *31*, 161.
2. Grigg, R.; Teasdale, A.; Sridharan, V. *Tetrahedron Lett.* **1991**, *32*, 3859.

3. Sakamoto, T.; Yasuhara, A.; Kondo, Y.; Yamanaka, H. *Heterocycles* **1993**, *36*, 2597.
4. Iyoda, M.; Miura, M.i; Sasaki, S.; Kabir, S. M. H.; Kuwatani, Y.; Yoshida, M. *Heterocycles* **1997**, *38*, 4581.
5. Fukuyama, Y.; Yaso, H.; Nakamura, K.; Kodama, M. *Tetrahedron Lett.* **1999**, *40*, 105.
6. Iwaki, T.; Yasuhara, A.; Sakamoto, T. *J. Chem. Soc., Perkin Trans. 1* **1999**, 1505.
7. Fukuyama, Y.; Yaso, H.; Mori, T.; Takahashi, H.; Minami, H.; Kodama, M *Heterocycles* **2001**, *54*, 259.

Stobbe condensation

Condensation of diethyl succinate and its derivatives with carbonyl compounds in the presence of a base.

References

1. Stobbe, H. *Ber. Dtsch. Chem. Ges.* **1893**, *26*, 2312.
2. El-Rayyes, N. R.; Al-Salman, Mrs. N. A. *J. Heterocycl. Chem.* **1976**, *13*, 285.
3. Baghos, V. B.; Nasr, F. H.; Gindy, M. *Helv. Chim. Acta* **1979**, *62*, 90.
4. Baghos, V. B.; Doss, S. H.; Eskander, E. F. *Org. Prep. Proced. Int.* **1993**, *25*, 301.
5. Moldvai, I.; Temesvari-Major, E.; Balazs, M.; Gacs-Baitz, E.; Egyed, O.; Szantay, C. *J. Chem. Res., (S)* **1999**, 3018.
6. Moldvai, I.; Temesvari-Major, E.; Gacs-Baitz, E.; Egyed, O.; Gomory, A.; Nyulaszi, L.; Szantay, C. *Heterocycles* **2001**, *53*, 759.
7. Liu, J.; Brooks, N. R. *Org. Lett.* **2002**, *4*, 3521.
8. Moldvai, I.; Temesvari-Major, E.; Incze, M.; Platthy, T.; Gacs-Baitz, E.; Szantay, C. *Heterocycles* **2003**, *60*, 309.

Stollé synthesis

Acid-catalyzed oxindole formation from aniline and α-chlorocarboxylic acid chloride.

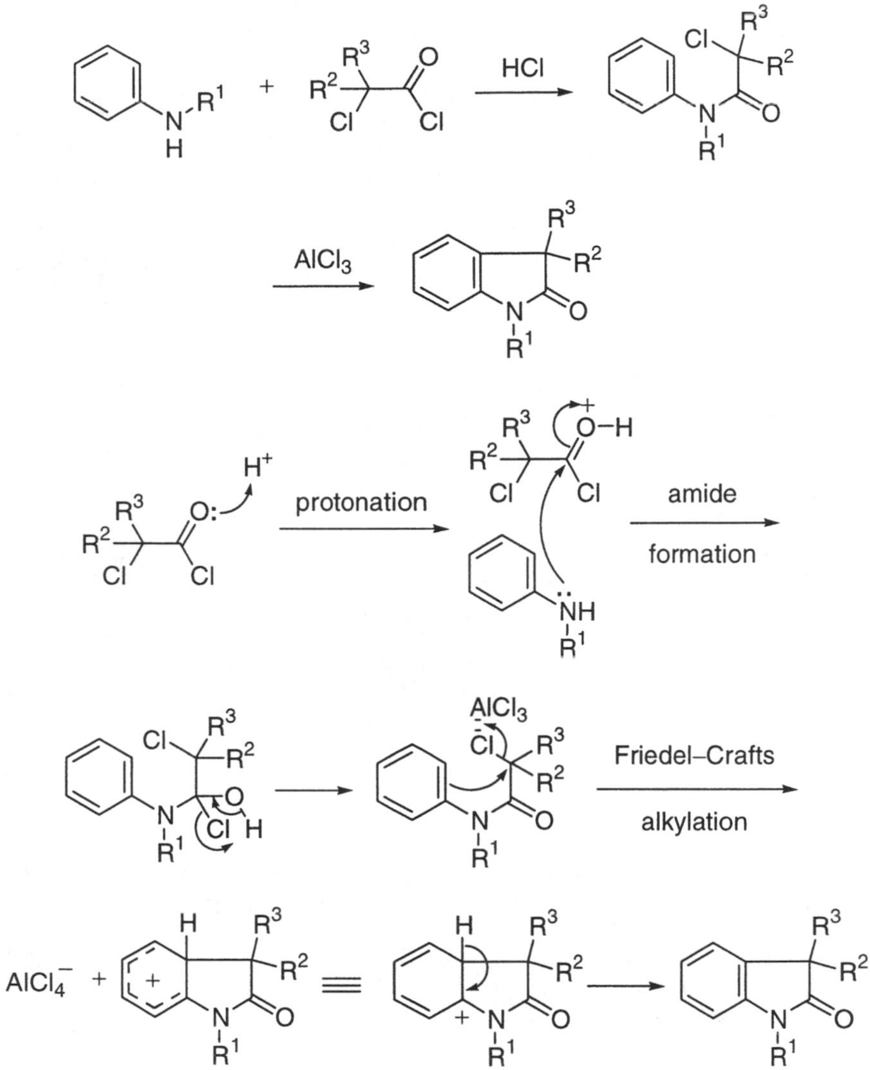

References

1. Stollé, R. *Ber. Dtsch. Chem. Ges.* **1913**, *46*, 3915.
2. Stollé, R. *Ber. Dtsch. Chem. Ges.* **1914**, *47*, 2120.
3. Przheval'skii, N. M.; Grandberg, I. I. *Khim. Geterotsikl. Soedin.* **1982**, 940.

398

Stork enamine reaction

A variant of the Robinson annulation, where bulky amines such as pyrrolidine are used, making the conjugate addition to MVK take place at the less hindered side of two possible enamines.

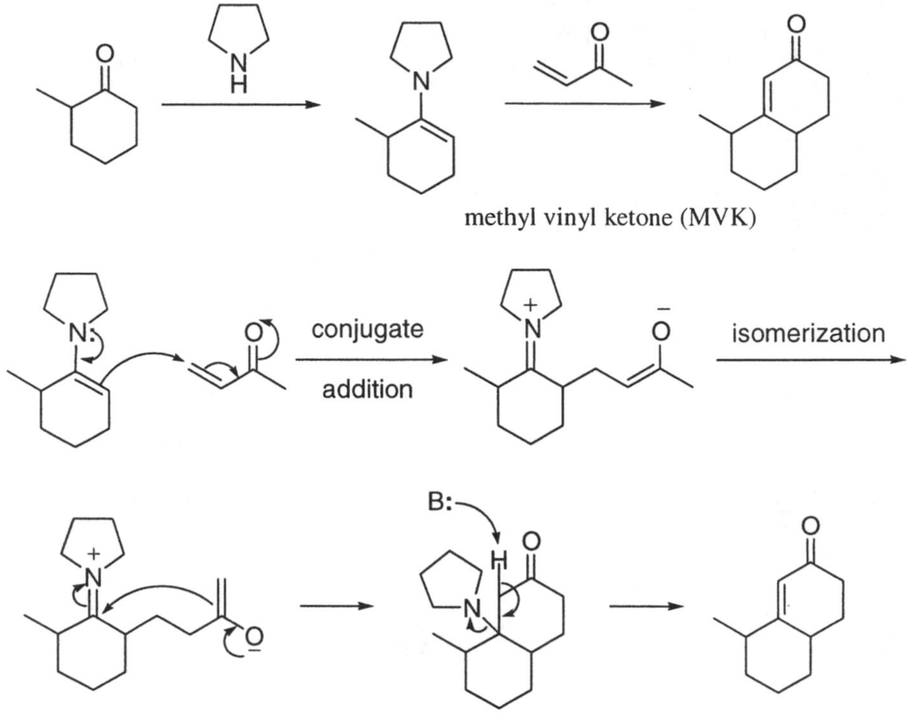

methyl vinyl ketone (MVK)

conjugate addition

isomerization

B:

References

1. Stork, G.; Terrell, R.; Szmuszkovicz, J. *J. Am. Chem. Soc.* **1954**, *76*, 2029.
2. *Enamines: Synthesis, Structure, and Reactions;* Cook, A. G., Ed.; Dekker: New York, **1969**, 514. (Review).
3. Autrey, R. L.; Tahk, F. C. *Tetrahedron* **1968**, *24*, 3337.
4. Hickmott, P. W. *Tetrahedron* **1982**, *38*, 1975.
5. Szablewski, M. *J. Org. Chem.* **1994**, *59*, 954.
6. Hammadi, M.; Villemin, D. *Synth. Commun.* **1996**, *26*, 2901.
7. Bridge, C. F.; O'Hagan, D. *J. Fluorine Chem.* **1997**, *82*, 21.
8. Li, J. J.; Trivedi, B. K.; Rubin, J. R.; Roth, B. D. *Tetrahedron Lett.* **1998**, *39*, 6111.
9. Yehia, N. A. M.; Polborn, K.; Muller, T. J. J. *Tetrahedron Lett.* **2002**, *43*, 6907.
10. Kesel, A. *Biochem. Biophys. Res. Commun.* **2003**, *300*, 793.

Strecker amino acid synthesis

Sodium cyanide-promoted condensation of aldehyde and amine to afford α-amino nitrile, which may be hydrolyzed to α-amino acid.

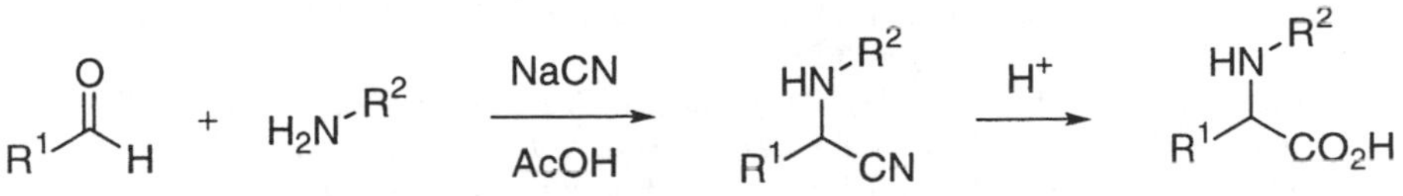
NaCN
AcOH
H+

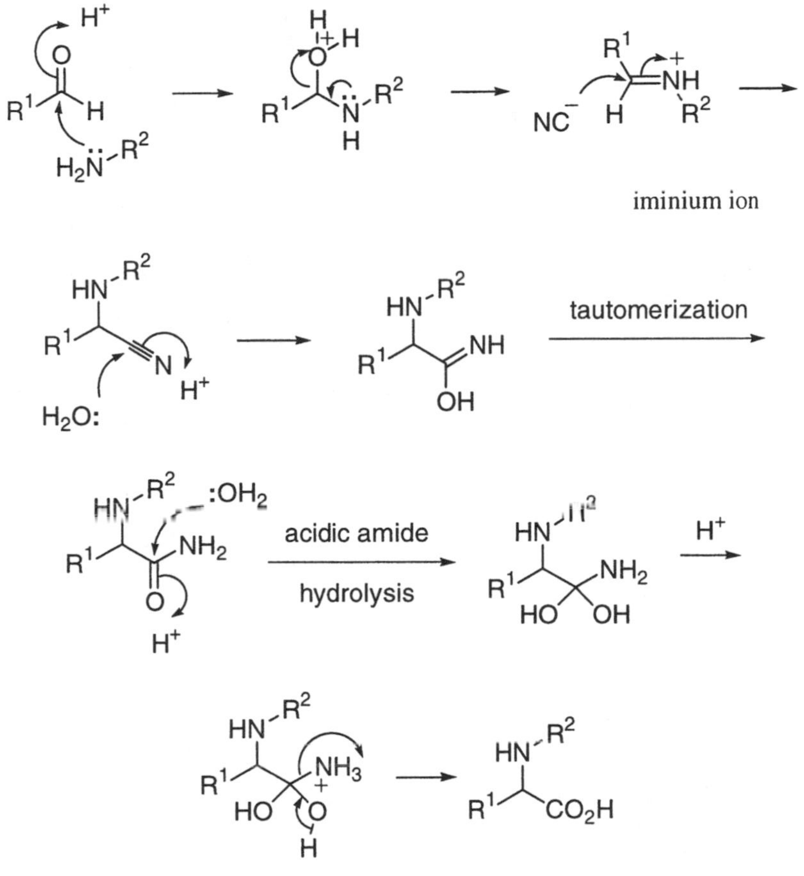
iminium ion
tautomerization
acidic amide
hydrolysis
H+

References

1. Strecker, A. *Justus Liebigs Ann. Chem.* **1850**, *75*, 27.
2. Chakraborty, T. K.; Hussain, K. A; Reddy, G. V. *Tetrahedron* **1995**, *51*, 9179.
3. Iyer, M. S.; Gigstad, K. M.; Namdev, N. D.; Lipton, M. *J. Am. Chem. Soc.* **1996**, *118*, 4910.

400

4. Iyer, M. S.; Gigstad, K. M.; Namdev, N. D.; Lipton, M. *Amino Acids* **1996**, *11*, 259.
5. Mori, A.; Inoue, S. *Compr. Asymmetric Catal. I-III* **1999**, *2*, 983. (Review).
6. Ishitani, H.; Komiyama, S.; Hasegawa, Y.; Kobayashi, S. *J. Am. Chem. Soc.* **2000**, *122*, 762.
7. Wede, J.; Volk, Franz-J.; Frahm, A. W. *Tetrahedron: Asymmetry* **2000**, *11*, 3231.
8. Davis, F. A.; Lee, S.; Zhang, H.; Fanelli, D. L. *J. Org. Chem.* **2000**, *65*, 8704.
9. Ding, K.; Ma, D. *Tetrahedron* **2001**, *57*, 6361.
10. Matrumoto, K.; Kim, J. C.; Hayashi, N.; Jenner, G. *Tetrahedron Lett.* **2002**, *43*, 9167.
11. Jenner, G.; Salem, R. B.; Kim, J. C.; Matsumoto, K. *Tetrahedron Lett.* **2003**, *44*, 447.
12. Volk, F. J.; Wagner, M.; Frahm, A. W. *Tetrahedron: Asymmetry* **2003**, *14*, 497.

Suzuki coupling

Palladium-catalyzed cross-coupling reaction of organoboranes with organic halides, triflates, *etc.* in the presence of a base (transmetallation is reluctant to occur without the activating effect of a base). For the catalytic cycle, see Kumada coupling on page 234.

$$R-X \ + \ R^1-B(R^2)_2 \ \xrightarrow[\text{NaOR}^3]{L_2Pd(0)} \ R-R^1$$

$$R-X \ + \ L_2Pd(0) \ \xrightarrow[\text{addition}]{\text{oxidative}} \ \underset{L}{\overset{R}{\diagdown}}Pd\underset{X}{\overset{L}{\diagup}}$$

$$R^1-B(R^2)_2 \ \xrightarrow[\text{addition of base}]{\text{NaOR}^3} \ \underset{-}{R^1-\overset{OR_3}{\underset{|}{B}}(R^2)_2}$$

$$\underset{L}{\overset{R}{\diagdown}}Pd\underset{X}{\overset{L}{\diagup}} \ + \ \underset{-}{R^1-\overset{OR^3}{\underset{|}{B}}(R^2)_2} \ \xrightarrow[\text{isomerization}]{\text{transmetallation}}$$

$$R^3O-B(R^2)_2 \ + \ \underset{H}{\overset{L}{\diagdown}}Pd\underset{R^1}{\overset{L}{\diagup}} \ \xrightarrow[\text{elimination}]{\text{reductive}} \ R-R^1 \ + \ L_2Pd(0)$$

References

1. Miyaura, N.; Suzuki, A. *Chem. Rev.* **1995**, *95*, 2457. (Review).
2. Suzuki, A. In *Metal-catalyzed Cross-coupling Reactions*; Diederich, F.; Stang, P. J., Eds.; Wiley–VCH: Weinhein, Germany, **1998**, 49–97. (Review).
3. Stanforth, S. P. *Tetrahedron* **1998**, *54*, 263. (Review).
4. Li, J. J. *Alkaloids: Chem. Biol. Perspect.* **1999**, *14*, 437. (Review).
5. Groger, H. *J. Prakt. Chem.* **2000**, *342*, 334.
6. Franzen, R. *Can. J. Chem.* **2000**, *78*, 957.
7. LeBlond, C. R.; Andrews, A. T.; Sun, Y.; Sowa, J. R., Jr. *Org. Lett.* **2001**, *3*, 1557.
8. Collier, P. N.; Campbell, A. D.; Patel, I.; Raynham, T. M.; Taylor, R. J. K. *J. Org. Chem.* **2002**, *67*, 1802.
9. Urawa, Y.; Ogura, K. *Tetrahedron Lett.* **2003**, *44*, 271.

Swern oxidation

Oxidation of alcohols to the corresponding carbonyl compounds using $(COCl)_2$, DMSO, and quenching with Et_3N.

$CO_2\uparrow$ + $CO\uparrow$ +

$Et_3N \bullet HCl\downarrow$ +

sulfur ylide

$R^1\overset{O}{\underset{}{\diagup}}R^2$ + $(CH_3)_2S\uparrow$

References

1. Huang, S. L.; Omura, K.; Swern, D. *J. Org. Chem.* **1976**, *41*, 3329.
2. Huang, S. L.; Omura, K.; Swern, D. *Synthesis* **1978**, *4*, 297.
3. Mancuso, A. J.; Huang, S.-L.; Swern, D. *J. Org. Chem.* **1978**, *43*, 2480.
4. Tidwell, T. T. *Org. React.* **1990**, *39*, 297. (Review).
5. Nakajima, N.; Ubukata, M. *Tetrahedron Lett.* **1997**, *38*, 2099.
6. Harris, J. M.; Liu, Y.; Chai, S.; Andrews, M. D.; Vederas, J. C. *J. Org. Chem.* **1998**, *63*, 2407.
7. Bailey, P. D.; Cochrane, P. J.; Irvine, F.; Morgan, K. M.; Pearson, D. P. J.; Veal, K. T. *Tetrahedron Lett.* **1999**, *40*, 4593.
8. Rodriguez, A.; Nomen, M.; Spur, B. W.; Godfroid, J. J. *Tetrahedron Lett.* **1999**, *40*, 5161.

9. Dupont, J.; Bemish, R. J.; McCarthy, K. E.; Payne, E. R.; Pollard, E. B.; Ripin, D. H. B.; Vanderplas, B. C.; Watrous, R. M. *Tetrahedron Lett.* **2001**, *42*, 1453.
10. Nishide, K.; Ohsugi, S.-i.; Fudesaka, M.; Kodama, S.; Node, M. *Tetrahedron Lett.* **2002**, *43*, 5177. (New odorless protocols).
11. Firouzabadi, H.; Hassani, H.; Hazarkhani, H. *Phosphorus, Sulfur Silicon Related Elements* **2003**, *178*, 149.

Tamao–Kumada oxidation

Oxidation of alkyl fluorosilanes to the corresponding alcohols.
Cf. Fleming oxidation.

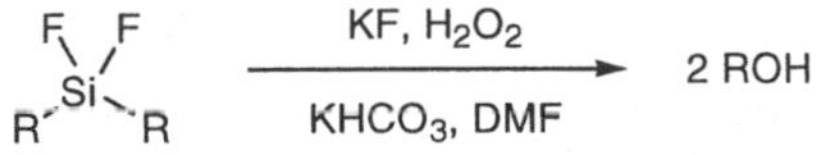

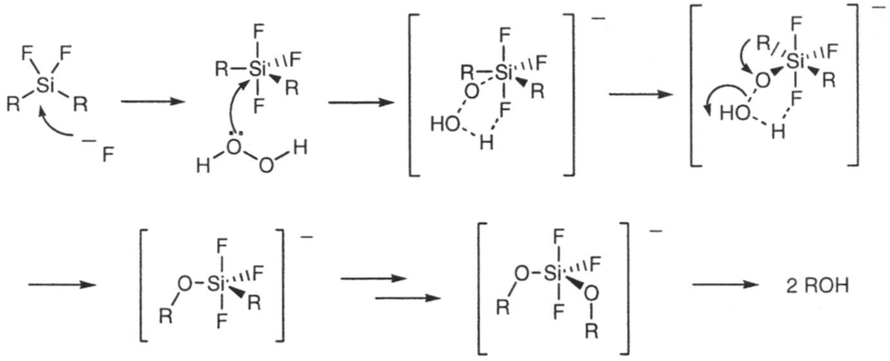

References

1. Tamao, K.; Ishida, N.; Kumada, M. *J. Org. Chem.* **1983**, *48*, 2120.
2. Kim, S.; Emeric, G.; Fuchs, P. L. *J. Org. Chem.* **1992**, *57*, 7362.
3. Jones, G. R.; Landais, Y. *Tetrahedron* **1996**, *52*, 7599.
4. Hunt, J. A.; Roush, W. R. *J. Org. Chem.* **1997**, *62*, 1112.
5. Knölker, H.-J.; Jones, P. G.; Wanzl, G. *Synlett* **1997**, 613.
6. Studer, A.; Steen, H. *Chem.--Eur. J.* **1999**, *5*, 759.
7. Barrett, A. G. M.; Head, J.; Smith, M. L.; Stock, N. S.; White, A. J. P.; Williams, D. J. *J. Org. Chem.* **1999**, *64*, 6005.

Tebbe olefination (Petasis alkenylation)

$$Cp_2Ti\underset{Cl}{\overset{}{\diagdown}}Al(CH_3)_2 + \underset{R}{\overset{O}{\|}}_{R^1} \longrightarrow \underset{R}{\overset{}{\diagup}}_{R^1} + Cp_2Ti=O + ClAl(CH_3)_2$$

Tebbe's reagent

$$Cp_2TiCl_2 + 2\ Al(CH_3)_3 \xrightarrow{\text{transmetallation}} Cp_2Ti\underset{CH_3}{\overset{H}{\diagdown}} \xrightarrow[\text{abstraction}]{\alpha\text{-hydride}}$$

titanocene dichloride

$$CH_4\uparrow + Cp_2Ti=CH_2 \xrightarrow[\text{coordination}]{Cl-Al(CH_3)_2} Cp_2Ti\underset{Cl}{\overset{}{\diagdown}}Al(CH_3)_2$$

titanocene methylidene   Tebbe's reagent

$$Cp_2Ti\underset{Cl}{\overset{}{\diagdown}}Al(CH_3)_2 \underset{\text{dissociation}}{\rightleftharpoons} Cl-Al(CH_3)_2 + \underset{O}{\overset{Cp_2Ti=CH_2}{\diagdown}}\overset{R^1}{\underset{R}{}}$$

$$\xrightarrow[\text{cycloaddition}]{[2+2]} \underset{\substack{O \\ R}}{\overset{Cp_2Ti-CH_2}{\diagup}}R^1 \xrightarrow[\text{cycloaddition}]{\text{retro-}[2+2]} \underset{R}{\overset{O}{\|}}_{R^1} + Cp_2Ti=O$$

oxatitanacyclobutane   Formation of the strong Ti=O is the driving force.

The Petasis reagent (Me_2TiCp_2, dimethyltitanocene) undergoes similar olefination reactions with ketones and aldehydes. The originally proposed mechanism [3] was very different from that of Tebbe olefination. However, later experimental data seem to suggest that both Petasis and Tebbe olefination share the same mechanism, i.e. the carbene mechanism involving a four-membered titanium oxide ring intermediate [6].

References

1. Tebbe, F. N.; Parshall, G. W.; Reddy, G. S. *J. Am. Chem. Soc.* **1978**, *100*, 3611.
2. Chou, T. S.; Huang, S. B. *Tetrahedron Lett.* **1983**, *24*, 2169.
3. Petasis, N. A.; Bzowej, E. I. *J. Am. Chem. Soc.* **1990**, *112*, 6392.
4. Schioett, B.; Joergensen, K. A. *J. Chem. Soc., Dalton Trans.* **1993**, 337.
5. Nicolaou, K. C.; Postema, M. H. D.; Claiborne, C. F. *J. Am. Chem. Soc.* **1996**, *118*, 1565.

6. Hughes, D. L.; Payack, J. F.; Cai, D.; Verhoeven, T. R.; Reider, P. J. *Organometallics* **1996**, *15*, 663.
7. Godage, H. Y.; Fairbanks, A. J. *Tetrahedron Lett.* **2000**, *41*, 7589.
8. Hartley, R. C.; McKiernan, G. J. *Perkin 1* **2002**, 2763–2793. (Review).
9. Jung, M. E.; Pontillo, J. *Tetrahedron* **2003**, *59*, 2729.

Thorpe–Ziegler reaction

The intramolecular version of the Thorpe reaction.

References

1. Baron, H.; Remfry, F. G. P.; Thorpe, Y. F. *J. Chem. Soc.* **1904**, *85*, 1726.
2. Rodriguez-Hahn, L.; Parra M., M.; Martinez, M. *Synth. Commun.* **1984**, *14*, 967.
3. Yakovlev, M. Yu.; Kadushkin, A. V.; Solov'eva, N. P.; Granik, V. G. *Heterocycl. Commun.* **1998**, *4*, 245.
4. Curran, D. P.; Liu, W. *Synlett* **1999**, 117.
5. Dansou, B.; Pichon, C.; Dhal, R.; Brown, E.; Mille, S. *Eur. J. Org. Chem.* **2000**, 1527.
6. Kovacs, L. *Molecules* **2000**, *5*, 127.
7. Gutschow, M.; Powers, J. C. *J. Heterocycl. Chem.* **2001**, *38*, 419.
8. Keller, L.; Dumas, F.; Pizzonero, M.; d'Angelo, J.; Morgant, G.; Nguyen-Huy, D. *Tetrahedron Lett.* **2002**, *43*, 3225.
9. Malassene, R.; Toupet, L.; Hurvois, J.-P.; Moinet, C. *Synlett* **2002**, 895.
10. Malassene, R.; Vanquelef, E.; Toupet, L.; Hurvois, J.-P.; Moinet, C. *Org. Biomol. Chem.* **2003**, *1*, 547.

Tiemann rearrangement

Treatment of amidoximes, derived from amides and hydroxylamine, with benzenesulfonyl chloride and water leads to ureas.

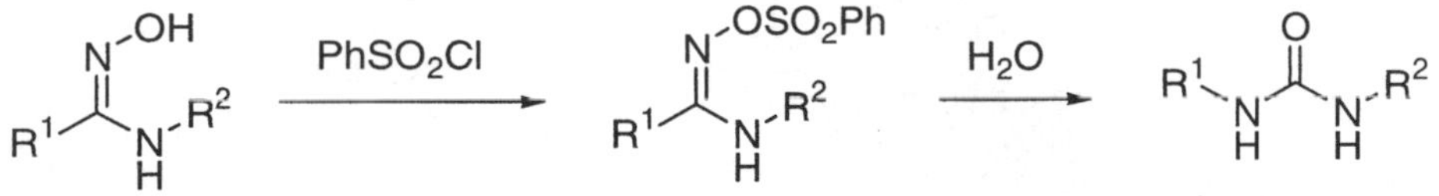

amidoxime   urea

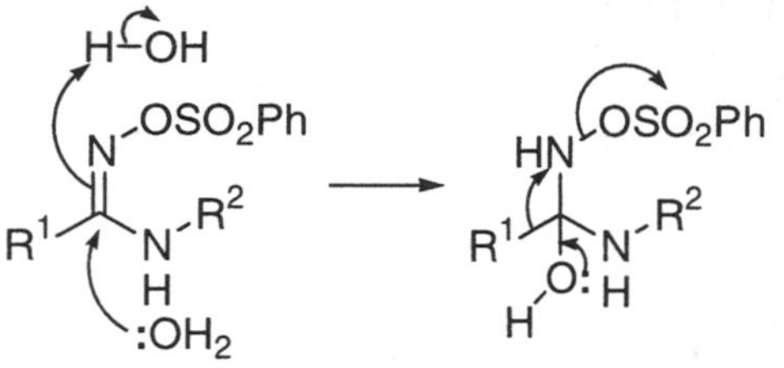

The substituent *anti* to the leaving group ($^-$OSO$_2$Ph) migrates.

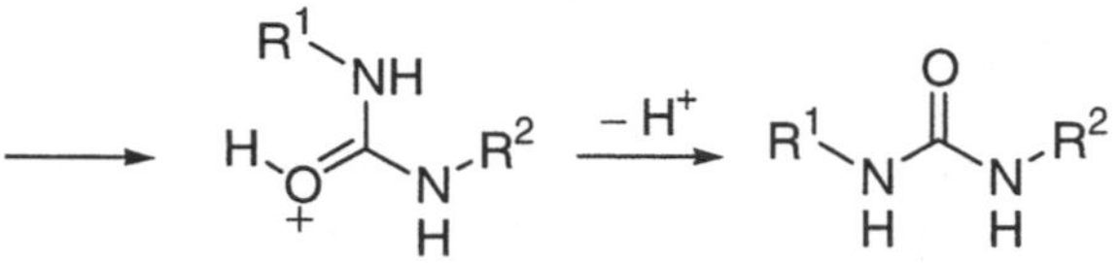

References

1. Tiemann, F. *Ber. Dtsch. Chem. Ges.* **1891**, *24*, 4162.
2. Garapon, J.; Sillion, B.; Bonnier, J. M. *Tetrahedron Lett.* **1970**, 4905.
3. Adams, G. W.; Bowie, J. H.; Hayes, R. N.; Gross, M. L. *J. Chem. Soc., Perkin Trans. 2* **1992**, 897.
4. Bakunov, S. A.; Rukavishnikov, A. V.; Tkachev, A. V. *Synthesis* **2000**, 1148.
5. Richter, R.; Tucker, B.; Ulrich, H. *J. Org. Chem.* **1983**, *48*, 1694.
6. Eichinger, P. C. H.; Dua, S.; Bowie, J. H. *Int. J. Mass Spectrom. Ion Proc.* **1994**, *133*, 1.

Tiffeneau–Demjanov rearrangement

Carbocation rearrangement of β-aminoalcohols *via* diazotization to afford carbonyl compounds.

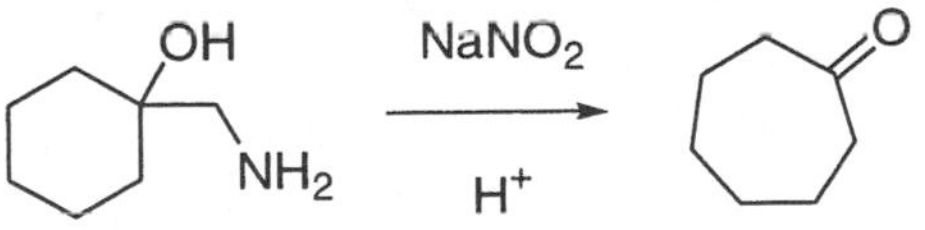

Step 1, Generation of N_2O_3

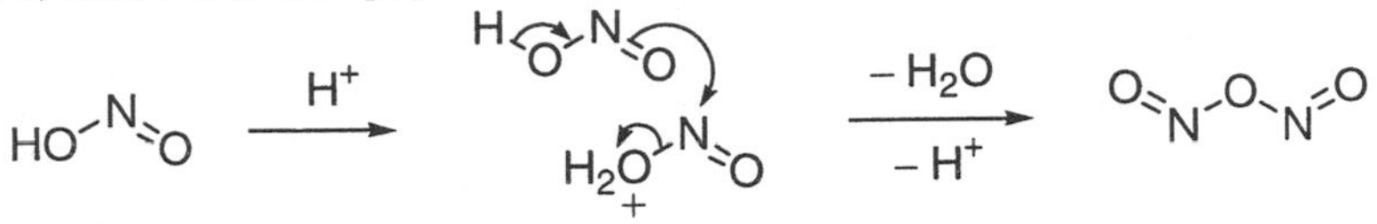

Step 2, Transformation of amine to diazonium salt

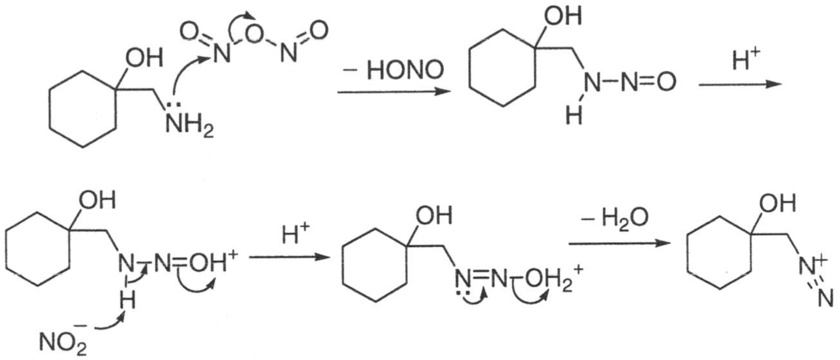

Step 3, Ring-expansion *via* rearrangement

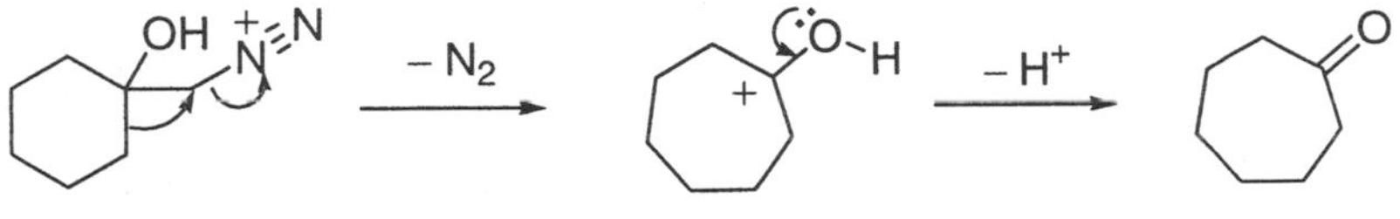

References

1. Tiffeneau, M.; Weil, P.; Tehoubar, B. *Compt. Rend* **1937**, *205*, 54.
2. Smith, P. A. S.; Baer, D. R. *Org. React.* **1960**, *11*, 157. (Review).
3. Jones, J. B.; Price, P. *J. Chem. Soc., Chem. Commun.* **1969**, 1478.
4. Parham, W. E.; Roosevelt, C. S. *J. Org. Chem.* **1972**, *37*, 1975.
5. McKinney, M. A.; Patel, P. P. *J. Org. Chem.* **1973**, *38*, 4059.
6. Jones, J. B.; Price, P. *Tetrahedron* **1973**, *29*, 1941.
7. Dave, V.; Stothers, J. B.; Warnhoff, E. W. *Can. J. Chem.* **1979**, *57*, 1557.

410

8. Haffer, G.; Eder, U.; Neef, G.; Sauer, G.; Wiechert, R. *Justus Liebigs Ann. Chem.* **1981**, 425.
9. Thomas, R. C.; Fritzen, E. L. *J. Antibiotics* **1988**, *41*, 1445.
10. Stern, A. G.; Nickon, A. *J. Org. Chem.* **1992**, *57,* 5342.
11. Fattori, D.; Henry, S.; Vogel, P. *Tetrahedron* **1993**, *49*, 1649.
12. Houdai, T.; Matsuoka, S.; Murata, M.; Satake, M.; Ota, S.; Oshima, Y.; Rhodes, L. L. *Tetrahedron* **2001**, *57*, 5551.

Tishchenko reaction

Esters from the corresponding aldehydes and Al(OEt)$_3$, which serves as a homogeneous catalyst.

References

1. Tishchenko, V. *J. Russ. Phys. Chem. Soc.* **1906**, *38*, 355.
2. Saegusa, T.; Ueshima, T.; Kitagawa, S. *Bull. Chem. Soc. Jpn.* **1969**, *42*, 248.
3. Ogata, Y.; Kishi, I. *Tetrahedron* **1969**, *25*, 929.
4. Berberich, H.; Roesky, P. W. *Angew. Chem., Int. Ed.* **1998**, *37*, 1569.
5. Lu, L.; Chang, H.-Y.; Fang, J.-M. *J. Org. Chem.* **1999**, *64*, 843.
6. Mascarenhas, C.; Duffey, M. O.; Liu, S.-Y.; Morken, J. P. *Org. Lett.* **1999**, *1*, 1427.
7. Bideau, F. L.; Coradin, T.; Gourier, D.; Hénique, J.; Samuel, E. *Tetrahedron Lett.* **2000**, *41*, 5215.
8. Toermaekangas, O. P.; Koskinen, A. M. P. *Org. Process Res. Dev.* **2001**, *5*, 421.
9. Chang, C.-P.; Hon, Y.-S. *Huaxue* **2002**, *60*, 561. (Review).
10. Shirakawa, S.; Takai, J.; Sasaki, K.; Miura, T.; Maruoka, K. *Heterocycles* **2003**, *59*, 57.

412

Tollens reaction

Condensation of carbonyl compounds possessing an α-hydrogen with formaldehyde in the presence of base.

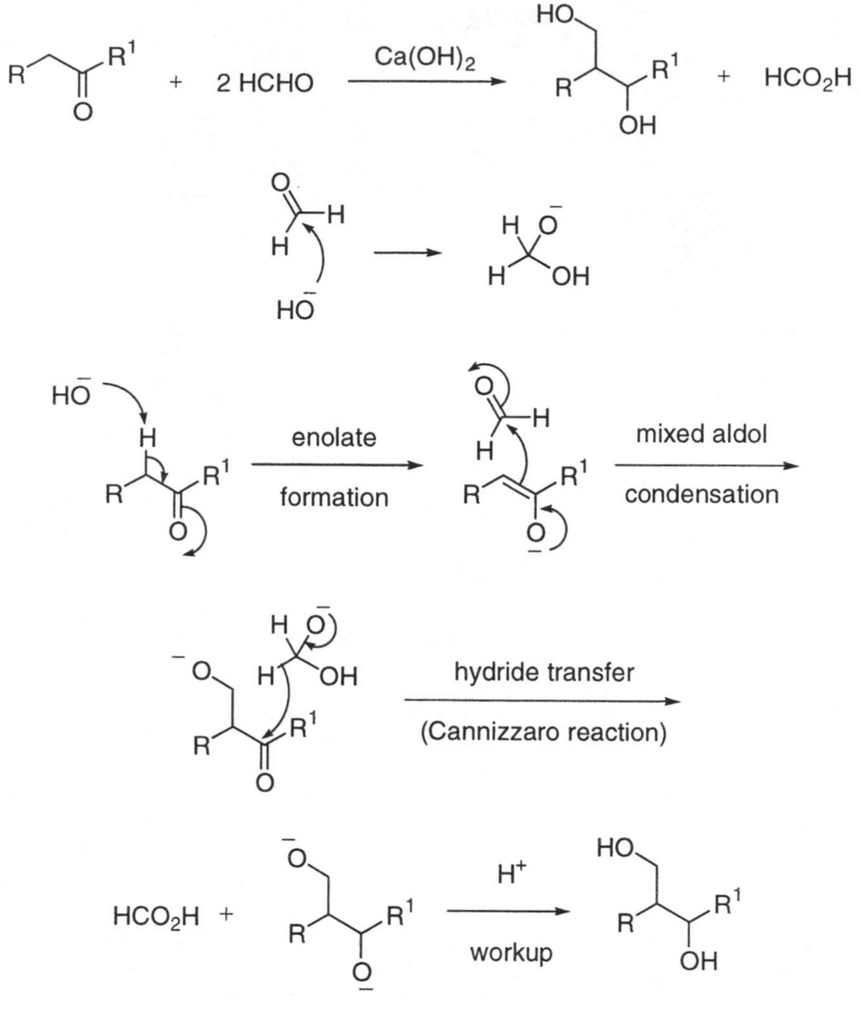

R—CH₂—C(=O)—R¹ + 2 HCHO →(Ca(OH)₂)→ R—CH(CH₂OH)—CH(OH)—R¹ + HCO₂H

enolate formation

mixed aldol condensation

hydride transfer

(Cannizzaro reaction)

HCO₂H + →(H⁺, workup)→

References

1. Parry-Jones, R.; Kumar, *J. Educ. Chem.* **1985**, *22*, 114.
2. Jenkins, I. D. *J. Chem. Educ.* **1987**, *64*, 164.
3. Munoz, S.; Gokel, G. W. *J. Am. Chem. Soc.* **1993**, *115*, 4899.
4. Yin, Y.; Li, Z.-Y.; Zhong, Z.; Gates, B.; Xia, Y.; Venkateswaran, S. *J. Materials Chem.* **2002**, *12*, 522.

5. Breedlove, C. H.; Softy, John. *J. College Sci. Teaching* **1983**, *12*, 281.
6. Huang, S.; Mau, A. W. H. *J. Phys. Chem. B* **2003**, *107*, 3455.

Tsuji–Trost reaction

Palladium-catalyzed allylation using nucleophiles.

π-allyl complex

References

1. Tsuji, J.; Takahashi, H.; Morikawa, M. *Tetrahedron Lett.* **1965**, 4387.
2. Tsuji, J. *Acc. Chem. Res.* **1969**, *2*, 144. (Review).
3. Godleski, S. A. In *Comprehensive Organic Synthesis;* Trost, B. M.; Fleming, I., eds.; *Vol. 4.* Chapter 3.3. Pergamon: Oxford, **1991**. (Review).
4. Bolitt, V.; Chaguir, B.; Sinou, D. *Tetrahedron Lett.* **1992**, *33*, 2481.
5. Moreno-Mañas, M.; Pleixats, R. In *Advances in Heterocyclic Chemistry;* Katritzky, A. R., ed.; Academic Press: San Diego, **1996**, *66*, 73. (Review).
6. Tietze, L. F.; Nordmann, G. *Eur. J. Org. Chem.* **2001**, 3247.
7. Sato, Y.; Yoshino, T.; Mori, M. *Org. Lett.* **2003**, *5*, 31.
8. Page, P. C. B.; Heaney, H.; Reignier, S.; Rassias, G. A. *Synlett* **2003**, 22.

Ueno–Stork cyclization

Substituted tetrahydrofuran from radical cyclization of δ-bromoolefin.

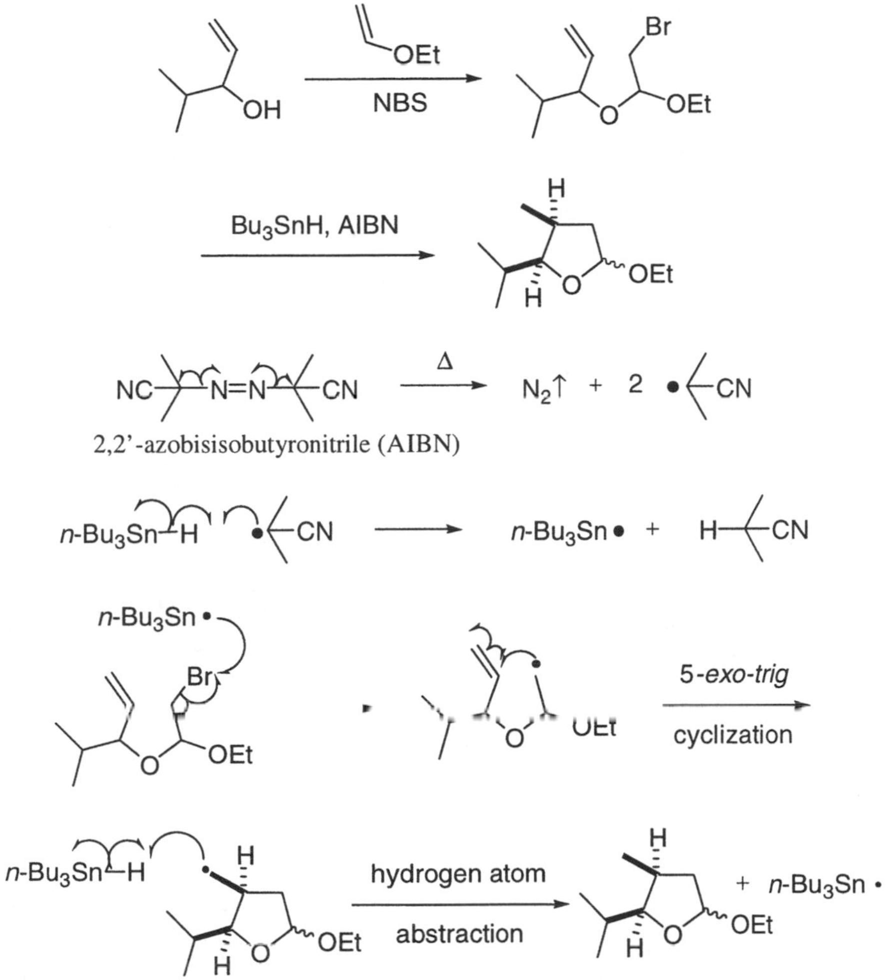

2,2'-azobisisobutyronitrile (AIBN)

5-*exo-trig*

cyclization

hydrogen atom

abstraction

References

1. Ueno, Y.; Chino, K.; Watanabe, M.; Moriya, O.; Okawara, M. *J. Am. Chem. Soc.* **1982**, *104*, 5564.
2. Stork, G.; Mook, R.; Biller, S. A.; Rychnovsky, S. D. *J. Am. Chem. Soc.* **1983**, *105*, 3741.
3. Villar, F.; Renaud, P. *Tetrahedron Lett.* **1998**, *39*, 8655.
4. Villar, F.; Andrey, O.; Renaud, P. *Tetrahedron Lett.* **1999**, *40*, 3375.
5. Villar, F.; Equey, O.; Renaud, P. *Org. Lett.* **2000**, *2*, 1061.

Ugi reaction

Four-component condensation (4CC) of carboxylic acids, *C*-isocyanides, amines, and oxo compounds to afford peptides. *Cf.* Passerini reaction.

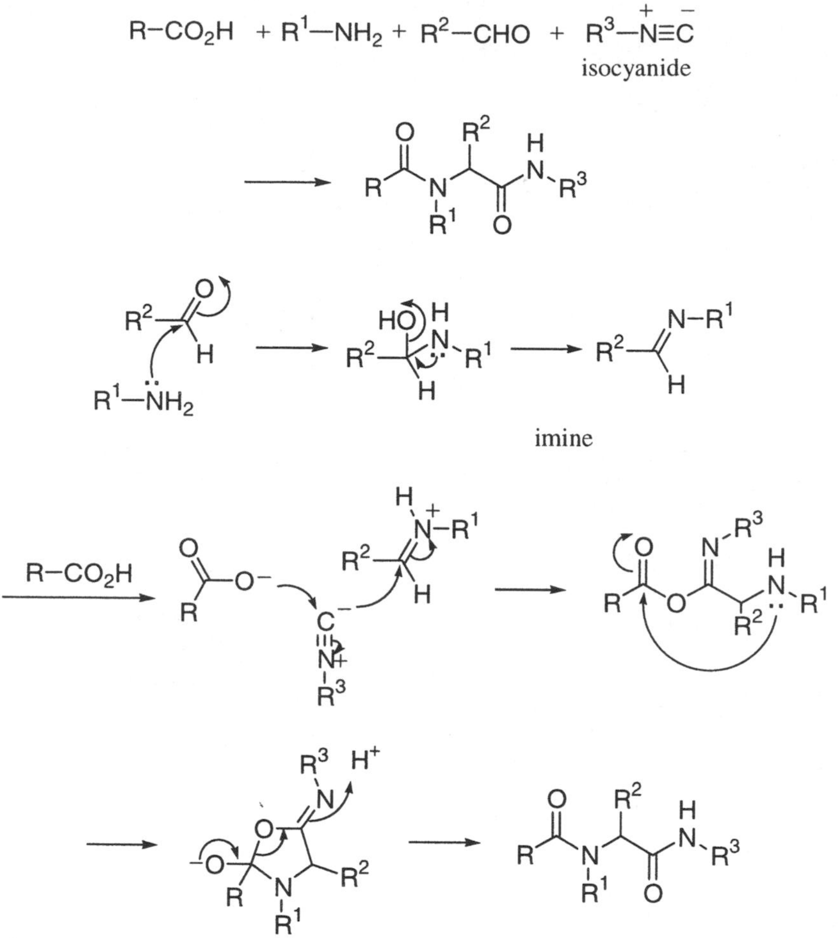

$$R-CO_2H \ + \ R^1-NH_2 \ + \ R^2-CHO \ + \ R^3-\overset{+}{N}\overset{-}{\equiv}C$$

isocyanide

imine

References

1. Ugi, I. *Angew. Chem., Int. Ed. Engl.* **1962**, *1*, 8.
2. Skorna, G.; Ugi, I. *Chem. Ber.* **1979**, *112*, 776.
3. Hoyng, C. F.; Patel, A. D. *Tetrahedron Lett.* **1980**, *21*, 4795.
4. Ugi, I.; Lohberger, S.; Karl, R. In *Comprehensive Organic Synthesis*; Trost, B. M.; Fleming, I., Eds.; Pergamon: Oxford, **1991**, *Vol. 2*, 1083. (Review).
5. Dömling, A.; Ugi, I. *Angew. Chem., Int. Ed.* **2000**, *39*, 3168. (Review).
6. Ugi, I. *Pure Appl. Chem.* **2001**, *73*, 187. (Review).

7. Zimmer, R.; Ziemer, A.; Grunner, M.; Brüdgam, I.; Hartl, H.; Reissig, H.-U. *Synthesis* **2001**, 1649.

8. Kennedy, A. L.; Fryer, A. M.; Josey, J. A. *Org. Lett.* **2002**, *4*, 1167.

9. Baldoli, C.; Maiorana, S.; Licandro, E.; Zinzalia, G.; Perdicchia, D. *Org. Lett.* **2002**, *4*, 4341.

10. Portlock, D. E.; Ostaszewski, R.; Naskar, D.; West, L. *Tetrahedron Lett.* **2003**, *44*, 603.

11. Beck, B.; Larbig, G.; Mejat, B.; Magnin-Lachaux, M.; Picard, A.; Herdtweck, E.; Docmling, A. *Org. Lett.* **2003**, *5*, 1047.

12. Hebach, C.; Kazmaier, U. *Chem. Commun.* **2003**, 596.

418

Ullmann reaction

Homocoupling of aryl iodide in the presence of Cu.

The overall transformation of PhI to PhCuI is an oxidative addition process.

References

1.   Ullmann, F. *Justus Liebigs Ann. Chem.* **1904**, *332*, 38.
2.   Fanta, P. E. *Synthesis* **1974**, 9.
3.   Stark, L. M.; Lin, X.-F.; Flippin, L. A. *J. Org. Chem.* **2000**, *65*, 3227.
4.   Belfield, K. D.; Schafer, K. J.; Mourad, W.; Reinhardt, B. A. *J. Org. Chem.* **2000**, *65*, 4475.
5.   Venkatraman, S.; Li, C.-J. *Tetrahedron Lett.* **2000**, *41*, 4831.
6.   Farrar, J. M.; Sienkowska, M.; Kaszynski, P. *Synth. Commun.* **2000**, *30*, 4039.
7.   Ma, D.; Xia, C. *Org. Lett.* **2001**, *3*, 2583.
8.   Buck, E.; Song, Z. J.; Tschaen, D.; Dormer, P. G.; Reider, P. J. *Org. Lett.* **2002**, *4*, 1623.
9.   Hameurlaine, A.; Dehaen, W. *Tetrahedron Lett.* **2003**, *44*, 957.

Vilsmeier–Haack reaction

The Vilsmeier–Haack reagent, chloroiminium salt, is a weak electrophile, therefore, the Vilsmeier–Haack reaction works better with electron-rich carbocycles and heterocycles.

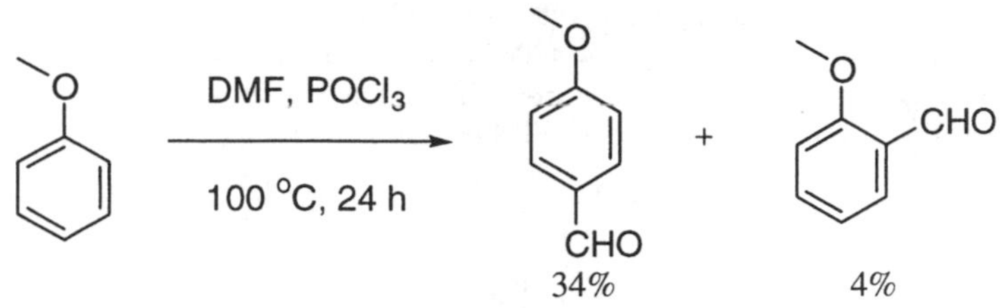

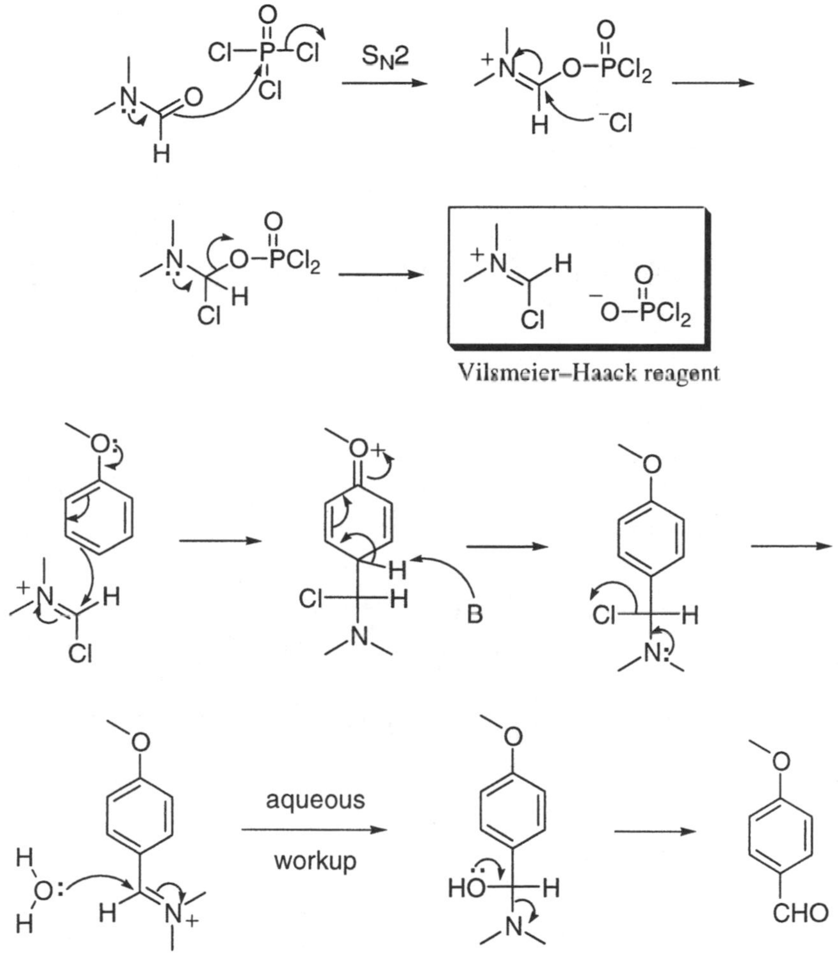

34%   4%

Vilsmeier–Haack reagent

420

References

1. Vilsmeier, A.; Haack, A. *Ber. Dtsch. Chem. Ges.* **1927**, *60*, 119.
2. Marson, C. M.; Giles, P. R. *Synthesis Using Vilsmeier Reagents* CRC Press, **1994**. (Review).
3. Seybold, G. *J. Prakt. Chem.* **1996**, *338,* 392–396 (Review).
4. Jones, G.; Stanforth, S. P. *Org. React.* **1997**, *49*, 1. (Review).
5. Ali, M. M.; Tasneem; Rajanna, K. C.; Sai Prakash, P. K. *Synlett* **2001**, 251.
6. Thomas, A. D.; Asokan, C. V. *J. Chem. Soc., Perkin Trans. 1* **2001**, 2583.
7. Tasneem, *Synlett* **2003**, 138. (Review of Vilsmeier–Haack reagent).

von Braun reaction

Treatment of tertiary amines with cyanogen bromide, resulting in a substituted cyanamide and alkyl halides.

$$R^1\text{-}\underset{R_2}{\overset{R}{N}} \quad \xrightarrow{\text{Br}-\text{CN}} \quad R^1\text{-}\underset{R_2}{\overset{CN}{N}} \quad + \quad \text{Br}-R$$

Cyanogen bromide (BrCN) is a *counterattack reagent*.

$$R^1\text{-}\underset{R_2}{\overset{R}{N:}} \xrightarrow{NC\text{-}Br} \xrightarrow{S_N2} \underset{R^1\text{-}\overset{+}{N}\text{-}R_2}{\overset{R}{\underset{}{}} \text{-}CN} \xrightarrow{S_N2} R^1\text{-}\underset{R_2}{\overset{CN}{N}} + \text{Br}-R$$

References

1. von Braun, J. *Ber. Dtsch. Chem. Ges.* **1907**, *40*, 3914.
2. Hageman, H. A. *Org. React.* **1953**, *7*, 198. (Review).
3. Nakahara, Y.; Niwaguchi, T.; Ishii, H. *Tetrahedron* **1977**, *33*, 1591.
4. Fodor, G.; Nagubandi, S. *Tetrahedron* **1980**, *36*, 1279.
5. Perni, R. B.; Gribble, G. W. *Org. Prep. Proced. Int.* **1980**, *15*, 297.
6. McLean, S.; Reynolds, W. F.; Zhu, X. *Can. J. Chem.* **1987**, *65*, 200.
7. Cooley, J. H.; Evain, E. J. *Synthesis* **1989**, 1.
8. Aguirre, J. M.; Alesso, E. N.; Ibanez, A. F.; Tombari, D. G.; Moltrasio Iglesias, G. Y. *J. Heterocycl. Chem.* **1989**, *26*, 25.
9. Laabs, S.; Scherrmann, A.; Sudau, A.; Diederich, M.; Kierig, C.; Nubbemeyer, U. *Synlett* **1999**, 25.
10. Chambert, S.; Thamosson, F.; Décout, J.-L. *J. Org. Chem.* **2002**, *67*, 1898.

von Richter reaction

Treatment of nitroarenes with cyanide to result in carboxylation at the *ortho* position of the nitro group.

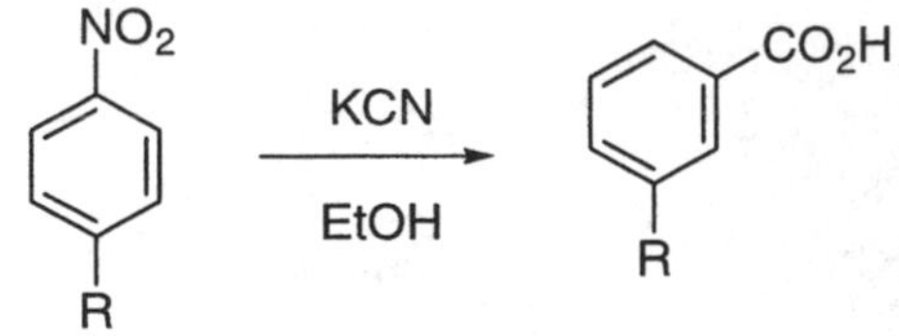

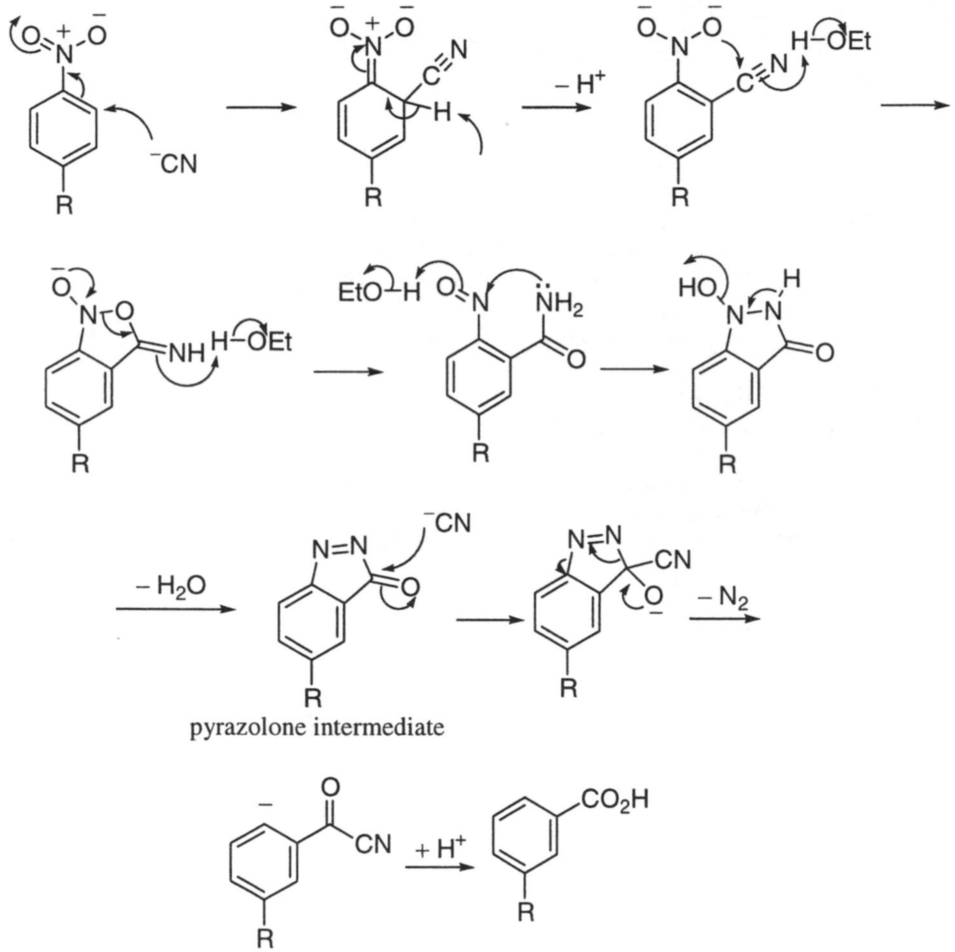

pyrazolone intermediate

References

1. von Richter, V. *Ber. Dtsch. Chem. Ges.* **1871**, *4*, 21, 459, 553.
2. Rosenblum, M. *J. Am. Chem. Soc.* **1960**, *82,* 3796.
3. Rogers, G. T.; Ulbricht, T. L. V. *Tetrahedron Lett.* **1968**, *23,* 1029.
4. Ellis, A. C.; Rae, I. D. *J. Chem. Soc., Chem. Commun.* **1977**, 152.
5. Tretyakov, E. V.; Knight, D. W.; Vasilevsky, S. F. *Heterocycl. Commun.* **1998**, *4*, 519.
6. Tretyakov, E. V.; Knight, D. W.; Vasilevsky, S. F. *J. Chem. Soc., Perkin Trans. 1* **1999**, 3721.
7. Brase, S.; Dahmen, S.; Heuts, J. *Tetrahedron Lett.* **1999**, *40*, 6201.

Wacker oxidation

Palladium-catalyzed oxidation of olefins to ketones.

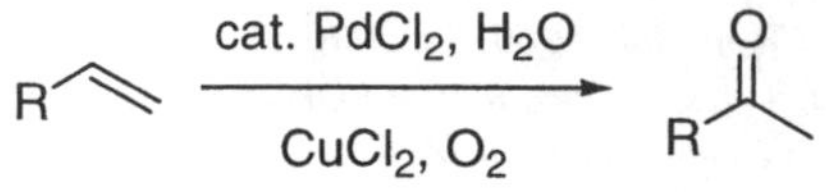

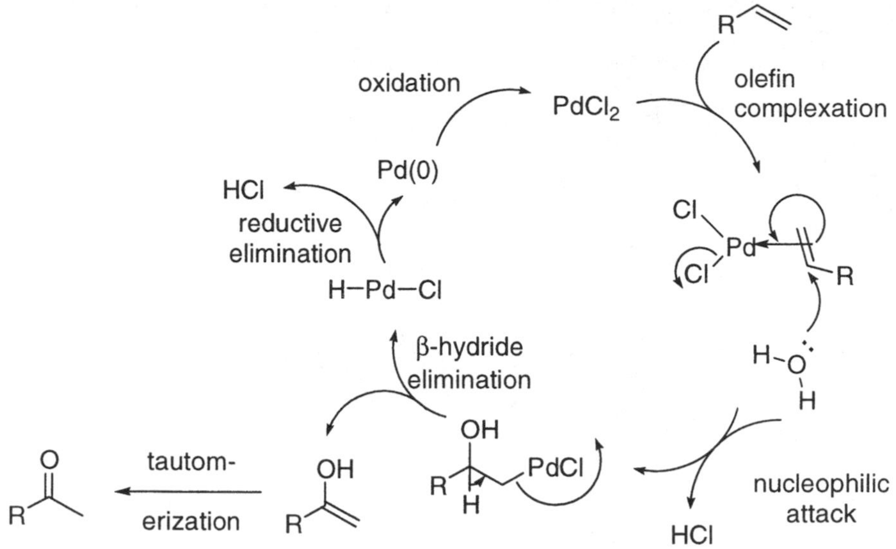

Regeneration of Pd(II):

$$Pd(0) + 2\ CuCl_2 \longrightarrow PdCl_2 + 2\ CuCl$$

Regeneration of Cu(II):

$$CuCl + O_2 \longrightarrow CuCl_2 + H_2O$$

References

1. Tsuji, J. *Synthesis* **1984**, 369. (Review).
2. Miller, D. G.; Wayner, Danial D. M. *J. Org. Chem.* **1990**, *55*, 2924.
3. Hegedus, L. S. *Transition Metals in the Synthesis of Complex Organic Molecule* **1994**, University Science Books: Mill Valley, CA, pp 199–208. (Review).
4. Hegedus, L. S. In *Comp. Org. Syn.* Trost, B. M.; Fleming, I., Eds.; Pergamon, **1991**, Vol. *4*, 552. (Review).
5. Tsuji, J. In *Comp. Org. Syn.* Trost, B. M.; Fleming, I., Eds.; Pergamon, **1991**, Vol. *7*, 449. (Review).

6. Kang, S.-K.; Jung, K.-Y.; Chung, J.-U.; Namkoong, E.-Y.; Kim, T.-H. *J. Org. Chem.* **1995**, *60*, 4678.
7. Feringa, B. L. *Transition Met. Org. Synth.* **1998**, *2*, 307. (Review).
8. Gaunt, M. J.; Yu, J.; Spencer, J. B. *Chem. Commun.* **2001**, 1844.
9. Barker, D.; Brimble, M. A.; McLeod, M.; Savage, G. P.; Wong, D. J. *J. Chem. Soc., Perkin Trans. 1*, **2002**, 924.
10. Thadani, A. N.; Rawal, V. H. *Org. Lett.* **2002**, *4,* 4321.
11. Choi, K.-M.; Mizugaki, T.; Ebitani, K.; Kaneda, K. *Chem. Lett.* **2003**, *32,* 180.
12. Takacs, J. M.; Jiang, X.-t. *Current Org. Chem.* **2003**, *7,* 369.

Wagner–Meerwein rearrangement

Acid-catalyzed alkyl group migration of alcohols to give more substituted olefins.

References

1. Wagner, G. *J. Russ. Phys. Chem. Soc.* **1899**, *31*, 690.
2. Hogeveen, H.; Van Kruchten, E. M. G. A. *Top. Curr. Chem.* **1979**, *80*, 89. (Review).
3. Martinez, A. G.; Vilar, E. T.; Fraile, A. G.; Fernandez, A. H.; De La Moya Cerero, S.; Jimenez, F. M. *Tetrahedron* **1998**, *54*, 4607.
4. Birladeanu, L. *J. Chem. Educ.* **2000**, *77*, 858.
5. Kobayashi, T.; Uchiyama, Y. *Perkin 1* **2000**, 2731.
6. Trost, B. M.; Yasukata, T. *J. Am. Chem. Soc.* **2001**, *123*, 7162.
7. Cerda-Garcia-Rojas, C. M.; Flores-Sandoval, C. A.; Roman, L. U.; Hernandez, J. D.; Joseph-Nathan, P. *Tetrahedron* **2002**, *58*, 1061.
8. Colombo, M. I.; Bohn, M. L.; Ruveda, E. A. *J. Chem. Educ.* **2002**, *79*, 484.
9. Roman, L. U.; Cerda-Garcia-Rojas, C. M.; Guzman, R.; Armenta, C.; Hernandez, J. D.; Joseph-Nathan, P. *J. Nat. Products* **2002**, *65*, 1540.
10. Garcia Martinez, A.; Teso Vilar, E.; Garcia Fraile, A.; Martinez-Ruiz, P. *Tetrahedron* **2003**, *59*, 1565.
11. Guizzardi, B.; Mella, M.; Fagnoni, M.; Albini, A. *J. Org. Chem.* **2003**, *68*, 1067.

Wallach rearrangement

Conversion of azoxy compounds to *p*-hydroxy azo compounds upon treatment with acid.

References

1. Wallach, O.; Belli, L. *Ber. Dtsch. Chem. Ges.* **1880**, *13*, 525.
2. Cichon, L. *Wiad. Chem.* **1966**, *20*, 641.
3. Buncel, E.; Keum, S. R.; *J. Chem. Soc., Chem. Commun.* **1983**, 578.
4. Shine, H. J.; Subotkowski, W.; Gruszecka, E. *Can. J. Chem.* **1986**, *64*, 1108.
5. Okano, T. *Kikan Kagaku Sosetsu* **1998**, *37*, 130.
6. Hattori, H. *Kikan Kagaku Sosetsu* **1999**, *41*, 46.
7. Lalitha, A.; Pitchumani, K.; Srinivasan, C. *J. Mol. Catal. A: Chem.* **2000**, *162*, 429.

Weinreb amide

N-Methoxy-*N*-methyl-amide. Due to the chelation effect, nucleophilic addition of an organometallic reagent adds only once to give ketone, whereas normal amides would have led to double addition to afford tertiary alcohol.

stable

References

1. Nahm, S.; Weinreb, S. M. *Tetrahedron Lett.* **1981**, *22*, 3815.
2. Sibi, M. P. *Org. Prep. Proc. Int.* **1993**, *25*, 15.
3. Mentzel, M.; Hoffmann, H. M. R. *J. Prakt. Chem.* **1997**, *339*, 517.
4. Singh, J.; Satyamurthi, N.; Aidhen, I. S. *J. Prakt. Chem.* **2000**, *342*, 340.
5. McNulty, J.; Grunner, V.; Mao, J. *Tetrahedron Lett.* **2001**, *42*, 5609.
6. Conrad, R. M.; Grogan, M. J.; Bertozzi, C. R. *Org. Lett.* **2002**, *4*, 1359.
7. Davis, F. A.; Prasad, K. R.; Nolt, M. B.; Wu, Y. *Org. Lett.* **2003**, *5*, 925.
8. Ruiz, J.; Sotomayor, N.; Lete, E. *Org. Lett.* **2003**, *5*, 1115.

Weiss reaction

Synthesis of *cis*-bicyclo[3.3.0]octane-3,7-dione.

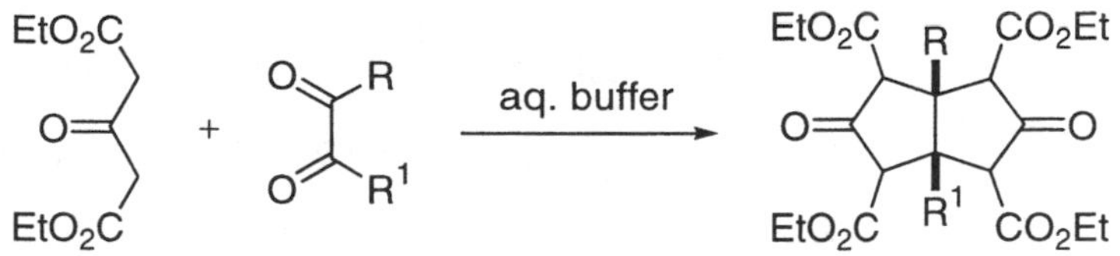

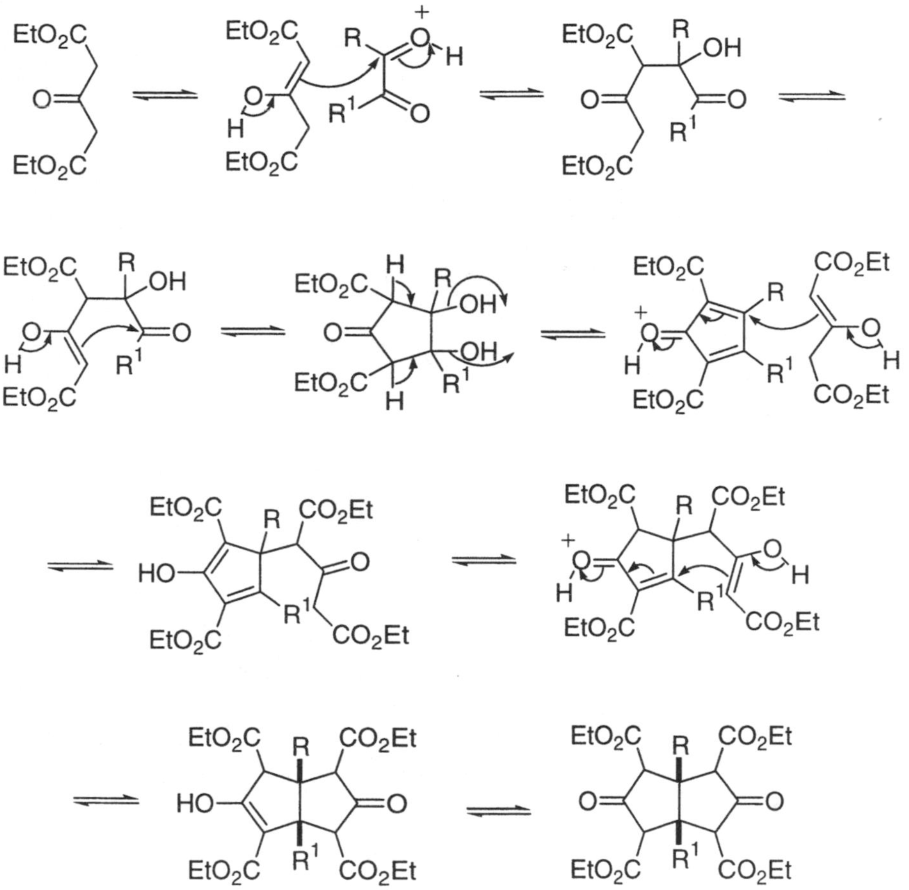

430

References

1. Weiss, U.; Edwards, J. M. *Tetrahedron Lett.* **1968**, *9,* 4885.
2. Gupta, A. K.; Fu, X.; Snydert, J. P.; Cook, J. M. *Tetrahedron* **1991**, *47*, 3665.
3. Reissig, H. U. *Org. Synth. Highlights* **1991**, 121. (Review).
4. Fu, X.; Cook, J. M. *Aldrichimica Acta* **1992**, *25*, 43. (Review).
5. Fu, X.; Kubiak, G.; Zhang, W.; Han, W.; Gupta, A. K.; Cook, J. M. *Tetrahedron* **1993**, *49*, 1511.
6. Van Ornum, S. G.; Li, J.; Kubiak, G. G.; Cook, J. M. *J. Chem. Soc., Perkin Trans. 1* **1997**, 3471.

Wenker aziridine synthesis

Aziridine synthesis *via* treatment of a β-amino alcohol with sulfuric acid to give β-aminoethyl sulfuric acid, which is subsequently treated with base.

References

1. Wenker, H. *J. Am. Chem. Soc.* **1935**, *57*, 2328.
2. Leighton, P. A.; Perkins, W. A.; Renquist, M. I. *J. Am. Chem. Soc.* **1947**, *69*, 1540.
3. Kaschelikar, D. V.; Fanta, P. E. *J. Am. Chem. Soc.* **1960**, *82*, 4927.
4. Kaschelikar, D. V.; Fanta, P. E. *J. Am. Chem. Soc.* **1960**, *82*, 4930.
5. Allen, C. F. H.; Spangler, F. W.; Webster, E. R. *Org. Synth., Coll. Vol. IV,* **1963**, 433.
6. Fanta, P. E.; Walsh, E. N. *J. Org. Chem.* **1966**, *31*, 59.
7. Dermer, O. C.; Ham, G. E. *Ethyleneimine and Other Aziridines,* Academic Press: New York, **1969**. (Review).
8. Gaertner, V. R. *J. Org. Chem.* **1970**, *35*, 3952.
9. Brewster, K; Pinder, R. M. *J. Med. Chem.* **1972**, *15*, 1078.
10. Nakagawa, Y.; Tsuno, T.; Nakajima, K.; Iwai, M.; Kawai, H.; Okawa, K. *Bull. Chem. Soc. Jpn.* **1972**, *45*, 1162.
11. Deyrup, J. A. In *Small Ring Heterocycles*, Part 1, Hassner, A., ed.; Wiley-Interscience: New York, **1983**, 1–214. (Review).
12. Park, J.-i.; Tian, G.; Kim, D. H. *J. Org. Chem.* **2001**, *66*, 3696.
13. Xu, J. *Tetrahedron: Asymmetry* **2002**, *13*, 1129.

Wharton oxygen transposition reaction

Reduction of α,β-epoxy ketones by hydrazine to allylic alcohols.

References

1. Wharton, P. S.; Bohlen, D. H. *J. Org. Chem.* **1961**, *26*, 3615.
2. Wharton, P. S. *J. Org. Chem.* **1961**, *26*, 4781.
3. Caine, D. *Org. Prep. Proced. Int.* **1988**, *20*, 1.
4. Dupuy, C.; Luche, J. L. *Tetrahedron* **1989**, *45*, 3437.
5. Di Filippo, M.; Fezza, F.; Izzo, I.; De Riccardis, F.; Sodano, G. *Eur. J. Org. Chem.* **2000**, 3247.

Willgerodt–Kindler reaction

Conversion of ketones to the corresponding thioamide and/or ammonium salt.

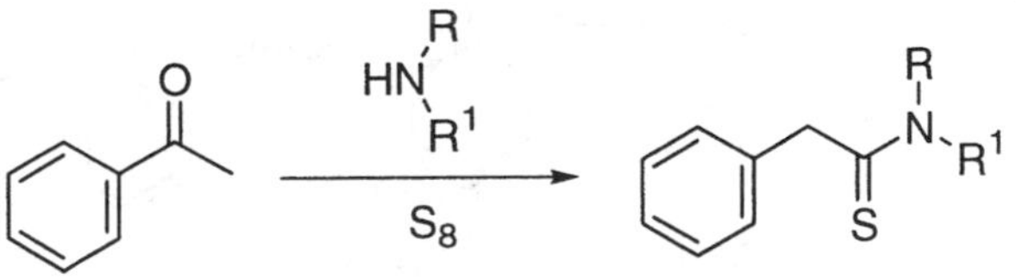

thioamide

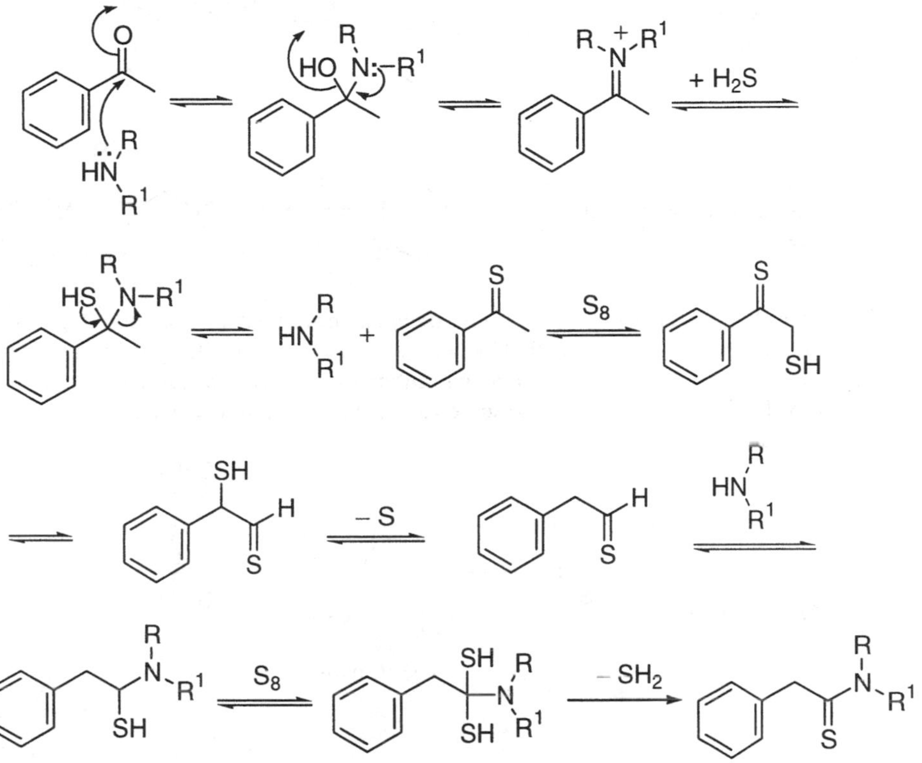

A slightly different mechanism has also been proposed:

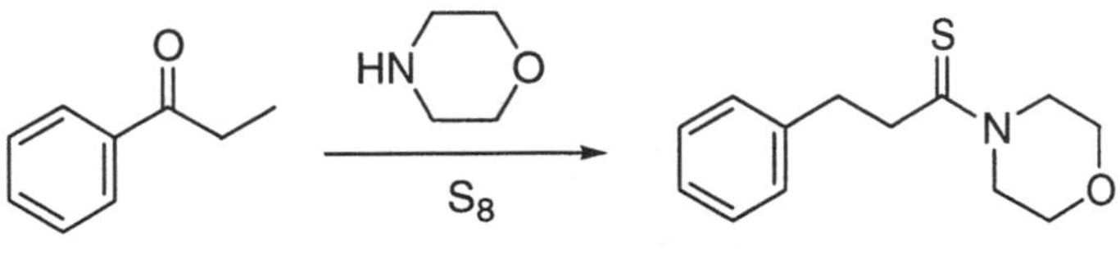

thioamide

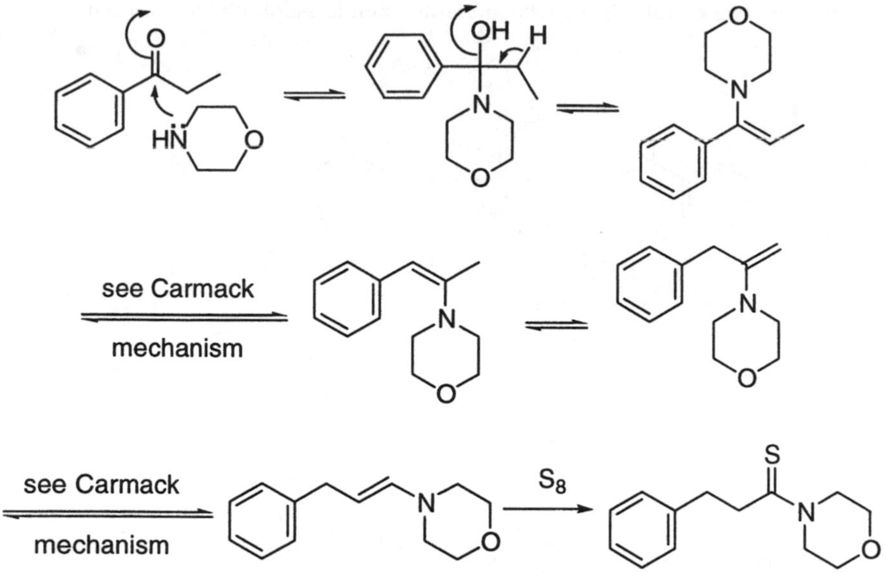

In Carmack's mechanism [5], the most unusual movement of a carbonyl group from methylene carbon to methylene carbon was proposed to go through an intricate pathway *via* a highly reactive intermediate with a sulfur-containing heterocyclic ring. The sulfenamide serves as the isomerization catalyst:

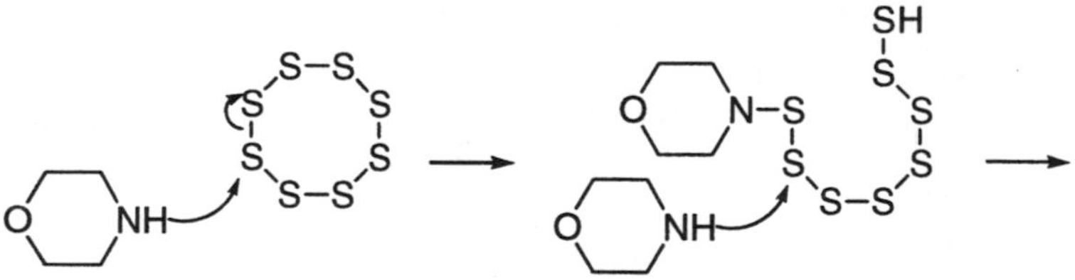

sulfenamide

thiirene

436

References

1. Willgerodt, C. *Ber. Dtsch. Chem. Ges.* **1887**, *20*, 2467.
2. Schneller, S. W. *Int. J. Sulfur Chem. B* **1972**, *7*, 155.
3. Schneller, S. W. *Int. J. Sulfur Chem.* **1973**, *8*, 485.
4. Schneller, S. W. *Int. J. Sulfur Chem.* **1976**, *8*, 579.
5. Carmack, M. *J. Heterocycl. Chem.* **1989**, *26*, 1319.
6. You, Q.; Zhou, H.; Wang, Q.; Lei, X. *Org. Prep. Proced. Int.* **1991**, *23*, 435.
7. Chatterjea, J. N.; Singh, R. P.; Ojha, N.; Prasad, R. *J. Inst. Chem. (India)* **1998**, *70*, 108.
8. Moghaddam, F. M.; Ghaffarzadeh, M.; Dakamin, M. G. *J. Chem. Res., (S)* **2000**, 228.
9. Poupaert, J. H.; Bouinidane, K.; Renard, M.; Lambert, D. M.; Isa, M. *Org. Prep. Proced. Int.* **2001**, *33*, 335.
10. Alam, M. M.; Adapa, S. R. *Synth. Commun.* **2003**, *33*, 59.

Williamson ether synthesis

Ether from the alkylation of alkoxides by alkyl halides.

References

1. Williamson, A. W. *J. Chem. Soc.* **1852**, *4,* 229.
2. Dermer, O. C. *Chem. Rev.* **1934**, *14,* 385. (Review).
3. Smith, R. G.; Vanterpool, A.; Kulak, H. J. *Can. J. Chem.* **1969**, *47,* 2015.
4. Freedman, H. H.; Dubois, R. A. *Tetrahedron Lett.* **1975**, *16,* 3251.
5. Hamada, Y.; Kato, N.; Kakamu, Y.; Shioiri, T. *Chem. Pharm. Bull.* **1981**, *29,* 2246.
6. Jursic, B. *Tetrahedron* **1988**, *44,* 6677.
7. Tan, S. N.; Dryfe, R. A.; Girault, H. H. *Helv. Chim. Acta* **1994**, *77,* 231.
8. Silva, A. L.; Quiroz, B.; Maldonado, L. A. *Tetrahedron Lett.* **1998**, *39,* 2055.
9. Weissberg, A.; Dahan, A.; Portnoy, M. *J. Comb. Chem.* **2001**, *3,* 154.
10. Peng, Y.; Song, G. *Green Chem.* **2002**, *4,* 349.
11. Stabile, R. G.; Dicks, A. P. *J. Chem. Educ.* **2003**, *80,* 313.

438

Wittig reaction

Olefination of carbonyls using phosphorus ylides.

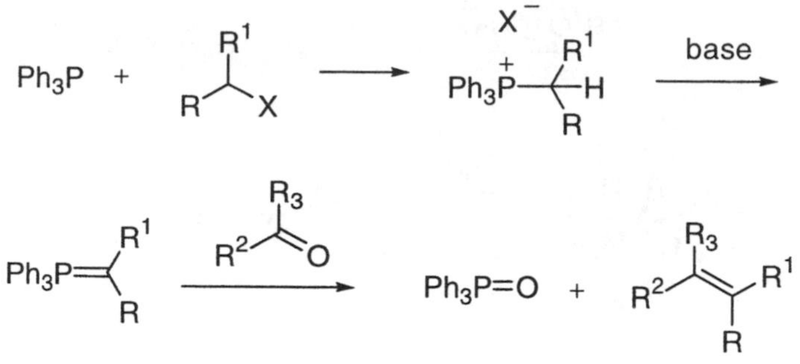

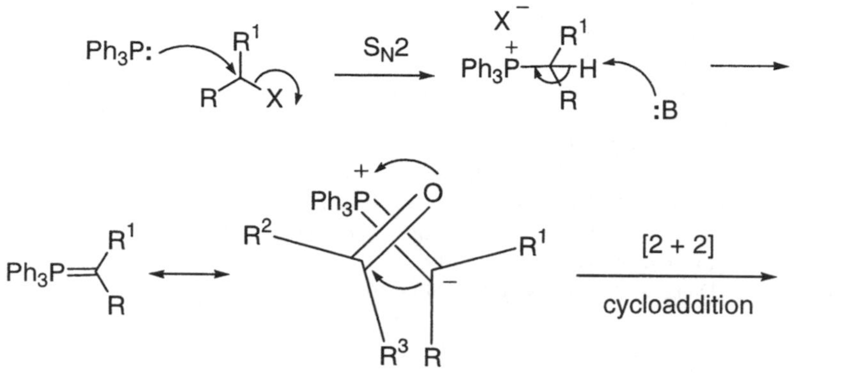

"puckered" transition state, irreversible and concerted

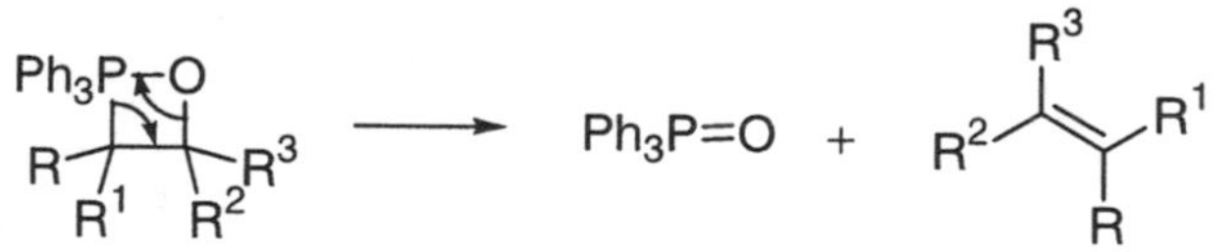

oxaphosphetane intermediate

References

1. Wittig, G.; Schöllkopf, U. *Ber. Dtsch. Chem. Ges.* **1954**, *87*, 1318.
2. Murphy, P. J.; Brennan, J. *Chem. Soc. Rev.* **1988**, *17*, 1–30. (Review).
3. Maryanoff, B. E.; Reitz, A. B. *Chem. Rev.* **1988**, *89*, 863–927. (Review).
4. Vedejs, E.; Peterson, M. J. *Top. Stereochem.* **1994**, *21*, 1. (Review).
5. Heron, B. M. *Heterocycles* **1995**, *41*, 2357.
6. Murphy, P. J.; Lee, S. E. *J. Chem. Soc., Perkin Trans. 1* **1999**, 3049.
7. De Luca, L.; Giacomelli, G.; Porcheddu, A. *Org. Lett.* **2002**, *4*, 533.
8. Blackburn, L.; Kanno, H.; Taylor, R. J. K. *Tetrahedron Lett.* **2003**, *44*, 115.

[1,2]-Wittig rearrangement

Treatment of ethers with alkyl lithium results in alcohols.

The radical mechanism is also possible as radical intermediates have been identified.

References

1. Wittig, G.; Löhmann, L. *Justus Liebigs Ann. Chem.* **1942**, *550*, 260.
2. Hoffmann, R. W. *Angew. Chem.* **1979**, *91*, 625.
3. Tomooka, K.; Yamamoto, H.; Nakai, T. *Justus Liebigs Ann. Chem.* **1997**, 1275.
4. Maleczka, R. E., Jr.; Geng, F. *J. Am. Chem. Soc.* **1998**, *120*, 8551.
5. Tomooka, K.; Kikuchi, M.; Igawa, K.; Suzuki, M.; Keong, P.-H.; Nakai, T. *Angew. Chem., Int. Ed.* **2000**, *39*, 4502.
6. Katritzky, A. R.; Fang, Y. *Heterocycles* **2000**, *53*, 1783.
7. Kitagawa, O.; Momose, S.-i.; Yamada, Y.; Shiro, M.; Taguchi, T. *Tetrahedron Lett.* **2001**, *42*, 4865.
8. Barluenga, J.; Fañanás, F. J.; Sanz, R.; Trabada, M. *Org. Lett.* **2002**, *4*, 1587.
9. Lemiègre, L.; Regnier, T.; Combret, J.-C.; Maddaluno, J. *Tetrahedron Lett.* **2003**, *44*, 373.

[2,3]-Wittig rearrangement

Transformation of allyl ethers into homoallylic alcohols by treatment with base. Also known as Still–Wittig rearrangement.

R^1 = alkynyl, alkenyl, Ph, COR, CN.

References

1. Cast, J.; Stevens, T. S.; Holmes, J. *J. Chem. Soc.* **1960**, 3521.
2. Nakai, T.; Mikami, K. *Org. React.* **1994**, *46*, 105. (Review).
3. Bertrand, P.; Gesson, J.-P.; Renoux, B.; Tranoy, I. *Tetrahedron Lett.* **1995**, *36*, 4073.
4. Maleczka, R. E., Jr.; Geng, F. *Org. Lett.* **1999**, *1*, 1111.
5. Tsubuki, M.; Kamata, T.; Nakatani, M.; Yamazaki, K.; Matsui, T.; Honda, T. *Tetrahedron: Asymmetry* **2000**, *11*, 4725.
6. Itoh, T.; Kudo, K. *Tetrahedron Lett.* **2001**, *42*, 1317.
7. Pévet, I.; Meyer, C.; Cossy, J. *Tetrahedron Lett.* **2001**, *42*, 5215.
8. Anderson, J. C.; Skerratt, S. *Perkin 1* **2002**, 2871.
9. McGowan, G. *Au. J. Chem.* **2002**, *55*, 799.
10. Schaudt, M.; Blechert, S. *J. Org. Chem.* **2003**, *68*, 2913.

Wohl–Ziegler reaction

Radical-initiated allylic bromination using NBS, and catalytic AIBN as initiator or NBS under photolysis.

Initiation:

2,2'-azobisisobutyronitrile (AIBN)

Propagation:

The succinimidyl radical now is available for the next cycle of the radical chain reaction

References

1. Wohl, A. *Ber. Dtsch. Chem. Ges.* **1919**, *52*, 51.
2. Ziegler, K.; *et al.* **1942**, *551*, 80.
3. Wolfe, S.; Awang, D. V. C. *Can. J. Chem.* **1971**, *49*, 1384.
4. Ito, I.; Ueda, T. *Chem. Pharm. Bull.* **1975**, *23*, 1646.
5. Pennanen, S. I. *Heterocycles* **1978**, *9*, 1047.
6. Rose, U. *J. Heterocycl. Chem.* **1991**, *28*, 2005.
7. Allen, J. G.; Danishefsky, S. J. *J. Am. Chem. Soc.* **2001**, *123*, 351.

442

8. Jeong, I. H.; Park, Y. S.; Chung, M. W.; Kim, B. T. *Synth. Commun.* **2001**, *31*, 2261.
9. Detterbeck, R.; Hesse, M. *Tetrahedron Lett.* **2002**, *43*, 4609.
10. Stevens, C. V.; Van Heecke, G.; Barbero, C.; Patora, K.; De Kimpe, N.; Verhe, R. *Synlett* **2002**, 1089.

Wolff rearrangement

One carbon homologation *via* the intermediacy of α-diazoketone and ketene.

α-diazoketone   ketene intermediate

α-ketocarbene

Treatment of the ketene with water would give the corresponding homologated carboxylic acid.

References

1. Wolff, L. *Justus Liebigs Ann. Chem.* **1912**, *394*, 25.
2. Meier, H.; Zeller, K. P. *Angew. Chem.* **1975**, *87*, 52.
3. Podlech, J.; Linder, M. R. *J. Org. Chem.* **1997**, *62*, 5873.
4. Wang, J.; Hou, Y. *J. Chem. Soc., Perkin Trans. 1* **1998**, 1919.
5. Müller, A.; Vogt, C.; Sewald, N. *Synthesis* **1998**, 837.
6. Lee, Y. R.; Suk, J. Y.; Kim, B. S. *Tetrahedron Lett.* **1999**, *40*, 8219.
7. Tilekar, J. N.; Patil, N. T.; Dhavale, D. D. *Synthesis* **2000**, 395.
8. Yang, H.; Foster, K.; Stephenson, C. R. J.; Brown, W.; Roberts, E. *Org. Lett.* **2000**, *2*, 2177.
9. Xu, J.; Zhang, Q.; Chen, L.; Chen, H. *J. Chem. Soc., Perkin Trans. 1* **2001**, 2266.
10. Kirmse, W. *Eu. J. Org. Chem.* **2002**, 2193. (Review).
11. Bogdanova, A.; Popik, V. V. *J. Am. Chem. Soc.* **2003**, *125*, 1456.
12. Julian, R. R.; May, J. A.; Stoltz, B. M.; Beauchamp, J. L. *J. Am. Chem. Soc.* **2003**, *125*, 4478.

444

Wolff–Kishner reduction

Carbonyl reduction to methylene using basic hydrazine.

$$R \underset{R^1}{\overset{O}{\|}} \xrightarrow[\text{NaOH, reflux}]{NH_2NH_2} R \frown R^1$$

References

1. Kishner, N. *J. Russ. Phys. Chem. Soc.* **1911**, *43*, 582.
2. Wolff, L. *Justus Liebigs Ann. Chem.* **1912**, *394*, 86.
3. Huang-Minlong Modification, Huang Minlong *J. Am. Chem. Soc.* **1946**, *68*, 2487.
4. Todd, D. *Org. React.* **1948**, *4*, 378. (Review).
5. Cram, D. J.; Sahyun, M. R. V.; Knox, G. R. *J. Am. Chem. Soc.* **1962**, *84*, 1734.
6. Szmant, H. H. *Angew. Chem., Int. Ed. Engl.* **1968**, *7*, 120.
7. Murray, R. K., Jr.; Babiak, K. A. *J. Org. Chem.* **1973**, *38*, 2556.
8. Akhila, A.; Banthorpe, D. V. *Indian J. Chem.* **1980**, *19B*, 998.
9. Bosch, J.; Moral, M.; Rubiralta, M. *Heterocycles* **1983**, *20*, 509.
10. Taber, D. F.; Stachel, S. J. *Tetrahedron Lett.* **1992**, *33*, 903.
11. Gadhwal, S.; Baruah, M.; Sandhu, J. S. *Synlett* **1999**, 1573.
12. Eisenbraun, E. J.; Payne, K. W.; Bymaster, J. S. *Ind. Eng. Chem. Res.* **2000**, *39*, 1119.
13. Szendi, Z.; Forgo, P.; Tasi, G.; Bocskei, Z.; Nyerges, L.; Sweet, F. *Steroids* **2002**, *67*, 31.
14. Chattopadhyay, S.; Banerjee, S. K.; Mitra, A. K. *J. Indian Chem. Soc.* **2002**, *79*, 906.
15. Bashore, C. G.; Samardjiev, I. J.; Bordner, J.; Coe, J. W. *J. Am. Chem. Soc.* **2003**, *125*, 3268.

Woodward *cis*-dihydroxylation

Cf. Prévost *trans*-dihydroxylation.

cyclic iodonium ion intermediate

neighboring group assistance

References

1. Woodward, R. B.; Brutcher, F. V. *J. Am. Chem. Soc.* **1958**, *80*, 209.
2. Mangoni, L.; Dovinola, V. *Tetrahedron Lett.* **1969**, *10*, 5235.
3. Kamano, Y.; Pettit, G. R.; Tozawa, M.; Komeichi, Y.; Inoue, M. *J. Org. Chem.* **1975**, *40*, 2136.
4. Brimble, M. A.; Nairn, M. R. *J. Org. Chem.* **1996**, *61*, 4801.
5. Hamm, S.; Hennig, L.; Findeisen, M.; Muller, D.; Welzel, P. *Tetrahedron* **2000**, *56*, 1345.

Wurtz reaction

Caron-carbon bond formation from the treatment of alkyl halides and sodium or magnesium metals.

$$R-X \xrightarrow{\text{Na(0)}} R-R + NaX$$

$$R-X \xrightarrow{\text{Na(0)}} R^- \ Na^+ + NaX$$

Ionic mechanism,

$$\xrightarrow{S_N2} R-R + X^-$$

Radical mechanism,

$$R-X \xrightarrow{R^- \ Na^+} NaX + 2R\bullet \longrightarrow R-R$$

References

1. Wurtz, A. *Justus Liebigs Ann. Chem.* **1855**, *96*, 364.
2. Connor, D. S.; Wilson, E. R. *Tetrahedron Lett.* **1967**, *8*, 4925.
3. Kwa, T. L.; Boelhouwer, C. *Tetrahedron* **1969**, *25*, 5771.
4. Garst, J. F.; Cox, R. H. *J. Am. Chem. Soc.* **1970**, *92*, 6389.
5. Hobbs, C. F.; Hamman, W. C. *J. Org. Chem.* **1970**, *35*, 4188.
6. Garst, J. F.; Hart, P. W. *J. Chem. Soc., Chem. Commun.* **1975**, 215.
7. Miyahara, Y.; Shiraishi, T.; Inazu, T.; Yoshino, T. *Bull. Chem. Soc. Jpn.* **1979**, *52*, 953.
8. Nenfield, R. E.; Cragg, R. H.; Jones, R. G.; Swain, A. C. *Nature* **1991**, *353*, 340.
9. HariPrasad, S.; Nagendrappa, G. *Indian J. Chem.* **1997**, *36B*, 1016.
10. Ceylan, M.; Budak, Y. *J. Chem. Res. (S)* **2002**, 416.
11. Banno, T.; Hayakawa, Y.; Umeno, M. *J. Organometallic Chem.* **2002**, *653*, 288. (Review).

Yamada coupling reagent

The use of diethyl phosphoryl cyanide (diethyl cyanophosphonate) for the activation of carboxylic acids.

References

1. Yamada, S.; Takeuchi, Y. *Tetrahedron Lett.* **1971**, *12*, 3595.
2. Yamada, S.-i.; Kasai, Y.; Shioiri, T. *Tetrahedron Lett.* **1973**, *14*, 1595.
3. Yokoyama, Y.; Shioiri, T.; Yamada, S. *Chem. Pharm. Bull.* **1977**, *25*, 2423.
4. Shioiri, T.; Hamada, Y. *J. Org. Chem.* **1978**, *43*, 3631.
5. Kato, N.; Hamada, Y.; Shioiri, T. *Chem. Pharm. Bull.* **1984**, *32*, 3323.
6. Hamada, Y.; Mizuno, A.; Ohno, T.; Shioiri, T. *Chem. Pharm. Bull.* **1984**, *32*, 3683.
7. Guzman, A.; Diaz, E. *Synth. Commun.* **1997**, *27*, 3035.
8. Mizuno, M.; Shioiri, T. *Tetrahedron Lett.* **1998**, *39*, 9209.
9. Elmore, C. S.; Dean, D. C.; Zhang, Y.; Gibson, C.; Jenkins, H.; Jones, A. N.; Melillo, D. G. *J. Labeled Compounds Radiopharm.* **2002**, *45*, 29.

Yamaguchi esterification

Esterification using 2,4,6-trichlorobenzoyl chloride (Yamaguchi reagent).

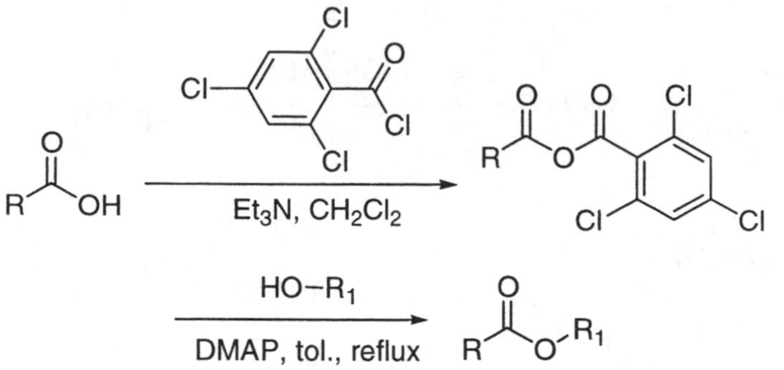

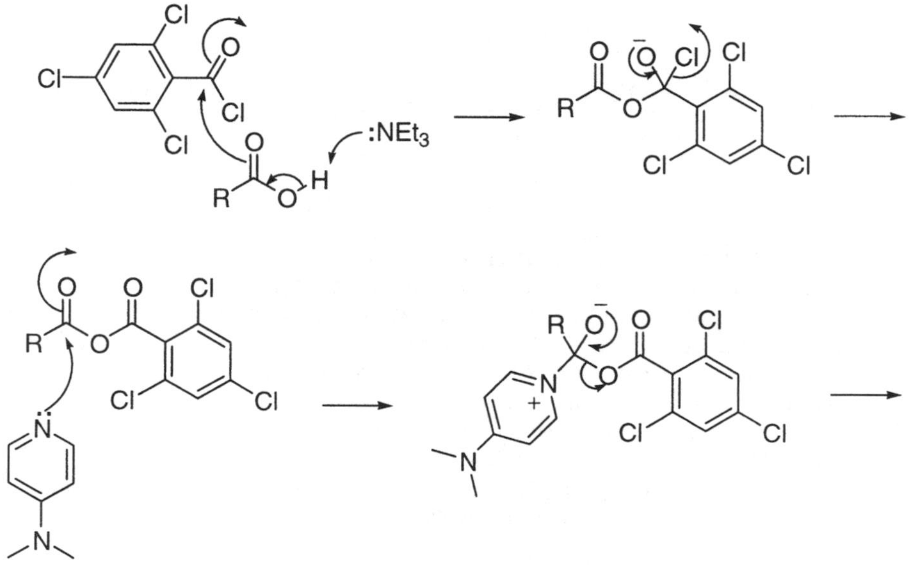

DMAP (Dimethylaminopyridine)

Steric hindrance of the chloro substituents blocks attack of the other carbonyl.

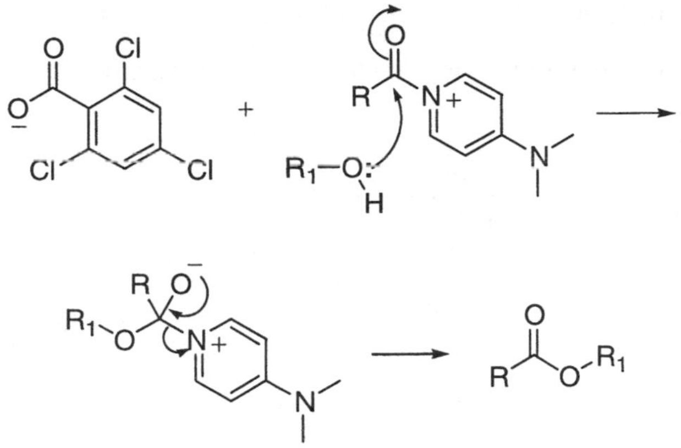

References

1. Inanaga, J.; Hirata, K.; Saeki, H.; Katsuki, T.; Yamaguchi, M. *Bull. Chem. Soc. Jpn.* **1979**, *52*, 1989.
2. Kawanami, Y.; Dainobu, Y.; Inanaga, J.; Katsuki, T.; Yamaguchi, M. *Bull. Chem. Soc. Jpn.* **1981**, *54*, 943.
3. Bartra, M.; Vilarrasa, J. *J. Org. Chem.* **1991**, *56*, 5132.
4. Richardson, T.; Rychnovsky, S. D. *Tetrahedron* **1999**, *55*, 8977.
5. Berger, M.; Mulzer, J. *J. Am. Chem. Soc.* **1999**, *121*, 8393.
6. Paterson, I.; Chen, D. Y.-K.; Acena, J. L.; Franklin, A. S. *Org. Lett.* **2000**, *2*, 1513.
7. Hamelin, O.; Wang, Y.; Depres, J.-P.; Greene, A. E. *Angew. Chem., Int. Ed.* **2000**, *39*, 4314.

450

Zaitsev elimination

E2 thermodynamic elimination to give the more substituted olefin.

major   minor

Hofmann elimination, on the other hand, furnishes the least highly substituted olefins.

References

Zaitsev elimination

1. Brown, H. C.; Wheeler, O. H. *J. Am. Chem. Soc.* **1956**, *78*, 2199.
2. Elrod, D. W.; Maggiora, G. M.; Trenary, R. G. *Tetrahedron Comput. Methodol.* **1990**, *3*, 163.
3. Reinecke, M. G.; Smith, W. B. *J. Chem. Educ.* **1995**, *72*, 541.
4. Lewis, D. E. *Book of Abstracts, 214th ACS National Meeting*, Las Vegas, NV, September 7–11, (**1997**).

Hofmann elimination

1. Eubanks, J. R. I.; Sims, L. B.; Fry, A. *J. Am. Chem. Soc.* **1991**, *113*, 8821.
2. Bach, R. D.; Braden, M. L. *J. Org. Chem.* **1991**, *56*, 7194.
3. Lai, Y. H.; Eu, H. L. *J. Chem. Soc., Perkin Trans. 1* **1993**, 233.
4. Sepulveda-Arques, J.; Rosende, E. G.; Marmol, D. P.; Garcia, E. Z.; Yruretagoyena, B.; Ezquerra, J. *Monatsh. Chem.* **1993**, *124*, 323.
5. Woolhouse, A. D.; Gainsford, G. J.; Crump, D. R. *J. Heterocycl. Chem.* **1993**, *30*, 873.
6. Bhonsle, J. B. *Synth. Commun.* **1995**, *25*, 289.
7. Berkes, D.; Netchitailo, P.; Morel, J.; Decroix, B. *Synth. Commun.* **1998**, *28*, 949.
8. Jimenez, R. M.; Soltero, J. F. A.; Manriquez, R.; Lopez-Dellamary, F. A.Palacios, G.; Puig, J. E.; Morini, M.; Schulz, P. C. *Langmuir* **2002**, *18*, 3767.
9. Morphy, J. R.; Rankovic, Z.; York, M. *Tetrahedron Lett.* **2002**, *43*, 6413.
10. Hernandez-Maldonado, A. J.; Yang, R. T.; Chinn, D.; Munson, C. L. *Langmuir* **2003**, *19*, 2193.

Zincke reaction

N-Aryl or *N*-alkyl pyridinium salt synthesis *via* the ring-opening reaction of aniline and 1-(2,4-dinitrophenyl)pyridium chloride.

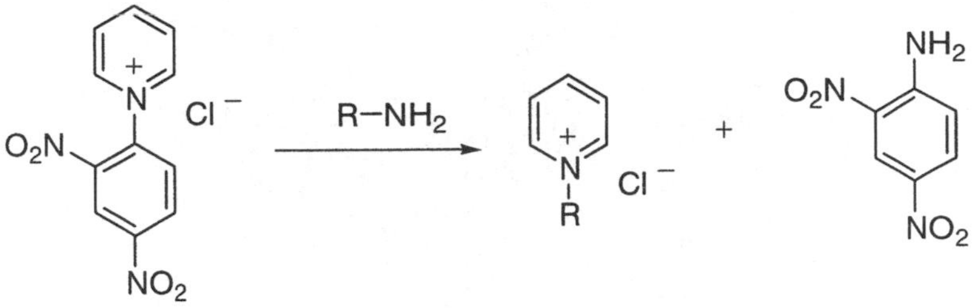

Zincke salt

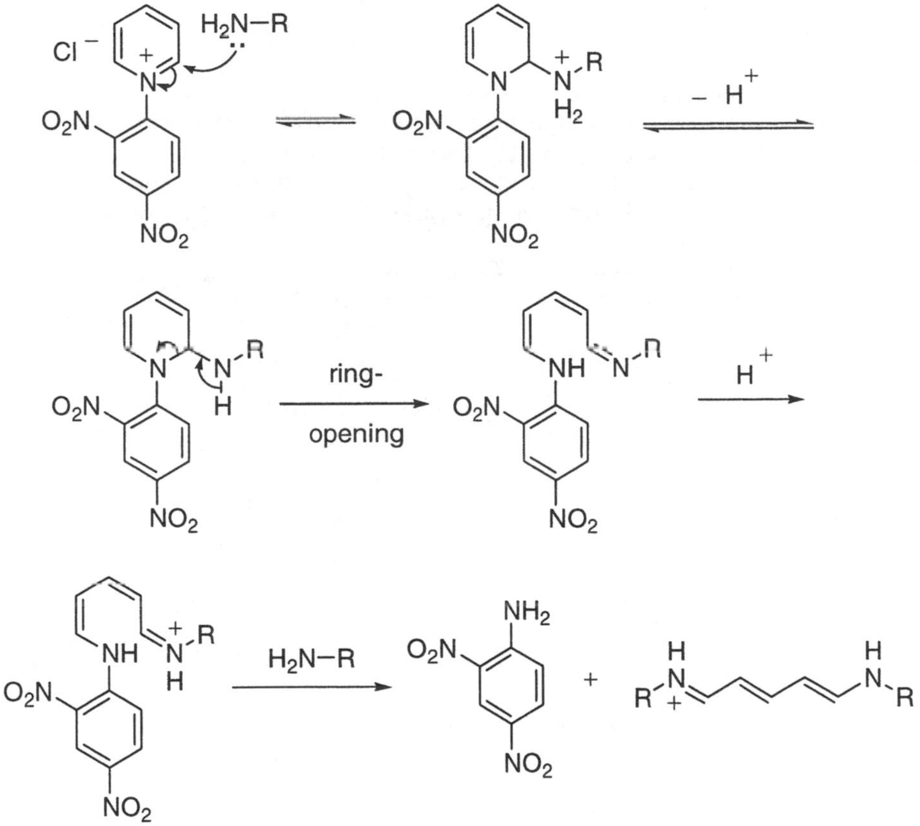

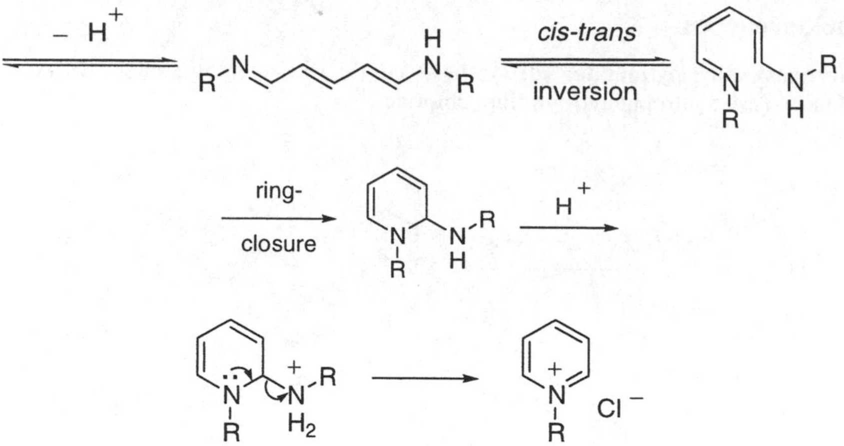

452

References

1. Zincke, T. *Justus Liebigs Ann. Chem.* **1903**, *330*, 361.
2. Marvell, E. N.; Caple, G.; Shahidi, I. *Tetrahedron Lett.* **1967**, *8*, 277.
3. Marvell, E. N.; Caple, G.; Shahidi, I. *J. Am. Chem. Soc.* **1970**, *92*, 5641.
4. Epszjun, J.; Lunt, E.; Katritzky, A. R. *Tetrahedron* **1970**, *26*, 1665. (Review).
5. de Gee, A. J.; Sep, W. J.; Verhoeven, J. W.; de Boer, T. J. *J. Chem. Soc., Perkin Trans. 1* **1974**, 676.
6. Eda, M.; Kurth, M. J. *J. Chem. Soc., Chem. Commun.* **2001**, 723.
7. Cheng, W.-C.; Kurth, M. J. *Org. Prep. Proced. Int.* **2002**, *34*, 585. (Review).

Zinin benzidine rearrangement (semidine rearrangement)

Acid-promoted rearrangement of hydrazobenzene to 4,4-diaminobiphenyl (benzidine) and 2,4-diaminobiphenyl.

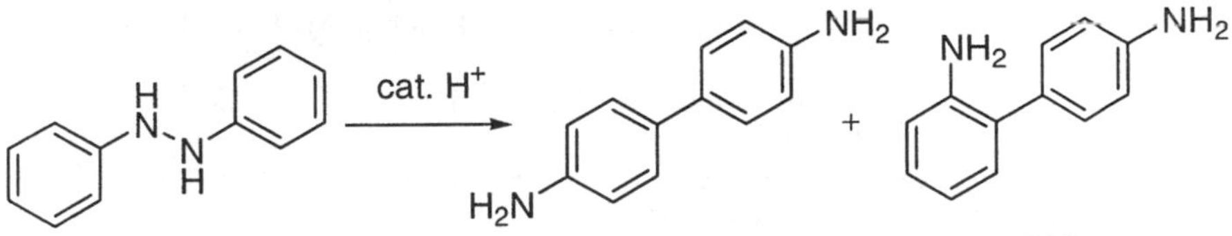

Hydrazobenzene   70% (benzidine)   30%

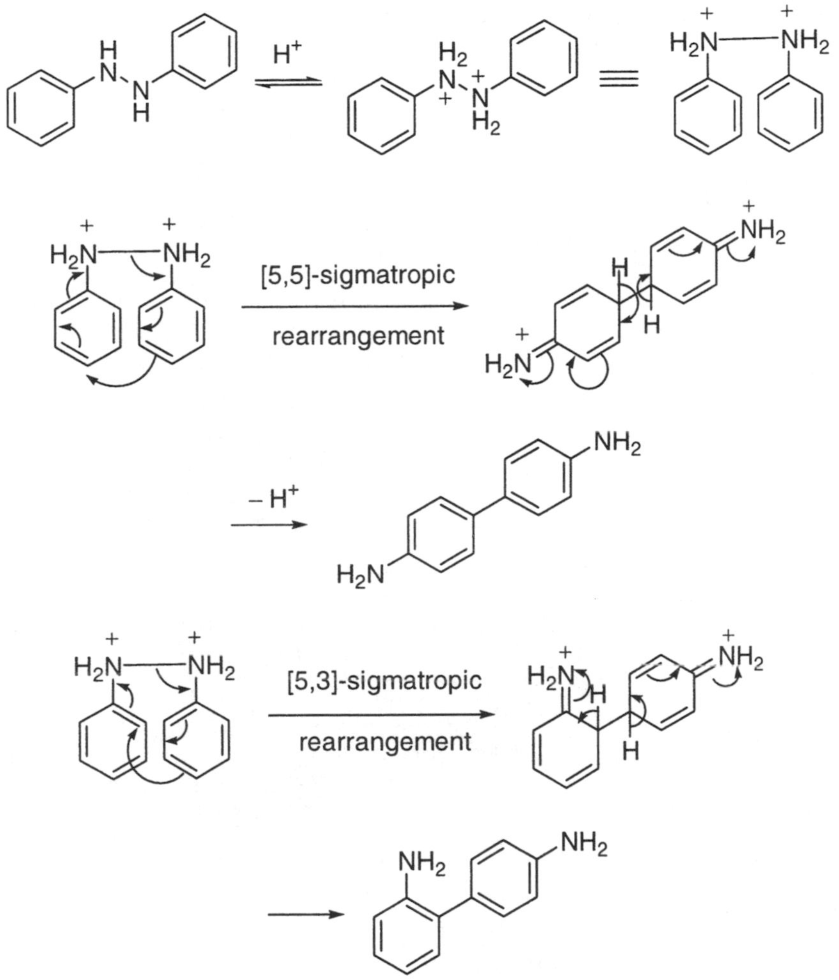

[5,5]-sigmatropic rearrangement

$-H^+$

[5,3]-sigmatropic rearrangement

454

References

1. Zinin, N. *J. Prakt. Chem.* **1845**, *36*, 93.
2. Shine, H. J.; Baldwin, C. M.; Harris, J. H. *Tetrahedron Lett.* **1968**, *9*, 977.
3. Banthorpe, D. V.; O'Sullivan, M. *J. Chem. Soc., Perkin Trans. 2* **1973**, 551.
4. Shine, H. J.; Zmuda, H.; Kwart, H.; Horgan, A. G.; Brechbiel, M. *J. Am. Chem. Soc.* **1982**, *104,* 5181.
5. Rhee, E. S.; Shine, H. J. *J. Am. Chem. Soc.* **1986**, *108,* 1000.
6. Shine, H. J. *J. Chem. Educ.* **1989**, *66*, 793.
7. Davies, C. J.; Heaton, B. T.; Jacob, C. *J. Chem. Soc., Chem. Commun.* **1995**, 1177.
8. Park, K. H.; Kang, J. S. *J. Org. Chem.* **1997**, *62,* 3794.
9. Buncel, E.; Cheon, K.-S. *J. Chem. Soc., Perkin Trans. 2* **1998**, 1241.
10. Buncel, E. *Can. J. Chem.* **2000**, *78*, 1251.
11. Benniston, A. C.; Clegg, W.; Harriman, A.; Harrington, R. W.; Li, P.; Sams, C. *Tetrahedron Lett.* **2003**, *44*, 2665.

Subject Index

462